Comparative Medicine

Erika Jensen-Jarolim
Editor

Comparative Medicine

Anatomy and Physiology

Editor
Erika Jensen-Jarolim
Comparative Medicine
Medical University Vienna and
University of Veterinary Medicine Vienna
Messerli Research Institute
Vienna, Austria

ISBN 978-3-7091-1558-9 ISBN 978-3-7091-1559-6 (eBook)
DOI 10.1007/978-3-7091-1559-6
Springer Wien Heidelberg New York Dordrecht London

Library of Congress Control Number: 2013956624

Printed on acid-free paper

Springer is part of Springer Science+Business Media (www.springer.com)

*To my unforgettable father.
Thank you for infecting your children with your lively interest in the animal kingdom and biology from our very beginning. Following your mandate, your grandchildren Annika, Sebastian, and Christopher are successfully contaminated.
Prosperous tree of life!*

Preface

In 2010 the Messerli Research Institute was founded as an interdisciplinary enterprise between the private Swiss Messerli Foundation and three universities: the University of Veterinary Medicine Vienna, Medical University Vienna, and University of Vienna. The intention was to investigate and improve all types of human–animal relationships and interactions.

In spite of the fact that human medicine and veterinary medicine share a long history of mutual fertilization, the number of interactions between these disciplines is still insufficient today. To foster the interdisciplinary dialogue for the sake of human and animal patients, the professorship for comparative medicine was embedded into the highly stimulating environment of the Messerli Research Institute.

In this first book, comparative medicine aligns head to head the anatomy and physiology of human and animal organisms to shape a solid basis for the understanding of conserved (or different) principles of life. It gives an outline on historic aspects and new identity of the discipline, as well as an update in juridical and ethical implications in the human–animal interaction.

We hope to enrich your reading with our contribution and thereby to contribute to the interdisciplinary dialogue.

Vienna, Austria Erika Jensen-Jarolim

Acknowledgement

The Editor would like to express her thanks to Mrs. Herta Messerli and the Messerli Foundation for making an interdisciplinary endeavour like this book possible. The authors truly enjoyed the constant exchange during the production phase.

Contents

List of Figures

List of Tables

Definition of Comparative Medicine: History and New Identity

1

Erika Jensen-Jarolim

Contents

Abstract

Most of today's knowledge on molecular mechanisms in mental and physical health and disease is significantly supported by animal studies. Innumerous nutritional and medical products such as vaccines were developed based on animal and human experiments, including heroic self-experiments, and on constant comparison of effects in different species during the history of medicine. The achieved medical standard changed our life expectations significantly by combatting or even eradicating many of previously deathly diseases. Directives for drug development for medical (Directive 2001/83/EC of the European Parliament 2001) and veterinary medicine (Directive 2001/82/EC of the European Parliament 2001) today comprise obligatory proof of concept and toxicity studies using different species of laboratory animals, followed by clinical studies in human or veterinary patients, where in fact the procedures do not differ substantially (Lombard 2007). However, comparative medicine

E. Jensen-Jarolim (✉)
Comparative Medicine, Messerli Research Institute, University of Veterinary Medicine Vienna, Medical University Vienna and University Vienna, Vienna, Austria
e-mail: erika.jensen-jarolim@meduniwien.ac.at

E. Jensen-Jarolim (ed.), *Comparative Medicine*,
DOI 10.1007/978-3-7091-1559-6_1, © Springer-Verlag Wien 2014

today aims to foster the constant exchange of know-how between human and veterinary medical disciplines and encompasses the treatment of animal patients in clinical studies as a modern means of drug development simultaneously following the 3R rule.

1.1 Introduction

Since the introduction of the 3R rule and its legislative implementation in 2010 (Directive of the European Parliament 2010), all researchers are forced even more to plan studies carefully according to replacement, reduction, and refinement of animal experiments (Editorial in Nature Immunol 2010) (see Chap. 14). This indicates an increased sensitivity in our society for ethical aspects and responsibility in animal and human experimentation. In vitro alternatives for animal studies allowing efficacy and nontoxicity testing at the same time, and reflecting all possible effects in a complex organism, are thus urgently searched for and supported (FRAME 2005). Several improvements have been achieved and include methods like cryo-conservation of embryos of transgenic animals or the development of in vitro 3D tissue models as an approximation to the real-life situation (Table 1.1).

We believe that the constantly improving electronic accessibility of scientific literature from the beginnings of "modern medicine" will strongly contribute to the 3Rs by preventing duplication of experiments.

There is, however, an agreement in the scientific community that it will not be possible to entirely replace animal models in the near future (Festing and Wilkinson 2007). The matter is emotionally discussed on both sides (Cressey 2011; Stephens 2011) and needs ethical reflection (see Chap. 15). Nevertheless, the overwhelming knowledge on molecular disease mechanisms and therapies has been gathered by previous, even historic, animal studies.

We are today convinced that comparative medicine can contribute to the 3Rs simultaneously by following its overall aim to improve the health of humans and animals.

Comparative medicine is a discipline with a broad knowledge on anatomy, physiology, and molecular pathophysiology for promoting health (Monath et al. 2010) and for the generation of novel therapies for humans as well as for animal patients (Schwabe 1968). According to the intriguing concept by the Comparative Oncology Trial Consortium of the National Cancer Institute, veterinary patients should be included early into clinical developments (Gordon et al. 2009). Thereby, animal numbers needed in preclinical studies would be reduced and simultaneously the animal patients provided with cutting-edge research. This strategy requires consequent "comparative" thinking and a constant interdisciplinary dialogue.

In the course of the formation of a new identity and redefinition of Comparative Medicine, it was obligatory to search the roots of this discipline in history. A single

Table 1.1 Alternative methods for animal experiments identified by the ECVAM (European Centre for the Validation of Alternative Methods) in 1999 (FRAME 2005)

1. Models, mannequins, and mechanical simulators
2. Films and interactive videos
3. Computer simulations and virtual reality systems
4. Self-experimentation and human studies
5. Plant experiments
6. Observational and field studies
7. In vitro studies on cell lines
8. Clinical practice

book could be identified specifically targeting the topic "History of Comparative medicine" by Lise Wilkinson which the author found concise and informative (Wilkenson 1992) and which will serve as "the red line" through this chapter. What was found is an amazing coevolution and cross-fertilization of medical, veterinary and later experimental medicine during the centuries. The developments were mostly driven by curiosity, by plagues of the time, and by the need of surrogate models for the human body. Animal cadavers were studied in order to get insights into anatomy and, later, physiology and pathophysiology of humans and other organisms (Adler 2004). Only few doctors were trained in medicine *and* veterinary medicine, and only in 1862 that the first professorship and chair dedicated to comparative medicine was established in France (Isensee 1845).

1.2 Medical Remedies Based on Comparative Observations Before Modern Medicine

In history the relation between humans and animals changed significantly several times depending on the culture (McClellan and Dorn 2006). In 20000–15000 BC, the animal-human relation between **Paleolithic men** and animals was that between the hunter and the hunted. The first interest in the health of herds may have risen, when humans followed wild sheep, cattle, and horses, in a hauling alliance with the canine packs from 12000 BC on. The interest of humans in the health condition of the herds raised when animals were domesticated, first around 9,000 BC in nomadic form, especially with sheep and goat, and when from 3700 to 3100 BC more solid associations between men and cattle or horses were built up. From this time point on, healthy herds represented food, survival, and wealth for man. It remains open whether primitive herbal remedies already were developed for both human and animal patients then.

Around 2000 BC, medicine and pharmacology had reached a high specialization level (Schwabe 1978) in ancient Egypt (Riviere and Papich 2009). With respect to animals, the major focus was given on bulls and cows, which were worshipped and treated by remedies as documented in the famous Kahun papyrus of the Lahun papyrus collection in the University College London (http://www.digitalegypt.ucl.ac.uk/lahun/papyri.html, Kahon papyrus: University College London). Even though it was a tradition to embalm human and animal bodies, Egyptian physicians had in fact restricted knowledge about the human anatomy.

In parallel, in **India** animal healing got highly developed due to the Hindu belief of reincarnation and awareness of all animals (Somvanshi 2006). As a consequence, Indian physicians were trained in animal care and treatment, with major focus on cattle, horses, and elephants. The *Charaka samhita* represents the oldest teaching book of Ayurveda medicine (Indian healing medicine, "Knowledge of life"), containing the *Haya-Ayurveda* for horses and the *Hasti Ayurveda* for elephants, from 300 to 500 AD.

The most important sources from **China** documenting herbal medicine for horses date back to 1122 BC. Horses were of impact for work and military defense and a symbol for aristocracy, and requiring specific prescriptions for equine diseases (Buell et al. 2010) (Fig. 1.1). For instance, isotherapy, was developed, i.e., transfer of diluted sweat of a domestic animal to diseased human.

The great Greek philosopher **Plato** (ca. 428–347 BC) taught the superiority and immortality of the human soul. In his opinion animal observations or even dissections were taboo, although the Greek god of medicine and healing, **Asklepios**, was believed to be treating man and animals (Wickkiser 2008). This is also the reason why the rod of Asklepios is used as a symbol of human medicine and pharmacy as well as veterinary medicine. Accordingly, **Hippocrates** (460–370 BC) found comparative pathology useful. In his opus *Airs, Waters, Places* he described case histories relevant for herds and the human population (Adams 1849). Hippocrates suggested that an approach to diagnosis and treatment should always be based on experience, observation, and logical reasoning. Later **Aristotle** (384–322 BC) described animal species and diseases and philosophized about mechanisms of interspecies transmission (Aristotle, translated by Farquharson 2004). He inspired **Erasistratus of Chios** (404–320 BC) and **Herophilus of Chalcedon** (ca. 330–255 BC) (Bay and Bay 2010) to open schools for comparative anatomy and physiology in Alexandria, where also cruel vivisection methods on prisoners have been practiced. Unfortunately, the burning of Alexandria's library has destroyed most of their opus.

The Roman erudite **Marcus Terentius Varro** (around 100 BC) attempted to create a universal encyclopedia on the most important disciplines, among them medicine (Disciplinarum libri IX, *or* Nine Books about the Disciplines) where he propagated the separation of human lepra groups based on his concept that *tiny invisible animals carried with the air* cause disease which enter the body through the mouth and the nostrils (Wilkenson 1992).

Aulus Cornelius Celsus (25 BC–50 AD) reported in his opus "De Medicina Libri Octo" experimental physiology based on his experience in dissecting and vivisecting many species. He also reported specific interventions, for instance, cupping, as a method to "*…drew out the poison from dog bites, especially if the dog was mad*" (Fooks et al. 2009).

The dissection of the human body remained no longer acceptable in Rome in **Claudius Galen**'s era (129–200 AD) (Lyons and Petrucelli 1978; Aufderheide

Fig. 1.1 Armée de terracotta à Xi'an—China. © [Delphimages]—Fotolia.com

2003). By his work and teachings, Galen's influence remained after several 100 years until the middle ages. He used vivisection as a method of dissections on living animals; the term "Galenic formulation" is today still used and indicated the composition of medical remedies. To circumvent the prohibition of human dissection, he dissected animal cadavers. He analyzed multiple species, ranging from pigs, sheep, cattle, cats, dogs, to weasels, bears, mice, and also a single elephant. The animals mostly used for anatomy studies, however, were Barbary macaques from which he concluded to human anatomy. This strategy had caused several persistent errors in the understanding of human anatomy in medical history.

In spite of his many different talents, **Publius Flavius Vegetius Magnus** (500 AD) has been long interpreted as a key person specifically for veterinary medicine due to his publication of *"Digestorum Artis Mulomedicinae libri,"* which was even much later, in the sixteenth century, published in Latin and German (Vegetius and Gargilius 1871).

1.3 From Middle Ages to Humanism and Renaissance, ca. 500–1600

In the Arabic world the famous Persian physician **Muhammad ibn Zakariyā Rāzī, in** short called **Rhazes** (865–925), wrote the first book on the contagiosity of measles and smallpox, al-Judari wa al-Hasbah (Modanlo 2008). As no dissection of humans was allowed for Muslim physicians, dissection of animals and comparative anatomy was pursued as a surrogate strategy.

However, at least in Europe, the antique period was followed by an era without significant progress in terms of science and medicine, and also dissections were prohibited (Aufderheide 2003). This was also due to the change from polytheistic belief as a basis of free philosophy and open scientific discussion to the monotheistic Christian belief.

In Europe, the Catholic Church was much empowered leading to aggressive expansion strategies, termed crusades, with tens of thousands of participants.

Traveling over several years, crusades were not only using horses but also accompanied by whole cattle trains for supply. Crusades prompted the transmission and outbreak of zoonotic diseases, anthrax and cattle plague across their routes. In parallel, natural catastrophes weakened the human population and typhus, influenza, smallpox, and bubonic plague were epidemic (Boccaccio, translated by Aldington 1930). The crusades enabled the import of the Oriental rat flea carrying the plague bacterium Yersinia pestis (Haensch et al. 2010). Although this all resulted in a demographic disaster, the contagious principle was recognized only vaguely and counterregulations like quarantine were hardly known (Mazzeo 1955).

As a result of these massive and deleterious plagues, however, the way of inoculation and transmission was thoroughly considered. **Albertus Magnus** (1206–1280) started studies on the anatomy and physiology of animals after joining the Dominican monastic order. In his work *Liber de animalibus*, he discussed human and animal plagues and even brought down the mechanism of inoculation to three possible reasons: (1) bites or injuries, (2) actual contact with diseased animals, or (3) respiration of air from the sick. His sphere of influence ranged between Paris and Cologne (Lindner 1962). He also specifically addressed dogs' health in *Practica canum* (Giese 2007).

Jordanus Rufus (around 1250) was supported by Frederic II, king of Sicily, and did profound research with special impact on the favorite species of his mentor, horses and falconry birds. He left detailed sketches on manipulation on diseased horses including the necessary mechanical apparatuses (Prévot 1991).

Knowledge and know-how on, for instance, quarantine measures gathered in the antique, had been lost at that time. However, the capacity of knowledge dissemination was largely improved by the invention of movable printing machines by **Johannes Gutenberg** around 1439 (Febvre and Martin 2010). From that time point on, official guidelines could be distributed much more effectively as broadsheets to the public, for instance, recommendations for common diseases of horses and cattle.

The Renaissance physician and philosopher **Girolamo Fracastoro** (1476–1553) from Verona made the significant observation that "germs" can transfer diseases and are species specific in animals or even plants (http://ocp.hul.harvard.edu/contagion/contributors.html 2013). **Andreas Vesalius** (1514–1564) practiced human dissections (Prioreschi 2001), whereas otherwise students still had to learn from animal dissections. Medical understanding was additionally facilitated by inventions, such as magnifying lenses by **Zacharias and Hans Janssen** around 1590, being improved to a focusing device in 1609 by **Galileo Galilei** (1564–1642), and finally to the first microscope by **Antonie van Leeuwenhoek** (1632–1723) (Nature 2009). He microscopically observed spermatozoa, bacteria, and sporozoa which he altogether termed *Animalcules*. Thus from today's point of view, he was an ultimate trigger of parasitology, microbiology, and bacteriology.

Within a short time, the microscope was applied for medicinal studies. **Athanasius Kircher** (1602–1680) reported in his book *Scrutinium Pestis* that in the blood of plague-infected patients *tiny worms* in tissues could be seen below the microscope, which affirmed his opinion that living organisms might transfer disease (Winkle 2005).

1.4 The Establishment of Colleges, Societies, and Congresses

Clearly, progress could be achieved faster with regular exchange of knowledge in a structured form, also allowing quality control of data instead of spontaneous, individual observations and opinions. In 1662 the basis for such was shaped with the foundation of the Royal Society in London. For the spreading of the scientific news, the classical journal *Philosophical Transactions* was established, which still exists today. **Robert Boyle** (1626–1691) published several key experiments there, among them the blood transfusion experiments from one animal to another, including from sheep into men, thus using human test subjects (Boyle 1666; Knight and Hunter 2007). Only rarely veterinarians achieved to publish their work in this journal. However, also other contemporary scientists were unlucky to do so.

The onset of the eighteenth century brought a new wave of plagues to Europe, among them bubonic plague (Haensch et al. 2010), smallpox, and cattle plague. Faster communication and access to knowledge fostered comparative approaches, especially with respect to the transfer and contagion of infections. The technology of transference, i.e., transfer of infections from human patients to animals and vice versa, was developed and may be regarded as experimental in vivo approaches. It had already been settled earlier in China (Temple and Needham 1986) and probably also in India (Dharampal 2000) that pox crusts might be used for therapy of the pocks and smallpox infections. There was even a market for buying smallpox crusts which were sniffed or eaten, possibly as the first prophylactic vaccine (Moore 1815). It is difficult to say whether this practice was developed de novo in the eighteenth century or whether it was overtaken from published work. However, **Emanuel Timone** (1665–1741) was the first to publish inoculation, also called grafting (Woodward 1714). This technique was widely practiced throughout the eighteenth century in Europe.

At this time, animal medicine was actually more or less nonexistent in Europe. This was changed with the cattle plagues in 1711 and in 1745–1757 (Cáceres 2011). **Bernardino Ramazzini** (1633–1714), from the faculty of Padua, and **Giovanni Maria Lancisi** (1654–1720), the pope's physician, were the first to acclaim the danger for the population from the plagues (Wilkinson 1984; Klaassen et al. 2011). But, any interest was in the beginning rejected by the arrogance of physicians who found it below their dignity to discuss cattle (Wilkenson 1992; Talbott 1970). However, **Mortimer Cromwell** (1693–1752), secretary of the Royal Society, raised the topic of plagues to an issue of national concern. The general strategies were quarantine, isolation of the diseased animals, fumigation of stables, and slaughter (Spinage 2003).

Erasmus Darwin (1731–1802) also was impacted by the tragedy of the plagues. He published thoughts on contagion in *Zoonomia* where he discusses infectious diseases in both humans and animals (Darwin 1796).

Possibly, these events together paved the way for the foundation of the first veterinary college in Lyon, France, in 1761, by **Claude Bourgelat** (1712–1779) (Isensee 1845; Grognier 1805). Bourgelat was a lawyer and an expert in horses and concerned about the devastating cattle plagues. Thus, like many personalities in the beginnings of veterinary medicine, he was a generalist. In fact, he invented "comparative pathobiology" before the veterinary profession (Mammerickx 1971). His major interest in horses biased the veterinary teaching and activities in Lyon during several years, but numerous colleagues from all over Europe visited Lyon at that time. When **Vicq d'Azyr** (1748–1794) (Isensee 1845) was appointed as the chair of comparative medicine in Lyon in 1782, he reoriented the institute's outlook more scientifically and towards comparative medicine, because he strongly believed in the unity of human and veterinary medicine. Equally important, he proposed '*. . .to carry out valuable experiments (in animals) which would be criminal if attempted in the treatment of human disease*'. According to Lise Wilkinson, these significant words actually introduced the use of animal models in medicinal research (Wilkenson 1992).

The foundation of other veterinary colleges followed, among them in Maison-Alfort, near Paris, which for the first time also comprised a farm.

Only 4 years after Lyon's foundation, the first veterinary school of Austria was initiated in Vienna by Empress Maria Theresa in 1765 (Mösle's Widow 1839). **Anton Hayne** (1786–1853) took first human surgical education but became later professor for animal medicine with focus on infectious diseases in Vienna (Hayne 1836, 1844). When Ignaz Semmelweis claimed his detection of puerperal fever and possible prevention by aseptic technique and hygiene, Hayne claimed priority for the discovery (Carter 1981).

Veterinary schools were founded and curricula developed not only in Austria but also in Sweden, Denmark, France, the Netherlands, and in Germany. When the Royal Veterinary College was founded in 1790 (http://www.rvc.ac.uk/About/Museums/Milestones.cfm, The Royal Veterinary College, University of London), some of the Lyon students decided to move to London. Among them **John Hunter** (1728–1793) moved back to London. He was originally trained in human anatomy and surgery, but had a great interest in comparative anatomy and animal physiology. He established a collection of more than 10,000 human and animal preparations and wax teaching models including exotic animals from the menagerie of the London Tower. The "Hunterian Collection" can still today be visited at the Royal College of Surgeons in London. His teaching on infectious diseases significantly encouraged next generations, including his student Edward Jenner, to tackle this topic.

Edward Jenner (1749–1823), who later should be coined the inventor of the smallpox vaccine (Jenner 1798; Riedel 2005), dissected many different species and had a general interest in diseases (Jenner, foreword Drewitt 1931). He should be recognized as a key person in comparative medicine, too (Fig. 1.2). Jenner turned first to rabies as an increasing problem in and outside London. He noted that previously infected dogs developed protection. He further introduced and developed animal models for rabies and proved that a healthy dog could be inoculated

Fig. 1.2 Edward Jenner. © [Georgios Kollidas]—Fotolia.com

with rabies using the saliva of a rabid dog. Transmission experiments were in general a tool at that time to understand the pathophysiology of infections. Jenner had also observed that people (milkers) who had been in contact with cowpox-infected cows did not develop vaccinia (smallpox, variola minor). Instead, they developed milder cowpox only (Jenner 1798). In 1796 Jenner made a historic experiment: He infected a healthy 8-year-old boy James Phipps first with lymph derived from a cowpox lesion of a milker's hand, followed by a challenge (variolation) with the smallpox virus. The boy developed a mild disease upon the cowpox infection, but was protected against the smallpox challenge. This indicated that cowpox could induce protective immunity against the severe smallpox infection and was a breakthrough in vaccinology and an important stimulator of immunology.

As a consequence of Jenner's inoculation experiments and many others, in animals and humans, the time became mature for the introduction of specific inoculation programs for the population. A commission on vaccinations of the Royal Academy of Medicine in France was established, being directed by **Jean-Baptiste Edouard Bousquet** (1794–1872), which elaborated guidelines on the advisability and frequency of vaccination and revaccination (Bazin 2011).

In Germany, **Robert Koch** (1843–1910), a country practitioner who had studied in Berlin and Göttingen, experimented with infections and become a truly notable personality in the history of epidemic diseases (http://ocp.hul.harvard.edu/contagion/contributors.html 2013; http://www.nobelprize.org/nobelprizes/medicine/

laureates/1905/. The Nobel Prize in Physiology or Medicine 1905; Robert Koch. Nobelprizeorg 1905). He discovered the anthrax bacillus, tuberculosis bacillus, and tuberculin, as well as Vibrio cholerae. Among experiments in many other species, he made infection experiments with young cattle that were not susceptible for the human strain of tuberculosis and proposed that a transmission of tuberculosis via cows' milk to humans and vice versa can be excluded (Koch 1903). This assumption, however, had to be corrected later (Raw 1906).

An important stimulator in the history of comparative medicine was the first veterinarians' conference in Hamburg organized by **John Gamgee** (1831–1894) in 1863, from which "The World Veterinary Association" evolved (Asch 2013). After the first conference in animal vaccination in London in 1880, **George Fleming** (1831–1901) proposed in *The Lancet* that chairs for comparative pathology should be established in all medical schools (Fleming 1871).

Rabies, sepsis, or putrid intoxication were further eminent burdens of that time. In 1803 the French physiologist **Francois Magendie** (1783–1855) inoculated a dog with saliva from a human rabies case and could by this animal experiment for the first time prove the interspecies transmission of rabies (Stahnisch 2003). Magendie was also interested in sepsis and experimented with putrid fish that was injected in animals. He observed that symptoms were elicited similar to yellow fever or typhus. Magendie introduced animal experiments with dogs and other species as a common physiological practice, notably in a time before anesthetics were invented 1846 (von den Driesch 2003). This prompted the foundation of the world's first anti-vivisection organization in the UK in 1875 (Pedersen 2002). **Pierre Victor Galtier** (1846–1908) in Lyons (Goret 1969), in 1879, used rabid saliva for inoculation of a sheep via the jugular vein resulting in *acquired immunity* (Galtier 1879, 1881). Comparative medicine was interpreted as experimental medicine at that time.

During this time a high knowledge of experimentation was gathered and many names of the experimental forerunners are still remembered: **Bernard Gaspard** (1788–1871) inoculated dogs, fox, pig, and lambs, thus carnivores and herbivores with lymph, blood, bile, and human urine, to test the prophylactic value of these treatments (Magendie 1822). **Rudolf Virchow** (1821–1902), the initiator of modern pathology, distinguished between pyemia and septicemia and thrombosis and embolism by dog studies. Thus he made key observations leading to precise medications based on animal experiments. In his *Address on the value of pathological experiments*, he fierily advocated the value of these experiments, comparing work on isolated tissues, animal studies, and vivisection, in the light of the formation of vegetarian and anti-vivisection movements (Virchow 1881).

The Danish physiologist **Peter Ludvig Panum** (1820–1885) described in dogs dose-dependent effects using putrid blood, flesh, brain, or human feces for experimentations (Kolmos 2006). **Friedrich August Johannes Löffler**, German bacteriologist and student of Koch, identified bacteria as a cause of diphteria, virus as cause of foot-and-mouth disease, Pseudomonas mallei causing glanders, and bacteria as the causes of swine erysipelas and plague (Biographie 1987). Thus scientists were regularly active in human and veterinary topics.

Also, anthrax infections were common, e.g., due to gunshot wounds providing anaerobic milieu (Swiderski 2004). **Pierre Rayer** in 1850 inoculated sheep with anthrax, whose blood was agglutinated. When he with his assistant **Casimir Davaine** (1812–1882) investigated the blood microscopically, he could see that it also contained small rod-shaped, nonmoving bodies, approximately twice the length of an erythrocyte. Typically, researchers did not only sacrifice animals for their studies but also human (poor) patients or criminals. Davaine, who also brought down animal experiments to a smaller laboratory scale (rabbits, rats, guinea pigs), inoculated a pregnant woman with anthrax blood who consequently died only 2 days later (Swiderski 2004). When she was dissected, her placenta was full of anthrax bacteria, whereas the fetus was completely clean. By this cruel experiment, Davaine detected the biological filter mechanism of the placental barrier. A series of filtration and precipitation experiments followed in order to be able to separate the infectious agent from the blood cells. **Robert Koch** followed the idea of a *contagium vivum*, a living organism that transfers disease. He focused on culturing techniques for which he became famous. He was the first to approve anthrax in culture in 1877 and demonstrated its spore formation and its transferability to animals (von den Driesch 2003). The Robert Koch Institute (RKI), founded in 1891, is today still a research and reference institution for monitoring infectious diseases in Germany. **Louis Pasteur** (1822–1895), among others, developed attenuation: Upon heating to 42–43 °C, no spore production took place, reducing the danger of infections during vaccination (Pasteur and Chamberland 2002). Others like **Pierre Paul Èmile Roux** (1853–1933) improved attenuation by potassium dichromate treatment of the vaccines. These developments were also very useful for other vaccines, like anthrax. The first anthrax vaccine for humans was introduced in 1954 only.

Inoculation experiments were also performed by researchers in comparative medicine: **Jean-Baptiste Auguste Chauveau** (1827–1917), a professor from the Lyon Veterinary School (Bazin 2011) did not only himself experiment on sepsis (Sanderson 1872) but also chaired a commission which evaluated comparative experiments in cattle, showing that the co-inoculation with smallpox and cowpox induced cowpox only, thereby proving species specificity of the infection. In contrast to Jenner's concept of cross-protection by cowpox inoculation against smallpox, Chauveau anticipated that smallpox itself could be attenuated by passage through these animals (Swiderski 2004). The commission, in 1864, hence decided for a human experiment in seven children from an orphanage. They were all inoculated with smallpox six times, passaged through cattle. Unfortunately, two developed confluent smallpox and the other five the more discrete form (Cookshank 1889).

A follower of Chauveau, **Jean Joseph Henri Toussaint** (1847–1890) cultivated fowl cholera bacteria in vitro and supplied it to Pasteur who finally developed the vaccine for this veterinary disease (Wilkenson 1992).

However, from the above it is clear that **Louis Pasteur** contributed significantly to vaccine development by numerous experiments (Fig. 1.3). He inoculated rabies to several species like rabbits, dogs, sheep, goats, and guinea pigs and found that

Fig. 1.3 Louis Pasteur, nineteenth-century scientist. © [Erica Guilane-Nachez]—Fotolia.com

especially brain tissue was highly infectious (Pasteur, interviewed by Illo 1996). At the International Congress of Medicine in Copenhagen, Pasteur announced that he had solved the rabies vaccine in rabbits. Much later, when his personally retained laboratory notebooks were finally archived in the Bibliothèque Nationale in Paris, they revealed that less of the reported animal experiments were actually documented than Pasteur had claimed (Geison 1995; Warner 1996). For instance, the famous and successful healing experiment in a 9-year-old patient, Joseph Meister, who had been bitten and infected by a rabid dog, was done with a specific rabies vaccine that had not been tested in dogs before. Anyway, the healing experiment contributed to the great international popularity of Pasteur. One should also remember that **Pierre Paul Émile Roux** (1853–1933) (Pasteur and Chamberland 2002) and others had contributed significantly to this development, however, without merit.

1.5 Expansion of Comparative Medicine from Europe to the USA

Hence, comparative medicine has paved the way to modern medicine. Especially in the fields of infectiology, immunology, and vaccination in the eighteenth–nineteenth century, great scientists and clinicians were quite used to "comparative thinking." However, overall coexistence or cooperation between medicine and veterinary medicine differed from country to country (Wilkenson 1992): In France there was coexistence and cooperation; in Britain old prejudices against veterinarians resulted in minimal cooperation between the disciplines; Italy sent veterinary students to the prestigious centers in France; in Germany the veterinary schools were characterized by an administrative uniformity; students from overseas went back from Europe to spread their knowledge to the USA.

Importantly, it was recognized during this period that clinical and experimental research must go hand in hand and that experimentally oriented institutions are urgently needed to accomplish what we would term translational research today. In Vienna the great patho-anatomist **Carl von Rokitansky** (1804–1878) triggered **Salomon Stricker** (1834–1898) to found the *Institute of Experimental Pathology* in 1873, which was from the beginning predominantly dedicated to laboratory research (Wyklicky 1985). In an attempt to adapt it to modern nomenclature, the institute was in 2000 renamed to Institute of Pathophysiolgy, and in 2010 during the leadership of the editor of this book renamed Institute of Pathophysiology and Allergy Research, indicating a still tight association to immunologic topics today.

William H. Welch (1850–1934) (Dhom 2001) had his training in Europe and strongly emphasized the European idea of medical research institutes directly linked to clinics. Being the first dean of the John Hopkins School of Medicine, Baltimore, Welch was also instrumental in formulating the policy for the foundation of the *Rockefeller Institute of Medical Research* in 1901 (Ackerknecht 2001), which was the first American equivalent to the Pasteur and Robert Koch Institutes in Europe and where he served as founding president. **Theobald Smith** (1859–1934) who had previously been appointed for comparative pathology at the veterinary school of Harvard became by Welch appointed professor and director for the Institute of Animal pathology at the Rockefeller Institute. The *Journal for Experimental Medicine* (*JEM*), printed by the Rockefeller University Press, is tightly connected with the Rockefeller Institute and still today belongs to the most prestigious scientific journals globally. The journal's webpage alludes to the alliance with comparative medicine by stating '*…The journal prioritizes studies on intact organisms and has made a commitment to publishing studies on human subjects*'.

Smith also was in personal contact with the Swiss **Karl Friedrich Meyer** (1884–1974), who was an exceptional scientist, eminent teacher, and "*walking encyclopedia*", active in the Berkeley University of California and at the Hooper Foundation for Medical Research. He followed "*…the tradition of Koch and Pasteur" and addressed medical problems in the fields of human and animal*

Table 1.2 Mission statements of international Departments of Comparative Medicine

Comparative Medicine at Stanford University
. . . advancing human and animal health through outstanding research, veterinary care and training naturally occurring and experimental animal models of human disease; biology and prevention of diseases of laboratory animals; research applications of animals
National Center for Research Resources: NCRR's "Division of Comparative Medicine helps meet the needs of biomedical researchers for high-quality, disease-free animals and specialized animal research facilities"
Comparative Medicine at Yale School of Medicine
. . .Research in the department is focused on modeling human diseases in laboratory animal models and diagnosing and treating disruptive diseases in laboratory animals
Comparative Medicine at UC Davis
. . . concept of "One Medicine" through interdisciplinary comparative medical research, teaching, and model development. The mission is to investigate the pathogenesis of human disease, using experimental animal models and naturally occurring animal diseases. . ..
Comparative Medicine at National Cancer Institute (NCI): The Comparative Oncology Trials Consortium (COTC)
. . .to improve the assessment of novel treatments for humans by treating pet animals-primarily cats and dogs -with naturally occurring cancer, giving these animals the benefit of cutting-edge research and therapeutics. . ..
Comparative Medicine at Messerli Research Institute, Vienna
. . .The chances of improving the fulfillment of the 3Rs (Replacement, Reduction, and Refinement) in medical research are realistic with the use of systematic comparative studies between humans and animals and the increasing encouragement of clinical studies in the veterinary field as an important alternative and supplement to studies using laboratory animals. . ..

diseases' (Sabin 1980). The following conveyed statement from Meyer indicates that the definition of comparative pathology and experimental medicine was indeed overlapping: '*You know, one must always get these infections either human or animal, into small laboratory animals. Then we can study them, because it's too expensive to study them in larger animals*'. Meyer being called the Pasteur of the twentieth century shall be the last personality in this search for the roots of comparative medicine (Elberg et al. 1976).

Nevertheless, the (hi)story of comparative medicine goes on at the experimental medical institutes, where it is used as a basic principle for evaluating pathophysiological mechanisms and fostering drug development. Even though comparative medicine has been long regarded as a branch of experimental medicine or as a branch of veterinary medicine, it has developed its own identity. At least most of the prestigious universities have established institutes for comparative medicine. The interpretation of the topic, however, is diverse as can be seen from their mission statements (Table 1.2). The Comparative Medicine chair at the Vienna Messerli Research Institute is a joint venture between the Medical University, Vienna, and the University of Veterinary Medicine, Vienna. Its mission specifically encompasses the 3Rs by including animals as patients into scientific studies (Russell and Burch 1959), a strategy acknowledged by the European Centre for the Validation of Alternative Methods (Table 1.1) (FRAME 2005). Thereby, drug development will be simultaneously speeded-up for progress in human and animal medicine.

Acknowledgements The author is especially obliged to Kerstin Weich, M.A., M.Sc, Unit of Ethics and Human-Animal Studies, Messerli Research Institute Vienna, Austria, and to Mag.phil Daniela Haarmann, Veterinary University Vienna, for critically reading the manuscript. Furthermore, the project was supported by Austrian Science Fund (FWF) projects SFB F4606-B19 and P23398-B11 and by the Messerli Foundation. Thanks to crossip communications, Vienna, Austria, for support in preparation of the figures.

References

Ackerknecht EH (2001) A short history of medicine. JHU Press, Baltimore

Adams F (1849) The genuine works of Hippocrates (350 B.C.), vol 1. William Wood and Company, New York, pp 73–75, p 88, p 165–166

Adler R (2004) Medical firsts: from Hippocrates to the human genome. Wiley, Hoboken

Aristotle, translated by Farquharson ASL (2004) On the gait of animals. Kessinger Publishing, Whitefish

Asch E (2013) A brief History of the World Veterinary Association. http://wwwworldvetorg/node/9106

Aufderheide AC (2003) The scientific study of mummies Cambridge. Cambridge University Press, Cambridge, p 608

Bay N, Bay B (2010) Greek anatomist herophilus: the father of anatomy. Anat Cell Biol 43:280–283

Bazin H (2011) Vaccination: a history from Lady Montagu to genetic engineering. John Libbey Eurotext, Montrouge, p 98

Neue deutsche Biographie (1987) Locherer—Maltza(h)n, vol 15. Bayerische Akademie der Wissenschaften, Berlin, p 33

Boccaccio G, translated by Aldington R (1930) The Decameron. Dell Publishing Co (1980); Kessinger Publishing (2005), p 384

Boyle R (1666) Trials proposed to be made for the improvement of the experiment of transfusing blood out of one live animal to another. Philos Trans 22:385–388

Buell P, May T, Ramey D (2010) Greek and Chinese horse medicine: déjà vu all over again. Sudhoffs Arch 94:31–56

Cáceres SB (2011) The long journey of cattle plague. Can Vet J 52:1140

Carter K (1981) Semmelweis and his predecessors. Med Hist 25:57–72

Cookshank EM (1889) Human small pox as a source of vaccine lymph. In: Lewis H (ed) History and pathology of vaccination. A critical inquiry, vol 2. HK Lewis, London, p 513

Cressey D (2011) Animal research: Battle scars Nearly one-quarter of biologists say they have been affected by animal activists. A Nature poll looks at the impact. Nature 470:452–453

Darwin E (1796) Zoonomia or the laws of organic life (In three parts), 4th edn, vol 1. Edward Earle (1818), US, p 213, p 463

Dharampal (2000) Chapters: operation of inoculation of the smallpox as performed in Bengall (Coult R, 1731): an account of the manner of inoculating for the smallpox in the East Indies (Holwell JZ, 1767) In: collected writings. Indian Science and Technology in the eighteenth century, vol 1. Other India Press Mapusa, Goa, India, p 149–179

Dhom G (2001) Geschichte der Histopathologie. Springer, New York (01.03.1982), p 500–502

Directive 2001/83/EC of the European Parliament and of the Council of 6 November 2001 on the Community code relating to medicinal products for human use (Consolidated version: 20/01/2011)

Directive 2001/82/EC of the European Parliament and of the council of 6 November 2001 on the community code relating to veterinary medical products. Official Journal L—311, 28/11/2004, p 1–66; as amended by Directive 2004/28/EC of the European Parliament and the Council of the 31 March 2004 amending Directive 2001/82/EC on the Community code relating to veterinary medical products. Official Journal L—136, 30/04/2004, p 58–84

Editorial. (2010) Reduce, Refine, Replace Nat Immunol 11:971
Elberg S, Schachter J, Foster L, Steele J (1976) Medical Research and Public Health, with Recollections. An Interview with Karl F Meyer, conducted by E T Daniel, Typoscript, 1961 and 1962: The Regents of the University of California, p 439
Verso World History Series (2010) Febvre L, Martin H. The coming of the book: the impact of printing, 1450–1800, 3rd edn. p 45
Festing S, Wilkinson R (2007) The ethics of animal research. Talking Point on the use of animals in scientific research. EMBO Rep 8:526–530
Fleming G (1871) Animal plagues: their history, nature, and prevention. London: Chapman and Hall
Fooks A, Johnson N, Rupprecht C (2009) Rabies. In: Barrett ADT, Stanberry LR (eds) Vaccines for biodefense and emerging and neglected diseases, 1st edn. Academic Press Elsevier, Oxford, p 611
FRAME (2005) Human microdosing reduces the number of animals required for pre-clinical pharmaceutical research. Altern Lab Anim 33:439
Galtier P (1879) Études sur la rage. Ann Med Vet 28:627–639
Galtier PV (1881) Les injections de virus rabique dans le torrent circulatoire ne provoquent pas l'éclosion de la rage et semblent conférer l'immunité. La rage peut être transmise par l'ingestion de la matière rabique. In: Comptes rendus de l'Académie des science, vol 93
Geison GL (1995) The private science of Louis Pasteur. Princeton University Press, Princeton
Giese M (2007) Ut canes pulcherrimos habeas…, die kynologische Hauptvorlage von Albertus Magnus De animalibus. In Kulturtransfer und Hofgesellschaft im Mittelalter: Wissenskultur am sizilianischen und kastilischen Hof im 13. Jahrhundert (Eds. Griebner G, Fried J). Oldenbourg Akademieverlag, p 239
Gordon I, Paoloni M, Mazcko C, Khanna C (2009) The comparative oncology trials consortium: using spontaneously occurring cancers in dogs to inform the cancer drug development pathway. PLoS Med 6(10):e1000161
Goret P (1969) An anniversary: the life and work of Pierre-Victor Galtier (1846–1908). Professor at the Ecole Vétérinaire de Lyon. Bull Acad Natl Med 153(3):75–77
Grognier LF (1805) Notice historique et raisonnée sur C Bourgelat. Huzard, Paris
Haensch S, Bianucci R, Signoli M, Rajerison M, Schultz M, Kacki S, Vermunt M, Weston D, Hurst D, Achtman M, Carniel E, Bramanti B (2010) Distinct Clones of Yersinia pestis Caused the Black Death. PLoS Pathog 6(10):e1001134
Hayne A (1836) Die Seuchen der nutzbaren Haussäugethiere, in Bezug ihrer Erkenntniss, Ursachen, Behandlung, Vorbauung durch therapeutische und veterinär-polizeyliche Mittel und Vergleichung mit den Krankheiten der Menschen. Leopold Grund, Vienna
Hayne A (1844) Handbuch über die besondere Krankheits- Erkenntnis- und Heilungslehre der sporadischen und seuchenartigen Krankheiten der nutzbaren Haustiere. Hayne, Vienna
http://ocp.hul.harvard.edu/contagion/contributors.html (2013) Contagion. Historic views on diseases and epidemics. Harvard University Library, Harvard
Isensee E (1845) Psychiatrie, Veterinärmedizin, Staatsarzneikunde, Medizinische Geographie und Statistik, Generalregister. In: Die Geschichte der Medicin und ihrer Hülfswissenschaften: Neuere & neueste Geschichte, vol 2. Liebmann, Berlin (Oct 7, 2011), p 1330
Jenner E (1798) Smallpox vaccine: an inquiry into the causes and effects of the variolæ vaccinæ: printed for the author by Sampson Low
Jenner E, foreword Drewitt FP (1931) The note book of Edward Jenner. In the possession of the Royal College of Physicians of London. Oxford University Press, Oxford
Klaassen Z, Chen J, Dixit V, Tubbs R, Shoja M, Loukas M (2011) Maria Lancisi (1654–1720): anatomist and papal physician. Clin Anat 24(7):802–806
Knight H, Hunter M (2007) Robert Boyle's memoirs for the natural history of human blood (1684): print, manuscript and the impact of Baconianism in seventeenth-century medical science. Medical History 51:145–165
Koch R (1903) Transmission of bovine tuberculosis to man. J Tuberc 5:41–55

Kolmos HJ (2006) Panum's studies on "putrid poison" 1856. An early description of endotoxin. Dan Med Bull 53(4):450–452
Lindner KE (1962) Von Falken, Hunden und Pferden. Deutsche Albertus-Magnus-Übersetzung aus der 1. Hälfte des 15. Jahrhunderts. Original title: Liber de animalibus. In: Quellen und Studien zur Geschichte der Jagd. Berlin: de Gruyter (vol 7, 8)
Lombard M, Pastoret P, Moulin A (2007) A brief history of vaccines and vaccination. Rev Sci Tech Off Int Epiz 26:29–48
Lyons AS, Petrucelli RJ (1978) Medicine: An illustrated history. Harry N. Abrams Inc., New York, p 399
Magendie FE (1822) Memoires physiologiques sur les maladies purulente et putrides, sur la vaceine. Paris. Journal de Physiologie Experimentale et pathologique (2):1–45
Mammerickx M (1971) Claude Bourgelat: avocat des vétérinaires. Bruxelles
Mazzeo M (1955) Sanitary assistance inspired by Christianity. III. Crusades; great epidemics (leprosy, plague, ergotism); hospital religious orders. Riv Stor Sci Mediche Nat 46(1):7–38
McClellan JI, Dorn H (2006) Science and technology in world history: an introduction, vol 1, 1st edn. JHU Press, London, pp 5–23
Modanlo H (2008) A Tribute to Zakariya Razi (865–925 AD), An Iranian Pioneer Scholar. Arch Iranian Med 11:673–677
Monath T, Kahn L, Kaplan B (2010) Introduction: one health perspective. ILAR J 51(3):193–198
Moore J (1815) The history of the small pox. Longman, Hurst, Rees, Orme and Brown, London, 219
Mösle's Widow Braumüller (1839) Picture of Vienna Containing a Historical Sketch of the Metropolis of Austria, a Complete Notice of All the Public Institutions (etc.) and a Short Description of the Most Picturesque Spots in the Vicinity with a Map of the Town and Suburbs. Mösle & Braumüller, Vienna
Nature (2009) Nature milestones in light microscopy. http://www.nature.com/milestones/milelight/index.html
Parliament E (2010) Directive 2010/63/EU of the European Parliament and of the council of 22 September 2010 on the protection of animals used for scientific purposes
Pasteur L, Chamberland R (2002) Classics of biology and medicine: summary report of the experiments conducted at Pouilly-le-Fort, near Melun, on the anthrax vaccination 1881. Yale J Biol Med 75(1):59–62
Pasteur L, interviewed by Illo J (1996) Pasteur and rabies: an interview of 1882. Med Hist 40 (3):373–377
Pedersen H (2002) Humane Education Animals and Alternatives in Laboratory Classes. Aspects, Attitudes, and Implications. In: Humanimal 4. Akademitryck AB, Edsbruk. Stockholm Stiftelsen Forskning utan djurförsök
Prévot B (1991) La Science du cheval au Moyen Age: Le Traité d'hippiatrie de Jordanus Rufus. Series: Collection Sapience, Klincksieck, Paris
Prioreschi P (2001) Determinants of the revival of dissection of the human body in the Middle Ages. Med Hypotheses 56(2):229–234
Raw N (1906) Human and bovine tuberculosis: the danger of infected milk. Br Med J 2 (2381):357–358
Riedel S (2005) Edward Jenner and the history of smallpox and vaccination. Proc (Bayl Univ Med Cent) 18(1):21–25
Riviere E, Papich ME (2009) Veterinary pharmacology and therapeutics: an introduction into the discipline, 9th edn. Wiley-Blackwell, Ames, IA, p 5
Russell WMS, Burch RL (1959) The sources, incidence, and removal of inhumanity. Methuen, London
Sabin AD (1980) Karl-Friedrich Meyer. Biogr Mem 52:269–332
Sanderson JB (1872) Criticisms of Dr. Chauveau of Lyons on the discussion at the pathological society on pyaemia. Br Med J 2(617):459–460
Schwabe C (1968) Animal diseases and world health. J Am Vet Med Assoc 153:1859–1863

Schwabe C (1978) Cattle, priests, and progress in medicine. University of Minnesota Press, Minneapolis
Somvanshi R (2006) Veterinary medicine and animal keeping in ancient India. Asian Agrihist 10:133–146
Spinage CA (2003) Cattle plague: a history. Publisher Kluwer Academic/Plenu, US
Stahnisch F (2003) Der Funktionsbegriff und seine methodologische Rolle im Forschungsprogramm des Experimentalphysiologen François Magendie (1783–1855) Inaugural-Dissertation zur Erlangung der medizinischen Doktorwürde des Fachbereichs Humanmedizin. Münster Hamburg London Universität Berlin
Stephens M (2011) Animal research: replacing the lab rat. Nat Immunol 471:449
Swiderski RM (2004) Anthrax: a history. McFarland, Jefferson, NC
Talbott J (1970) A biographical history of medicine: excerpts and essays on the men and their work. Grune & Stratton, New York, p 1211
Temple R, Needham JF (1986) The genius of China: 3,000 years of science, discovery, and invention. Simon and Schuster, New York, pp 135–136
Vegetius R, Gargilius M (1871) Digestorum Artis Mulomedicinae libri. (Ed Lommatzsch, E; 1871). Teubneri BG 1903, Leipzig, p 342
Virchow R (1881) An address on the value of pathological experiments. Br Med J 2 (1075):198–203
von den Driesch A (2003) Geschichte der Tiermedizin: 5000 Jahre Tierheilkunde: Schattauer Verlag
Warner JH (1996) Review on "The private Science of Louis Pasteur". Bull Hist Med 70:718–720
Wickkiser B (2008) Asklepios, medicine, and the politics of healing in fifth-century Greece. The John Hopkins University Press, Baltimore
Wilkenson L (1992) Animals & disease. An introduction to the history of comparative medicine. Cambridge University Press, Cambridge
Wilkinson L (1984) Rinderpest and mainstream infectious disease concepts in the eighteenth century. Med Hist 28(2):129–150
Winkle W (2005) Geißeln der Menschheit: Die Kulturgeschichte der Seuchen. Artemis & Winkler, 3rd edn. p 22
Woodward J (1714) Account of the procuring of the small pox by incision, or inoculation, as it has for some time been practiced in Constantinople. Being an extract of a Letter from Emanuel Timonius, Constantinople, December 1713. In: Hutton C, Shaw G, Pearson R (eds) The Philosophical Transactions of the Royal Society of London, 1713–1723 edn, vol 6
Wyklicky H (1985) History of the institute for general and experimental pathology of the University in Vienna. Wien Klin Wochenschr 97(8):346–349

Smallest Unit of Life: Cell Biology

2

Isabella Ellinger and Adolf Ellinger

Contents

Abstract

The **cell** is the smallest structural and functional unit of living organisms, which can exist on its own. Therefore, it is sometimes called the building block of life. Some organisms, such as bacteria or yeast, are unicellular—consisting only of a single cell—while others, for instance, mammalians, are multicellular.

I. Ellinger (✉)
Department of Pathophysiology and Allergy Research, Center for Pathophysiology, Infectiology and Immunology, Medical University Vienna, Vienna, Austria
e-mail: isabella.ellinger@meduniwien.ac.at

A. Ellinger
Department of Cell Biology and Ultrastructure Research, Center for Anatomy and Cell Biology, Medical University Vienna, Vienna, Austria
e-mail: adolf.ellinger@meduniwien.ac.at

E. Jensen-Jarolim (ed.), *Comparative Medicine*,
DOI 10.1007/978-3-7091-1559-6_2, © Springer-Verlag Wien 2014

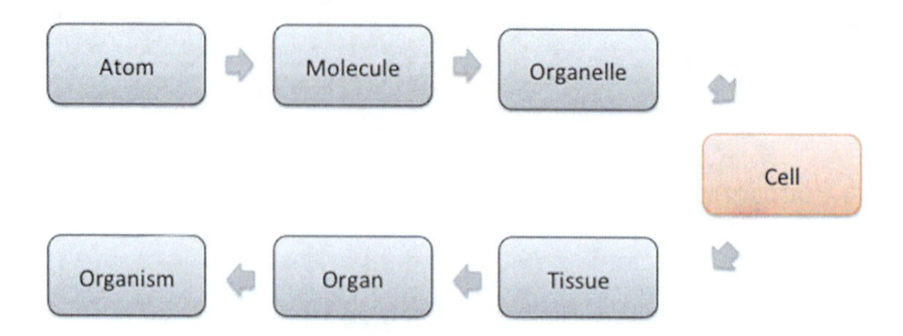

Fig. 2.1 The different levels of organization in multicellular organisms. The cell is highlighted in color and represents the smallest living biological structure

The human body is built from an estimated 100 trillion or 10^{14} cells. Such complex living systems have developed several levels of organization depending on each other, for example, organs, tissues, cells, and subcellular structures (Fig. 2.1). For the understanding of these biological systems, small units must be investigated at a time. The logical starting point for the examinations is the cell, since at the cellular level, all life is remarkably similar.

2.1 Introduction

The term **cell** comes from the Latin word ***cellula***, meaning a small room. This descriptive name for the smallest living biological structure was chosen by *Robert Hooke* in 1665 when he compared the cork cells he saw through his simple microscope to the small rooms monks lived in.

The **cell theory** developed in the middle of the nineteenth century by *Theodor Schwann*, *Matthias Jakob Schleiden*, and *Rudolf Virchow* states that all organisms are composed of at least one cell and all cells originate from preexisting ones (*Omnis cellula e cellula*). Vital functions of an organism take place within cells, and all cells contain the hereditary information necessary for regulating cell functions and for transmitting information to the next generation of cells.

Modern research in cell biology is based on integration of originally distinct research areas. It combines cytology, the initial way to study cells using morphological techniques with biochemistry and genetics/molecular biology to reveal the principal mechanisms of cells.

2.2 Cell Architecture

Generally, two kinds of cells are discerned, **eukaryotes** and **prokaryotes**, which developed from a common ancestor (Fig. 2.2). Most prokaryotes are unicellular organisms; they are classified into two large domains, Bacteria and Archaea. Eukaryotic organisms may be single cell or multicellular organisms. The four

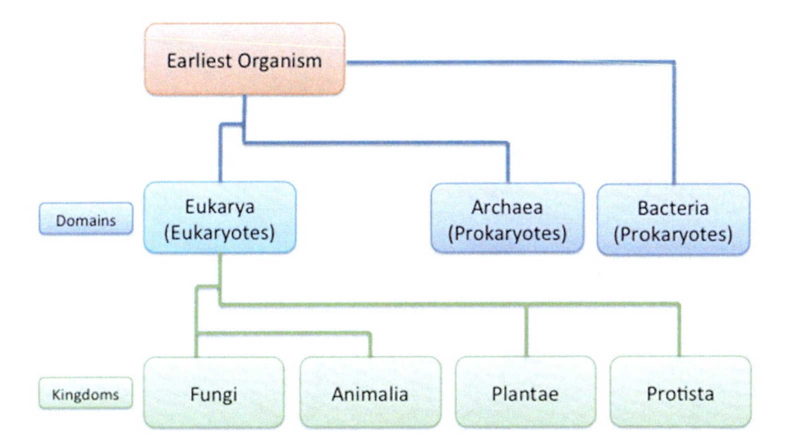

Fig. 2.2 The classification levels of organisms indicates that all eukaryotic organisms are found in the domain *Eukarya*, while in the other domains only prokaryotic cells exist. Humans as well as all other mammalian species belong to the kingdom *Animalia* and are multicellular organisms built from eukaryotic cells

kingdoms of eukaryotic organisms are Fungi, Plantae, Protista, and Animalia. To the latter group, all mammalian species belong, including humans.

Schematic depictions of prokaryotic and eukaryotic cells are indicated in Fig. 2.3. One principal distinction is that the genetic material of eukaryotes is contained within a nucleus. Furthermore, the genetic material is less structured in prokaryotes than in eukaryotes, where it is organized in chromosomes. Prokaryotes are usually much smaller than eukaryotic cells. This results in a higher surface-to-volume ratio, which enables a higher metabolic and growth rate and thereby a shorter generation cycle. Eukaryotes, in contrast, have various membrane-bound functional units, termed organelles. These subcellular structures help to compartmentalize the cells and to provide optimal conditions for various metabolic reactions. As a result, many distinct types of reactions can occur simultaneously in eukaryotes, thereby increasing cell efficiency. The majority of organelles are found in all types of eukaryotic cells, however, with cell-specific characteristics. Plant cells are characterized by additional organelles such as chloroplasts (responsible for photosynthesis) and vacuoles (fluid-filled organelles maintaining, e.g., the cell shape, and serving as dynamic waste baskets).

2.3 Eukaryotic Cell Differentiation, Structure and Size

Although, the principal structural units of all eukaryotic cells are similar (Fig. 2.3), more than 200 different cell types build up the adult human body. Their cell shapes vary considerably. Usually, the shape is typical for concrete cell types and represents the manifestation of the cell-type-specific function and state of differentiation. For example, the distinctive biconcave shape of red blood cells optimizes

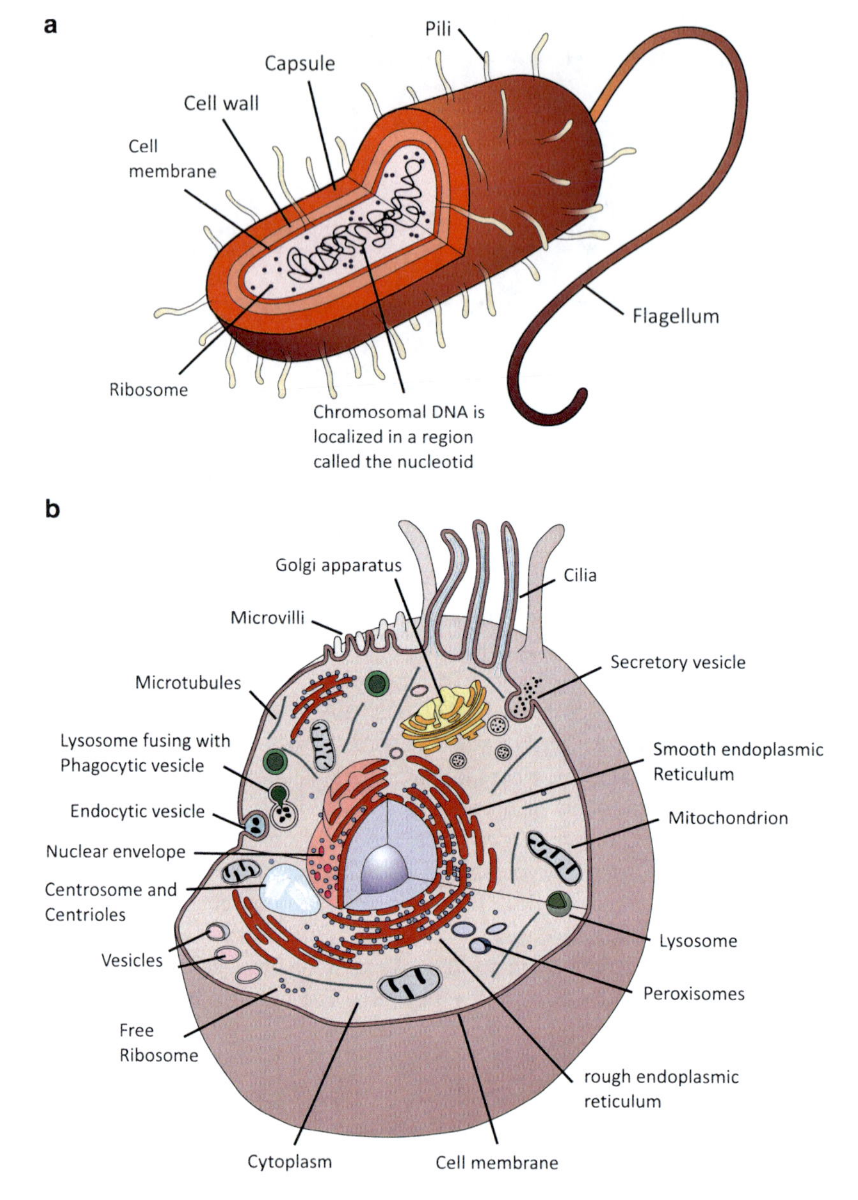

Fig. 2.3 Principal structures of (**a**) prokaryotic and (**b**) eukaryotic cells

their flow in large blood vessels, their remarkable flexibility helps them to squeeze even through tiny capillaries, and their surface-to-volume ratio is optimized for CO_2/O_2 exchange (Fig. 2.4). Epithelial cells line all inner and outer surfaces and cavities of the body insuring contact with and at the same time forming barriers against the environment. As a consequence, they are densely packed, with only

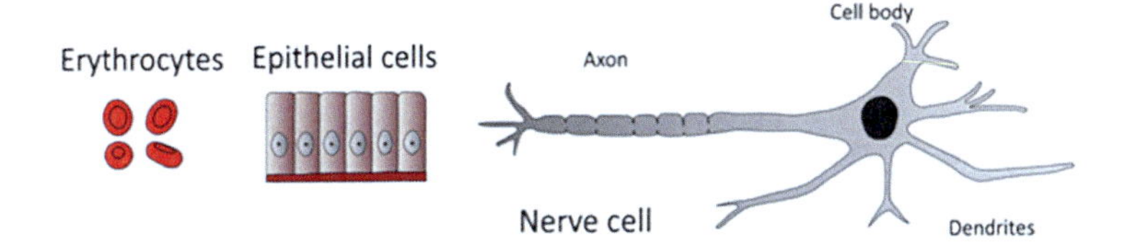

Fig. 2.4 The illustration shows three different phenotypes out of the 230 specialized human cell types; their function is reflected by their shape

narrow space between them. Entirely different, nerve cells, conducting impulses over long distances, exhibit multiple afferent processes (dendrites) and one efferent process (axon, neurite). Axons innervating muscle fibers in the limbs might reach lengths of 1 m in humans. A cell-type-specific shape, however, is not static, but changes depending on the stage of differentiation, the functional state, and signals obtained from the environment.

The multiple cell types in complex organisms such as humans are specialized members of a multicellular community. For the sake of this community, most cells have lost features that allow for their independent survival, but instead became experts for certain activities as a result of **cell differentiation**. Cell differentiation starts early in embryogenic development. In principle, all cells derived from the zygote have an identical genome (genotype). However, different genes can be activated in daughter cells, which results in the expression of cell-type-specific protein subsets and in cell-specific functions and shapes (phenotypes).

In humans, only the totipotent **stem cells** at the morula stage (see chapter on reproduction) have the potential to differentiate into any cell type of the organism (Fig. 2.5). Per definition, a stem cell is not finally differentiated and has the capability of unlimited self-renewal (as part of an organism), and upon cleavage, their daughter cells may either remain a stem cell or differentiate (asymmetric cleavage). Even in adults, multipotent stem cells exist (adult or somatic stem cells) and contribute to tissue homeostasis and repair. Major populations are the hematopoietic stem cells, which form all types of blood cells in the body; the mesenchymal stem cells, which, e.g., produce bone, cartilage, fat, and fibrous connective tissue; or the neural stem cells, generating the main phenotypes of the nervous system. The maintenance of the stem cell features relies, however, on their interaction with specific microenvironments called "stem cell niches." These niches regulate the division of the stem cells, ensuring on one hand their survival and protecting on the other hand the organism from exaggerated stem cell proliferation. Among the regulating niche factors are cell-cell interactions, cell-matrix interaction, oxygen tension, and absence or presence of certain metabolites.

The specialized, differentiated adult body cells were thought for long to represent the end point of the differentiation pathway. Research aimed to replace lost or damaged tissue by differentiation of embryonic or adult stem cells. Recent year's research now suggests that "reprogramming" of certain differentiated cell types into others could also be possible by a well-controlled process of genetic modification. This strategy may offer an additional possibility for tissue replacement.

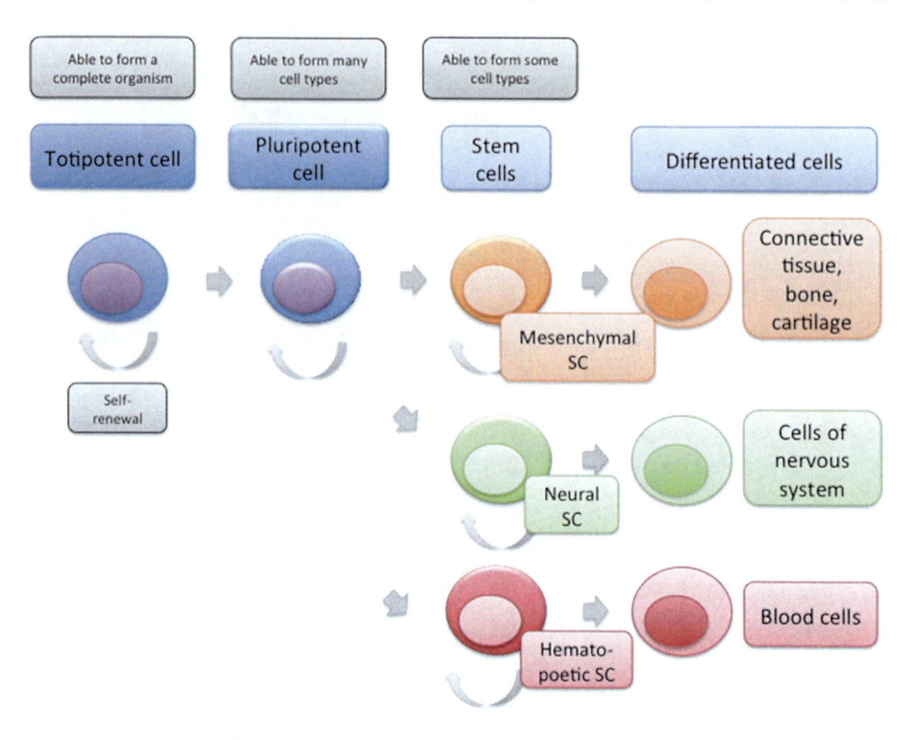

Fig. 2.5 Illustration of the steps of cell differentiation starting from totipotent eukaryotic cells, i.e., cells in the morula. Stem cells have the potential for self-renewal as well as differentiation, which is controlled by their microenvironments (stem cell niches)

Sizes and diameters of most eukaryotic cell types range from 5 to 20 μm (e.g., erythrocytes 7.5 μm, granulocytes 10–15 μm) up to 150–300 μm, which is the size of the oocytes in the female ovaries. Hepatocytes, an example of medium-sized cells, have a diameter of 20–40 μm. The relation of eukaryotic cells to other living organisms, structures, and smaller molecules is indicated in Fig. 2.6. Due to their small size, cells and their subcellular structures are invisible to the naked eye; their visual analysis requires the use of different types of microscopical techniques.

2.4 Important Molecules of Life

Besides all differences in function and form, all cells are built according to a common concept using similar molecules and metabolic processes. Usually, water, ions, and small organic molecules make up 75–80 % of the cell mass. More complex polymers (Fig. 2.7) serve for specific purposes. **Deoxyribonucleic acid** (DNA) a polymer of four nucleotides (deoxyadenosine monophosphate, deoxyguanosine monophosphate, deoxycytidine monophosphate, and deoxythymidine monophosphate) is used to

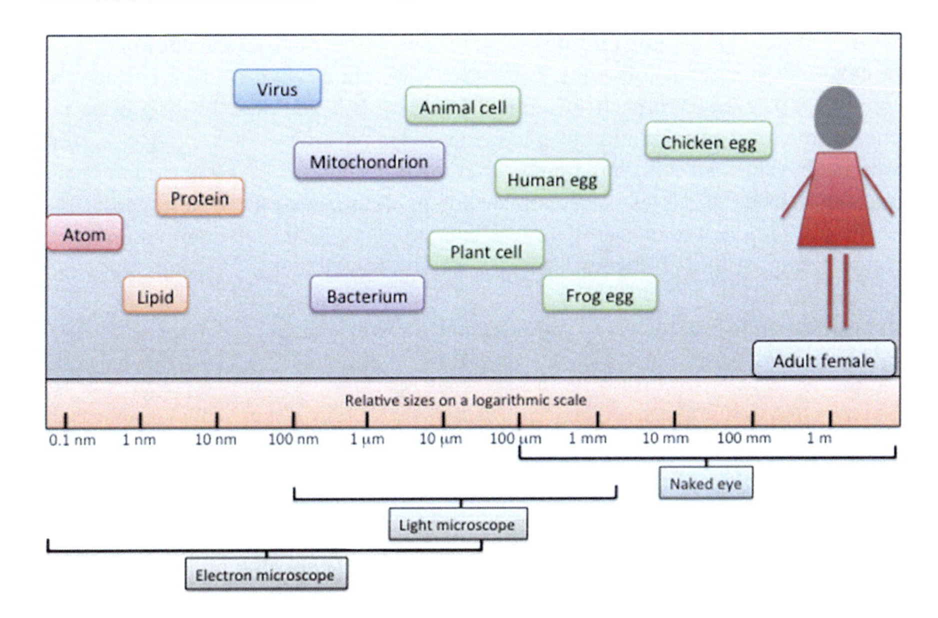

Fig. 2.6 Comparison of the sizes of eukaryotic cells (*green*) with subcellular elements, prokaryotic cells, and multicellular organisms

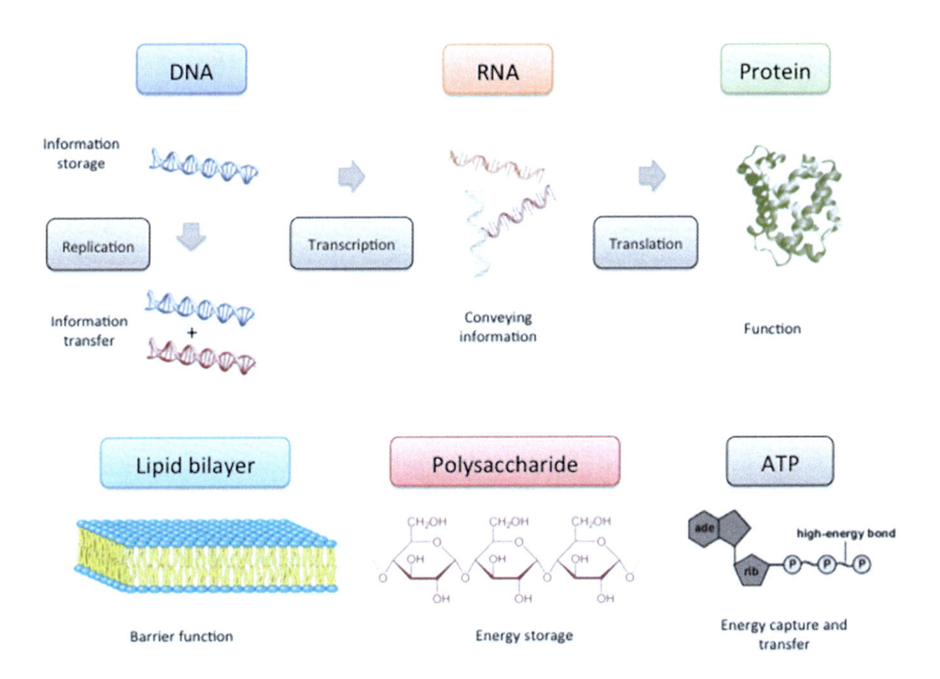

Fig. 2.7 Some major ubiquitously used cellular molecules. *DNA* deoxyribonucleic acid, *RNA* ribonucleic acid, *ATP* adenosine triphosphate, *ade* purine base adenine, *rib* pentose sugar ribose, *P* phosphate group

encode and store the genetic information in the nucleus and to pass the information to the next cell generation following *replication* and mitotic division (see chapter on reproduction). Various forms of **ribonucleic acids** (RNA) are used for the process of gene expression and regulation of gene expression. Messenger RNA (mRNA), for example, is a single-strand copy from a DNA made during *transcription*. After export from the nucleus, it serves as a template for **protein synthesis** from amino acids during *translation*. Proteins execute the myriad of cellular functions required to ensure living. **Lipid bilayers** separate cell compartments. Various types of **polysaccharides** (e.g., glycogen) are used to store energy, and molecules like **adenosine triphosphate** (ATP) are common tools to capture and transfer energy via high-energy bonds.

2.5 Cell Organelles in Animal Cells

Eukaryotic cells are divided into **nucleus** and **cytoplasm**, surrounded by the **plasma membrane**; the cytoplasm is further subdivided into the **cytosol**, the intracellular fluid, and **organelles** (see Fig. 2.3b). Analogous to organs, which are discrete functional units within an organism, **organelles** are compartments with specialized tasks within a eukaryotic cell. Most organelles are delimited by enclosing membranes, the nucleus being a prominent example. However, when organelles are defined by their tasks and specific functions, structures without the delimiting membrane, such as the ribosomes, the centrosome, or the cytoskeleton, may also be defined as organelles. The uniqueness of organelles is, at least partially, defined by their specific protein composition. Most cellular proteins are found only in one or a few compartments. The signals for this specific localization are defined by the amino acid sequence of the protein.

2.5.1 The Nucleus

The nucleus is the most obvious organelle in eukaryotic cells. It is enclosed by a double membrane (nuclear envelope) and communicates with the cytosol via numerous nuclear pore complexes. Within the nucleus the information to build all cellular components is stored. The memory medium is DNA (Fig. 2.7), which in humans has a total length of 2.3 m. The DNA consists of two complementary strains, which are known as the double helix. While for information storage only the coding strain is required, this helix structure increases stability. Within the DNA molecule, triplets of four bases (adenine, guanine, cytosine, thymine) code for each single amino acid. Twenty different amino acids are used to build up proteins.

The separation of the DNA from other subcellular structures became necessary when eukaryotes reached a higher level of differentiation and therefore increased their DNA content by up to 100-fold over prokaryotic cells. To prevent a tangle of the long DNA molecules, the DNA was not only separated within the nucleus but also split into smaller units (chromosomes) and highly compacted using, e.g., specific proteins such as histones.

The major processes which take place in the nucleus are *replication* of DNA, *transcription* of DNA sequences into mRNA molecules, and the *processing of mRNA* molecules, which are exported to the cytoplasm (see Fig. 2.7). *Translation* of mRNA into proteins is done outside the nucleus on specific organelles (ribosomes), which are preformed at distinct regions in the nucleus, the nucleoli. Ribosomal subunits are exported through nuclear pore complexes into the cytoplasm.

Following translation, proteins are imported into the various cell organelles by three main mechanisms: (1) Proteins moving from the cytosol into the nucleus are transported through **nuclear pores**. Pores function as selective gates, which actively transport macromolecules but also allow free diffusion of smaller molecules. (2) Proteins moving from the cytosol into the endoplasmic reticulum (ER), mitochondria, or peroxisomes are transported across organelle membranes by **protein translocators (chaperones)**. Finally, (3) proteins moving from one compartment to the next along the biosynthetic or endocytic pathway are transported via membrane-enclosed **transport vesicles**.

2.5.2 Membranes

All cells are surrounded by the **plasma membrane**, which on one hand serves as a boundary separating and protecting cells and on the other hand provides communication and exchange with the environment. The framework of the membranes is made up of **phospholipids**, amphipathic molecules that exhibit a hydrophilic head and a hydrophobic tail region, in which proteins/glycoproteins are distributed (Fig. 2.8). Based on the physicochemical properties of the membranes, the **fluid mosaic model** has been defined, emphasizing stability and high mobility at once; lipids and to a less extent proteins are highly mobile within the plane of the membrane, a basal feature for many cell functions.

The plasma membrane encloses the cell body. Endomembranes in the interior divide specific metabolic compartments (**organelles**) surrounded by the cytosol, a complex mixture of molecules in water. Membranes are composed of **lipids** (**phospholipids**, **cholesterol**, **glycolipids**) and **proteins** in a rough proportion of 2:1. However, there are great variations with extremes exemplified by mitochondrial cristae membranes (high protein content due to the enzymes of the respiratory chain) or nerve myelin sheets (high lipid content). The fluidity of membranes is defined by the length of the fatty acid chains, the amount of unsaturated bindings, and the amount of cholesterol molecules. Cholesterol stabilizes the membranes and reduces the fluidity.

Embedded proteins act as channels, protein pumps that move different molecules in and out of cells, enzymes, or linker proteins. The membrane is said to be **semipermeable**, in that it can either let a substance (molecules or ions) pass

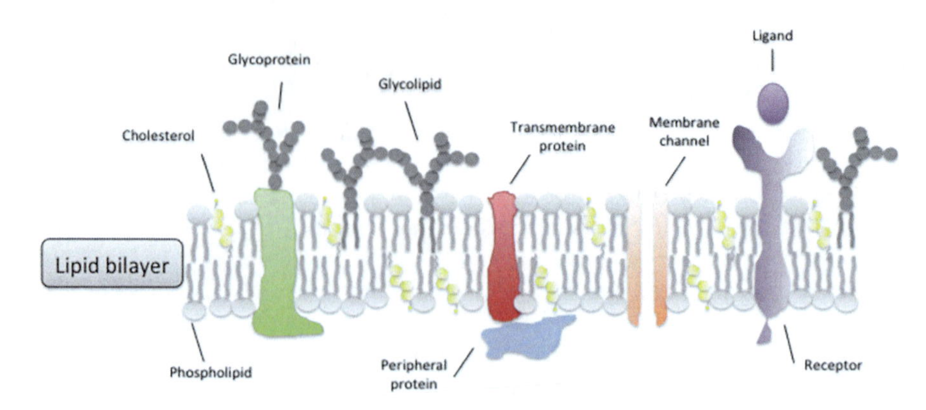

Fig. 2.8 Principal components of the cell membrane

through freely, pass through to a limited extent, or not pass through at all. Cell surface membranes also contain **receptor proteins** that allow cells to detect external signaling molecules such as hormones as well as nutrients (ligands).

Glycolipids mark the outer leaflet of the plasma membrane (directed to the extracellular space) and their oligosaccharide chains protrude from the cell surface. Together with the sugar chains of the **glycoproteins**, they form the **surface coat** (*glycocalyx*) covering the surface of all cells. These sugar chains hold a variety of functions, which range from receptor activity to cell recognition; they are major determinants of the blood group systems and responsible for surface protection of the intestinal or urinary tract.

2.5.3 Endocytosis and Endosomes

Endocytosis is the general term for the uptake of external materials into cells via formation of membrane pits and vesicles at the plasma membrane.

Uptake of particular substances (microorganisms, cells, cell fragments) is termed **phagocytosis** (cell eating) and is done by specific types of cells (**macrophages**, **granulocytes**). Phagocytosis plays a major role in the immune system, tissue remodeling, and cell renovation.

In contrast, all cells form endocytic vesicles with 50–150 nm in diameter (cell drinking). They either contain fluid with dissolved molecules, resulting in **fluid-phase endocytosis** or specific molecules, which are bound and taken up via receptors. The latter process is termed **receptor-mediated endocytosis**. In the first case molecules are taken up nonspecifically, according to their concentration in the extracellular fluid. On the contrary, receptor-mediated endocytosis is highly specific and regulated and enables the enrichment of molecules. In the course of vesicle formation, protein complexes are formed on the cytosolic face of the plasma membrane (coat formed by proteins such as **clathrin**, **coatomer proteins**, or **caveolin**), which enable the structural changes of the membrane during invagination, pinching off, and formation of vesicles. Adaptor protein complexes selectively link receptors and coat proteins. Many surface receptors have specific amino acid sequences in their cytoplasmic parts, which guide them into coated pits and allow efficient uptake of receptors and ligands. Resulting coated vesicles then rapidly shed their coat and fuse with early endosomes. Molecules taken up by receptor-mediated endocytosis are for instance nutrients such as low-density lipoproteins or iron-containing transferrin as well as aged molecules such as asialoglycoproteins.

Material taken up via endocytosis is further metabolized via **early endosomes**, **multivesicular bodies**, and **late endosomes**. The main function of endosomes is the **sorting** of molecules to different destinations within the cell (lysosome, cell surface). Hereby, the downregulation of pH via proton pumps (acidification) within the endosomes plays an important role. A gradual acidification of the endosomal lumen induces the release of receptors and bound ligands. While most of the receptors recycle back to the plasma membrane for reuse, most ligands end up in lysosomes for degradation and/or utilization in anabolic processes.

2.5.4 Cellular Degradation, Proteasomes and Lysosomes

Two major proteolytic systems contribute to the continuous removal of intracellular components.

The **ubiquitin/proteasome system** plays a major role in the maintenance of cellular homeostasis and protein quality control and in the regulation of essential cellular processes. The proteasome is a large cytosolic protein complex with proteolytic activity for degrading of unneeded or damaged proteins that have been marked for degradation by ubiquitination.

Lysosomes are membrane-bound organelles specialized for intracellular digestion of macromolecules. They contain ~40 different acid hydrolyses that operate best at a low pH (pH 4.5–5.0) and carry out controlled digestion of cellular and extracellular materials. Their unique membrane proteins are unusually highly glycosylated; thus their sugar residues facing the lumen prevent the enzymes inside from destroying the cell. The lysosome's membrane stabilizes the low pH by pumping in protons (H^+) from the cytosol via ATP-driven proton pumps. The digestive enzymes and lysosomal membrane proteins are synthesized in the **endoplasmic reticulum** (**ER**) and transported through the **Golgi apparatus** and the ***trans*-Golgi network** (**TGN**). Along this route, lysosomal proteins, after being tagged with a specific phosphorylated sugar group (mannose-6-phosphate—Man-6-P) are recognized by an appropriate receptor. They are sorted out from the biosynthetic secretory pathway and packaged into transport vesicles, which deliver their content via late endosomes to lysosomes.

2.5.5 The Biosynthetic Pathway and Associated Organelles: Endoplasmic Reticulum and Golgi Apparatus

The biosynthetic pathway defines the route of molecules from the site of protein synthesis to the site of their final destination. Starting point is the *transcription* of one strand of DNA into a complementary mRNA which in a next step is decoded to produce a polypeptide chain specified by the genetic code (*translation*, see Fig. 2.7). Translation is done at the **ribosomes** in the cytoplasm; these ribonucleoprotein aggregates are composed of a large and a small subunit, which are assembled in the nucleolus and exported into the cytoplasm. The subunits form the ribosome upon arrival of mRNA strands in the cytoplasm and start translation. Ribosomes either remain in the cytosol (**free ribosomes**) for translation of cytosolic, nuclear, peroxisomal, and mitochondrial proteins or attach to the endoplasmic reticulum and become **membrane-bound ribosomes**, thereby forming the **rough or granular endoplasmic reticulum** (rER). Here, membrane, lysosomal, and secretory proteins are synthesized.

Usually, **protein biosynthesis** starts on free ribosomes in the cytosol. The exceptions are a few mitochondrial proteins that are synthesized on ribosomes within the mitochondria. The subsequent fate of proteins depends on their amino acid sequences, which may contain sorting signals that direct the proteins to the organelle in which they are required. Proteins that lack this signal remain permanently in the cytosol.

2.5.5.1 Endoplasmic Reticulum

The **endoplasmic reticulum** is a membrane system participating in synthesizing, packaging, and processing of various molecules. It is an anastomosing network (reticulum) of cisterns, vesicles, and tubules. Transfer vesicles bud from specialized regions and deliver their contents to the Golgi apparatus for further processing and packaging. In mature cells the ER occurs in two forms: the rough ER and the smooth ER, respectively.

The **rough endoplasmic reticulum** synthesizes proteins for sequestration from the cytosol, including secretory proteins such as enzymes (e.g., digestive enzymes in pancreas acinar cells) or extracellular matrix molecules, proteins for insertion into the plasma membrane, and lysosomal enzymes. All these molecules have the signal sequence in common that binds to a cytoplasmic signal recognition particle (SRP). The SRP-polyribosome complex again is recognized by an ER-docking protein. Via attaching of the ribosome to the ER membrane, the nascent polypeptide chains are translocated into the rER lumen. Signal peptides are then cleaved and nascent proteins undergo folding with the aid of ER-located molecular chaperones. The latter also assist in quality control in that they retain misfolded or unassembled protein complexes. If modifications are unsuccessful, proteins are degraded. Another important posttranslational modification in the ER is core (initial) glycosylation of the growing polypeptide chain.

The **smooth endoplasmic reticulum (sER)** lacks ribosomes and thus appears smooth in the electron microscope. The tubular-vesicular membrane system contains enzymes important in lipid metabolism, steroid hormone synthesis, gluconeogenesis, and detoxification. It further plays a key role in regulating cytosolic calcium concentration by sequestering excess calcium. It is abundant in liver cells (hepatocytes), where it participates in glucose metabolism and drug detoxification. Specialized sER is found in striated muscle cells and regulates muscle contraction via sequestering and release of calcium ions.

2.5.5.2 Golgi Apparatus

The **Golgi apparatus** (**GA**) is the center of membrane flow and vesicle trafficking among organelles and has a key role in the final steps of secretion. The organelle is composed of stacks of flattened cisterns, vesicles, and vacuoles indicating the dynamic properties of the organelle. The *cis* face is usually close to the ER and is the entry of newly synthesized molecules. At the opposite side, the *trans* face, the TGN is the sorting, concentration, and packaging station for the completed molecules. Central functions of the GA are polysaccharide synthesis (terminal glycosylation), modification (e.g., sulfation of glycosaminoglycans, phosphorylation of lysosomal enzymes), and sorting of secretory products.

2.5.6 Peroxisomes

Peroxisomes are small membrane-bound organelles that use molecular oxygen to oxidize organic molecules and produce hydrogen peroxide (H_2O_2). With the aid of H_2O_2 and specific enzymes (e.g., catalase), peroxisomes oxidize and thereby detoxify various toxic substances such as alcohol. Oxidation reactions are also used to break down fatty acids in a process known as β-oxidation. The end products are then used in anabolic reactions. About 20 metabolic disorders caused by genetic anomalies result in defects in peroxisome function; range and severity of symptoms vary greatly, including biogenesis disorders and multi- or single-enzyme disorders. Among the different cell types, peroxisomes exhibit a high diversity with respect to their enzyme content and, additionally, they can adapt their function to altered environmental conditions.

2.5.7 Mitochondria

Mitochondria are membrane-bound organelles that carry out oxidative phosphorylation and produce most of the ATP (Fig. 2.7) of eukaryotic cells. The energy is generated by phosphorylation of pyruvate (derived from carbohydrates) and fatty acids (derived from lipids) in a multistep process. Mitochondria are found ubiquitously in almost all cell types, before all in cells and regions with elevated energy demand (hepatocytes, cardiac muscle cells, and proximal tubule cells of kidneys). Structural components are the outer mitochondrial membrane, the inner mitochondrial membrane (mostly infoldings in the form of cristae), the matrix, and the intermembrane space. The matrix, besides water and solutes, contains a variety of enzymes engaged in citric acid cycle and lipid oxidation, matrix granules (Calcium regulation), and mitochondrial ribosomes. Intercalated in the inner membrane are the components of the electron transport chain (respiratory chain). Mitochondria have an own circular DNA. This autonomous DNA comprises 37 genes: 13 for encoding proteins, 22 for transfer RNAs, and 2 for ribosomal RNAs. With the exception of chloroplasts, mitochondria are the only organelles in eukaryotic cells apart from the nucleus, which contain DNA. An explanation for the origin of the DNA is given by the endosymbiosis theory, which states that mitochondria as well as chloroplasts evolved from endosymbiotic bacteria.

2.5.8 Cytoskeleton

The cytoskeleton is a mesh of protein filaments in the cytoplasm of eukaryotic cells responsible for cell shape, stability, and the capacity for intracellular transport and cell movement. There are three major filament classes, microtubules, microfilaments (actin filaments), and intermediate filaments, which have a large subset of associated proteins. These proteins are responsible for the regulation of processes that initiate the nucleation of new filaments, their assembly and disassembly, cross-linking, stabilization, attaching to the plasma membrane, or severing into fragments.

Microtubules (MTs) are composed of protein subunits (α/β-tubulin heterodimers) and are polarized structures with a rapidly growing end and a decay end. They extend from microtubule-organizing centers (MTOC) close to the nucleus into the cell periphery and undergo rapid changes in length through changes in the balance between assembly and disassembly of the subunits into MTs (dynamic instability). Intracellular vesicular transport is mediated along MTs via the **motor proteins kinesin** and **dynein**, which in an energy-dependent process walk along MTs carrying their cargo (vesicles and content) from one organelle to another. The mitotic spindle apparatus is also built from MTs, which separate the sister chromatids during cell division (see mitosis, chapter on reproduction). MTs also provide a meshwork for organelle deployment and for cell polarity in, e.g., epithelial cells. Finally, MTs are major components of cell structures such as centrioles, cilia, or flagella, which move cells or move liquid over the surface of cells.

Actin filaments (AFs) are the thinnest and most flexible cytoskeletal element (6 nm). The filaments (F-actin/filamentous actin) are formed by polymerization of G (globular)-actin monomers. In some cell types they form rather stable arrays (e.g., muscle cells in association with myosin), while in most cell types, actin filaments are dynamic and repeatedly dissociate and reassemble. In addition to regulation of assembly and disassembly of G- ⇔ F-actin and vice versa, actin-binding proteins arrange microfilaments into networks and bundles (stress fibers), cross-link and attach them to the plasma membrane, and thus help to determine the shape and adhesive properties of cells. AFs are contractile, via interaction with **myosin**, the only actin-associated motor protein family. In muscle cells, myosin forms thick filaments

(A band of sarcomere). In non-muscle cells, myosin exists in a soluble form which binds to actin by its globular head; the free tail regions attaches to the plasma membrane and other cellular components to move these structures.

Various special formations of AFs in non-muscle cells exist. These are accumulations under the plasma membrane, called terminal web, parallel strands in the core of microvilli, ribbons in the cytoplasm of the leading edge of pseudopods, as well as a belt around the equator of dividing cells.

Intermediate filaments (IFs) are the most heterogeneous family with many tissue-specific forms found in the cytoplasm of animal cells. Their common function is to provide mechanical strength and to maintain cell shape. IFs may belong to the families of **nuclear lamins** (nuclear scaffold), **cytokeratins** (epithelial cells), **desmins** (muscle cells), **vimentin** (mesenchyme-derived cells), **glial fibrillary acidic proteins** (glial cells), and **neurofilaments** (neurons). This cell-type specificity, together with the stability and longevity of the single proteins, makes them particularly useful in the immunohistochemical determination of neoplastic cells. In most cell types, IFs form a network around the nucleus and extend throughout the cytoplasm attaching at specific regions of the plasma membranes (mechanical cell junctions: desmosome, hemidesmosome). They are specifically abundant in cells exposed to mechanical stress like muscle cells or keratinocytes of the skin.

Acknowledgement We gratefully acknowledge the support of Thomas Nardelli in preparing the artwork.

Further Readings

Alberts B, Johnson A, Lewis J, Raff M, Roberts K, Walter P (2008) Molecular biology of the cell, 5th edn. Garland Press, New York

Alberts B, Bray D, Hopkins K, Johnson A, Lewis J, Raff M, Roberts K, Walter P (2009) Essential cell biology, 3rd edn. Garland Press, New York

Lodish H, Berk A, Kaiser CA, Krieger M, Bretscher A, Ploegh H, Amon A, Scott MP (2013) Molecular cell biology, 7th edn. W.H. Freeman and Company, New York

3 Supporting Apparatus of Vertebrates: Skeleton and Bones

Wolfgang Sipos, Ursula Föger-Samwald, and Peter Pietschmann

Contents

Abstract

In vertebrates the skeleton consists of bones, cartilage, ligaments, and tendons and serves two major purposes: it is a supporting structure and a metabolic organ. With regard to the first aspect, the skeleton enables locomotion and supports the shape of the body. Moreover, bones protect internal organs such as the brain, heart, lung, and bone marrow. The bone marrow is an important site for the production of blood and immune cells. With regard to the metabolic function of the skeleton, the bone serves as an important storage organ for specific minerals. Very recent data also indicate that the bone is involved in the regulation of carbohydrate metabolism.

W. Sipos (✉)
Department of Farm Animals and Herd Management, University of Veterinary Medicine, Vienna, Austria
e-mail: Wolfgang.Sipos@vetmeduni.ac.at

U. Föger-Samwald • P. Pietschmann
Department of Pathophysiology and Allergy Research, Center for Pathophysiology, Infectiology and Immunology, Medical University Vienna, Vienna, Austria
e-mail: Ursula.Foeger-Samwald@meduniwien.ac.at; Peter.Pietschmann@meduniwien.ac.at

E. Jensen-Jarolim (ed.), *Comparative Medicine*,
DOI 10.1007/978-3-7091-1559-6_3, © Springer-Verlag Wien 2014

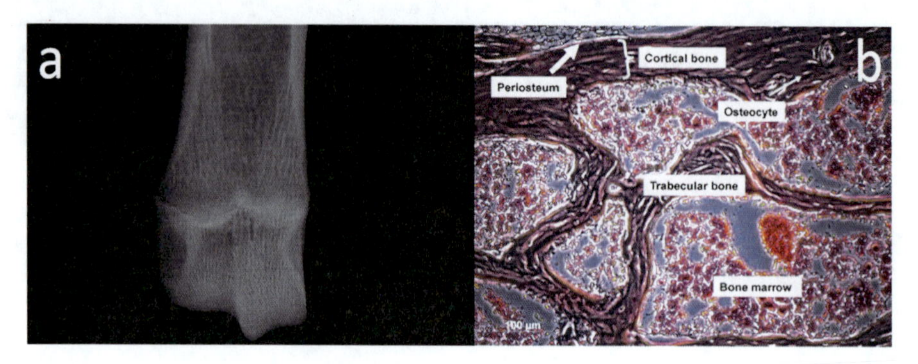

Fig. 3.1 *Micro-computed tomography* of the proximal part of the fourth metatarsal bone of *Sus scrofa* (Viscom microCT X8060II, resolution 18 μm; (**a**)). Cortical and trabecular bone can be clearly distinguished. (**b**) Hematoxylin-eosin stain of a murine vertebral body

3.1 Morphology and Chemistry of the Bone

Generally five types of bone are distinguished:

- Flat bones, such as the sternum, the skull, and the scapula.
- Long bones, e.g., femur, tibia, humerus, and finger bones.
- Short bones, e.g., wrist bones.
- Sesamoid bones: they are incorporated into tendons (such as the patella).
- Irregular bones, such as the vertebrae.

On a macroscopic level, cortical (compact) and trabecular (cancellous) bone can be distinguished. Seventy five percent of the skeleton is composed of cortical bone; 25 % of bone weight consists of trabecular bone. Generally, the outer parts of the bones are composed of cortical bone, whereas—at least in many locations—the inner part consists of trabecular bone (Fig. 3.1). Trabecular bone closely interacts with the bone marrow, and the metabolic activity of trabecular bone is much higher than that of cortical bone. The distribution of cortical versus trabecular bone varies according to the anatomical site. For instance, the shaft of long bones is almost exclusively composed of cortical bone, whereas vertebral bodies predominately consist of trabecular bone. The outer surface of the bone is covered by the periosteum, whereas the endosteum is located at the inner surface.

The bone is a very specialized form of connective tissue and is composed of a mineral phase and of an organic phase (bone matrix, bone cells). The mineral phase is predominately composed of hydroxyapatite $[Ca_{10}(PO_4)_6(OH)_2]$. In addition to calcium and phosphate, the bone also contains other anorganic components, such as carbonate, magnesium, and strontium. From a metabolic point of view, it is very important to consider that 99 % of body calcium, 85 % of body phosphate, and 50 % of body magnesium are stored in the bone. If blood calcium levels drop below the physiologic range, processes that activate the resorption of calcium from the bone are of great importance. Approximately 35 % of the bone is composed of bone matrix. The major component of bone matrix is type I collagen (other types of collagen are present in the bone as well but at minor amounts). Furthermore, bone matrix contains proteoglycans (aggrecan, decorin, hyaluronan) and glycoproteins (alkaline phosphatase, osteopontin, osteonectin, fibronectin, bone sialoprotein, osteocalcin). Further constituents of bone matrix are serum proteins, lipids, and water.

The bone is a highly structured material that—as a result of millions of years of evolution—ideally combines two properties: stiffness and elasticity.

3.2 Comparative Phylogeny of the Skeletal System

There is evidence of phylogenetic analogies to skeletal systems as early as in *Rhizopoda*, which are marine *Radiolaria* forming a porous central silica capsule. *Acantharia* in contrast use strontium sulfate. *Foraminifera* possess a porous lime body and thus serve as geologic indicator fossils. Sponge (*Porifera*) produce lime or silica spicula secreted by sclerocytes into a gelatinous intercellular substance. *Coelenterata* by means of forming lime skeletons enable the stock buildings of coral polyps. Whereas protozoa and parazoa developed quite simple, but nevertheless protective firm-substance structures, complex skeletal variations with diversified mechanical and physiologic capacities appeared during phylogeny of metazoan organisms.

Locomotory muscles in general need abutment to a skeletal system to be functional (see Chap. 4). Exceptions to this rule are hydroskeletons with a dermal muscle tube working against an incompressible liquid column. Several classes of animals use this hydraulic locomotion system. *Nematoda* posses a liquid column in a non-segmented visceral cavity, surrounded only by longitudinal muscles. Thus only serpentine movements are possible. *Annelida* possess a segmented visceral cavity and a cutaneous muscle tube consisting of longitudinal and circular muscles. The coelomic liquid forms the liquidous part of the hydroskeleton. Therefore, straight locomotion is possible through peristaltic movement. Hydroskeletons are also found in animals higher on the evolutionary tree, such as in *Arachnida* in which legs are moved hydraulically.

Nevertheless, complex and efficient movements are based on firm-substance skeletons, which can be divided into evolutionally older exoskeletons (as found in arthropods) and the generally phylogenetically younger endoskeletons in vertebrates. However, primitive endoskeletons can also be found in protozoa, sponge, and *Echinodermata* as mentioned above. Arthropods (primarily *Insecta*, *Crustacea*, and *Chelicerata*) possess a segmented body and extremities as well as a cuticular skeleton, which is the secreted product of the epidermis or hypodermis. The cuticula consists of the polysaccharide chitin (monomer: acetylglucosamine), which is embedded in a matrix of sclerotin. Joint membranes are only weakly sclerosed to enable locomotion. The epicuticula consists of a water-resistant layer of wax. In many *crustacea*, parts of the exoskeleton have increased firmness by incorporating calcium carbonate. Molluscs (*Gastropoda*, *Bivalvia*, and *Cephalopoda*) have three major anatomical parts, namely, the cephalopodium, the

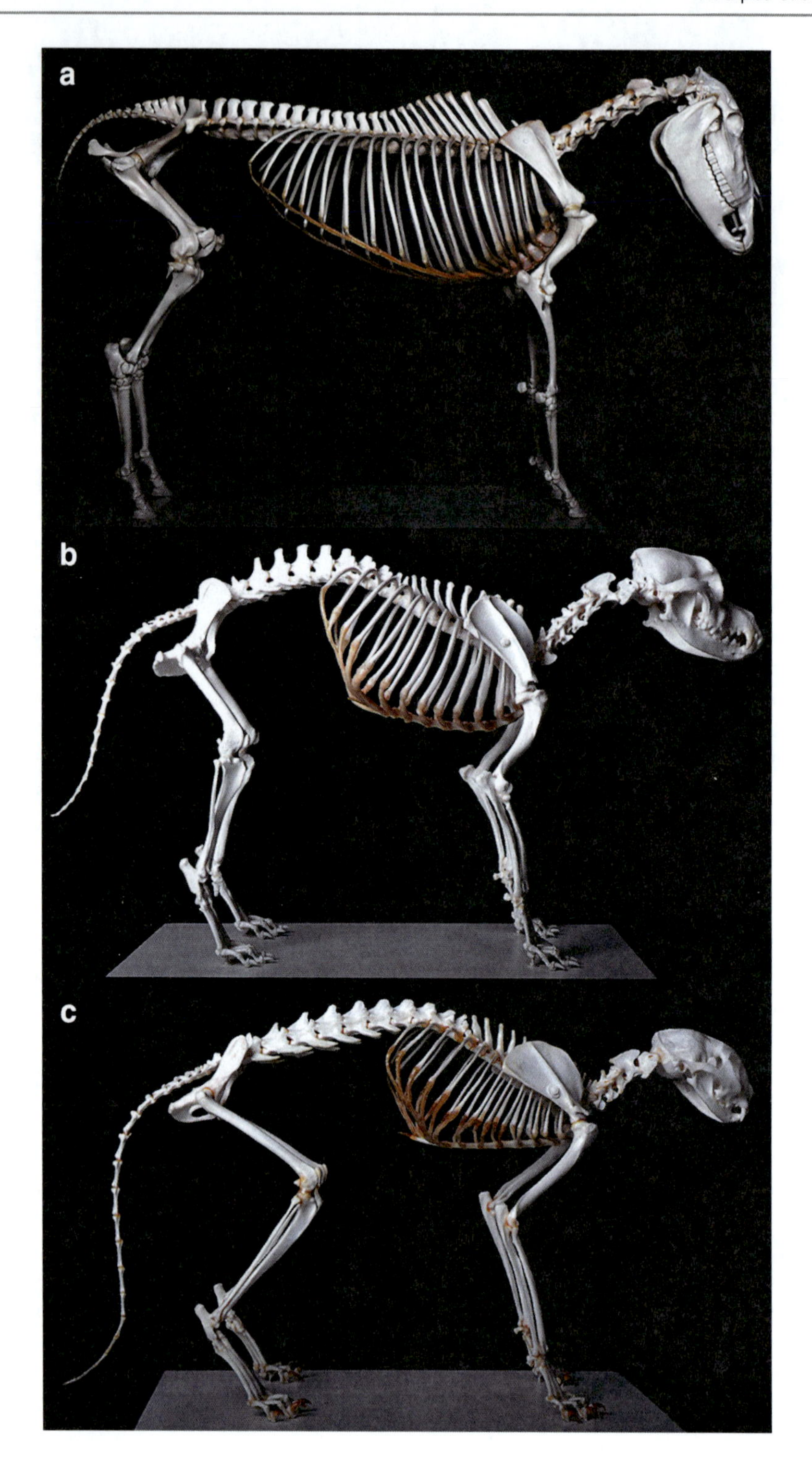
a
b
c

pallium, and the visceral complex. All molluscs, with the exception of the *Solenogastra*, have a shell formed by an inner pearl layer and an outer prism layer, which is protected by a conchiolin layer. *Echinodermata* possess an endoskeleton, which is based on mesodermal skeletal plates (sklerits) consisting of magnesium or calcium carbonate crystals.

The basic bearing and movement apparatus of *Chordata* consists of the chorda and longitudinal muscles, which are not attached to the chorda. In vertebrates, the chorda is surrounded by the segmented vertebral column and finally replaced by it. Muscles now attach to the axial skeleton. Skeletal muscle-based movement needs an arm formed by at least two bones and pairs of antagonistic muscles. Muscle efficiency is in part determined by the proportion of both parts of the arm. Therefore, runners have long extremities forming chains of arms, whose angle amplitudes increase from one joint to the following more distal joint. Several large ungulates, such as horses or deer, may serve as an example. In the course of their phylogeny, distal extremities have become disproportionally longer in dependence of increasing body height.

The endoskeleton of vertebrates consists of more than 200 bones (Fig. 3.2). The flexible axial skeleton is formed by a different number of vertebrae. Cartilaginous fish and snakes, for example, possess approximately 400 vertebrae. The mammalian vertebral column on the other hand relatively constantly consists of 7 cervical, 13 thoracal, 6–7 lumbar, 5–6 sacral, and a variable number of coccygeal vertebrae. Ribs, whose distal ends are connected by the sternum, function to protect the thoracal organs and to enable respiration starting with the tetrapods. The tension of the trunk of larger mammals is upheld by the so-called bowstring bridge-apparatus. Herein, the vertebral column plus the intervertebral muscles and ligamenta forms the bow, and the ventral abdominal muscles (especially the M. rectus abdominis) form the string.

The static and dynamic differences between the biped human skeleton and the skeletons of quadruped mammals are self-evident. Very important from the orthopedic point of view, mechanical forces on the vertebral column differ largely between these two basic skeletal types leading to different bone and joint diseases in quadruped mammals and humans. In contrast to biped humans or other mammals

Fig. 3.2 Comparison of skeletons. (**a**) The skeleton of a horse (in this case a Shetland pony), (**b**) a canine skeleton, and, finally, (**c**) a feline skeleton. These examples of different mammal skeletons are thought to demonstrate some differences of the bone apparatus, which evolved in adaption to different overall behavior and nutrition physiology. The horse is a herbivore and adopted to open grassland in arid climates. Therefore, horses are highly specialized runners. The skull resembles one of a typical herbivore with pronounced molars. Horses have elongated distal extremities with only the third metacarpal/metatarsal as well as digit left. As arduous runners horses possess a voluminous thoracal cavity to give place for the high-capacity lungs and heart. Cats are tiptoed hunters with a very flexible spine, which enables them to climb on trees and perform jumps from impressive heights. Their skulls and distal extremities are perfectly adapted to hunting and killing prey. With respect to locomotion the canine skeleton is somewhere in between. Dogs are persevering runners, which attack their prey when it is exhausted. Thus, dogs have long extremities and typically a deep thoracal cavity for the same reason as for horses (photographs are courtesy of Prof. Gerald Weißengruber, Institute of Anatomy, VMU, Vienna)

wearing their body exclusively or preferentially on their hind limbs (such as kangaroos), more than half of the body weight is sustained by the forelimbs in quadruped mammals.

Two pairs of extremities are found already in the oldest gnathostome fish. In tetrapods, pentadactyl pairs of extremities evolved. To enable extremities to move the trunk, they have to be connected to the axial skeleton. A complex shoulder girdle (starting with amphibians) is necessary especially when performing 3D movements with the forelimbs. The clavicula disappears in species whose limbs only perform pendular movements, such as is the case in larger herbivores. In mammals, the forelimb skeleton consists of the scapula, the humerus, then the radius and ulna, which can fuse to different extents in different species (especially in ungulates), the carpal and metacarpal bones (which can also fuse and/or be reduced in number in different species) as well as the digits. A reduction and partial fusion of digital radii as well as a simultaneous elongation of the distal extremities can be seen in ungulates as an adaption to fast locomotion as well as in bats as an adaption to flight. A shortening of digital radii can be observed in dragging (mole) or swimming (whales) vertebrates. A reduction of the number of extremities can be found as an adaption to serpentine or swimming locomotion (snakes, whales, and sirenia). The pelvic girdle, connecting the hind limbs to the trunk, consists of the ilium, the ischium, and the pubis. The mammalian hind limb consists of the femur plus patella (a sesamoid bone), the tibia and fibula (for which the same accounts as for the radius and ulna), and finally the tarsal, metatarsal, and digital bones.

The cranial skeleton houses the important larger organs of perception including the associated neurologic centers as well as the entrance to the gastrointestinal (GI) tract. The neurocranium forms the capsule for the organs of perception; the splanchnocranium consists of serial (branchiomere) skeletal clips, surrounding the cranial respiratory and GI tract; and the dermatocranium forms the bone carapace. Starting with tetrapods these three parts fuse now forming the syncranium with a complex jaw apparatus. Skull bones are differentiated into cranial bones (os occipitale, os presphenoidale, os basisphenoidale, os temporale, os frontale, os parietale, os interparietale, and os ethmoidale) and facial bones (os nasale, os lacrimale, os zygomaticum, maxilla, os incisivum, os palatinum, vomer, os pterygoideum, mandibula, os hyoideum, sinus paranasales).

3.3 Cells of the Bone

In the bone many different cell types can be found. These include osteoblasts, osteocytes, bone lining cells, osteoclasts, endothelial cells, fibroblasts, bone marrow stromal cells, adipocytes, bone macrophages, and various cell types that reside in the bone marrow including hematopoietic stem cells.

Osteoblasts are assumed to descend from mesenchymal stem cells. Further cells that belong to this lineage are adipocytes, chondroblasts, myocytes, and fibroblasts. Runx2 (runt-related transcription factor 2) and osterix are important transcription

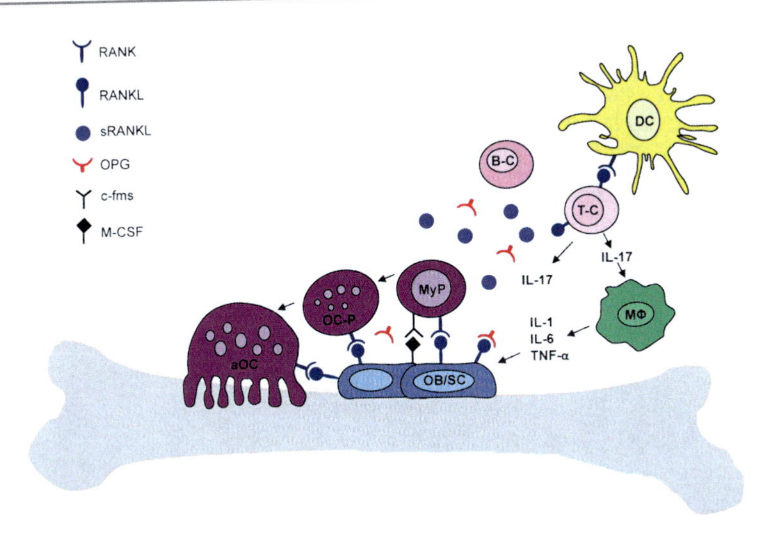

Fig. 3.3 *The RANK-RANKL-OPG axis* is the key regulatory pathway for osteoclastogenesis with the cytokine RANKL binding to its receptor RANK. OPG is a decoy receptor for RANKL and therefore inhibits RANKL action on osteoclast precursors/osteoclasts. *MyP* myeloid progenitor cell, *OC-P* osteoclast precursor, *aOC* activated osteoclast, *OB/SC* osteoblast/stromal cell, *MΦ* macrophage, *T-C* T cell, *B-C* B cell, *DC* dendritic cell (modified from Rauner et al. 2007)

factors that promote the differentiation of precursor cells towards osteoblasts. In contrast to the aforementioned factors, PPAR-gamma (peroxisome proliferator activator receptor gamma) inhibits osteoblast differentiation but promotes the differentiation of precursor cells towards adipocytes. In recent years the canonical Wnt β-catenin signaling pathway has been found to regulate the expression of genes important for osteoblasts such as Runx2. In order to activate this pathway, small molecules from the Wnt family bind to co-receptors (low-density lipoprotein-related protein 5/6, Frizzled) and thereby activate a signaling cascade. The binding of the Wnt molecules to their co-receptors is inhibited by antagonists such as sclerostin or dickkopf-1. Osteoblasts have two main functions: they are responsible for bone formation and regulate the generation of the bone degrading cell, the osteoclast. In the process of bone formation, first, bone matrix is synthesized; in a second step the organic components of the bone are mineralized. After the process of bone formation has been completed, the osteoblast either converts into a bone lining cell ("resting osteoblast") or further differentiates into an osteocyte. Osteocytes are cells that are completely embedded in the bone. Osteocytes communicate with each other via special canaliculi and form a three-dimensional network that is very important as a mechanosensor.

Osteoclasts are multinucleated cells that are able to resorb bone. Osteoclasts derive from myeloid precursor cells and are therefore related to monocytes, macrophages, and dendritic cells. In the process of osteoclast differentiation, several (mononuclear) osteoclast precursors fuse to multinucleated osteoclasts (Fig. 3.3). For the generation of osteoclasts two factors are essential: M-CSF (macrophage colony stimulating factor) and RANKL (receptor activator of nuclear factor κB ligand). RANKL is expressed by osteoblasts, bone marrow stromal cells, osteocytes, activated T cells, B cells, and endothelial cells. The receptor for RANKL (RANK) is expressed by osteoclast precursors, dendritic cells, and endothelial cells. Osteoprotegerin (OPG) is an antagonist of RANKL. Research conducted over the last decade has demonstrated that a huge number of cytokines, growth factors, and hormones that regulate the generation of osteoclasts do this by altering the RANKL/OPG ratio. For instance, proinflammatory cytokines such as TNF-α, interleukin (IL)-1, IL-6, and IL-17 activate the generation of osteoclasts by increasing the expression of RANKL. Thus many cells of the immune system (e.g., during inflammatory reactions) also have effects on bone metabolism. Interactions between the bone and the immune systems are currently investigated in the field of osteoimmunology. In contrast to most proinflammatory cytokines, 17β-estradiol decreases RANKL production, increases OPG expression, and consequently inhibits the generation of osteoclasts.

In order to dissolve the organic components of the bone, osteoclasts use several enzymes such as cathepsin K, collagenases, and matrix metalloproteases. Moreover, the osteoclast is able to create a very acidic environment at the resorption sites and thereby dissolves the anorganic components of the bone.

As already mentioned above, there are tight relationships between the bone and bone marrow. While in the past bone had been regarded as a mere mechanical protector of bone marrow, recent research revealed exciting reciprocal interactions between the two compartments. For instance, an osteoblastic niche for the long-term storage of hematopoietic stem cells has been identified. In birds, the bursa of Fabricius is the major anatomical site of B cell development. In mammals, bone marrow is considered to function as an equivalent of the bursa of Fabricius. In this context it is interesting to note that osteoblasts support the commitment and the differentiation of B cells from hematopoietic stem cells.

3.4 Bone Remodeling

In all vertebrates bone remodeling is a continuous lifelong process. The purpose of bone remodeling is to adapt bone tissue to alterations of mechanical loading; to regulate mineral homeostasis, in particular calcium and phosphate metabolism; and to repair micro-cracks. Bone remodeling occurs in microscopic compartments ("bone modeling units," BMUs). The process of bone remodeling is initiated by the osteoclasts and lasts approximately 2–4 weeks. After the osteoclasts have completed the phase of bone resorption and thereby created a so-called Howship's lacuna, osteoblasts appear on the scene and synthesize bone matrix. In a second step the newly synthesized bone matrix is mineralized. Upon the completion of bone formation, the osteoblasts transform into bone lining cells or osteocytes. The site where bone had been remodeled now enters a quiescent phase. Bone remodeling occurs both in the cortical and the trabecular compartment of the bone. Despite the fact that cortical bone is more abundant than trabecular bone, the rate of bone remodeling is much higher in trabecular bone. It has been estimated that in humans every bone site is remodeled approximately every 2–5 years. Under physiologic conditions the amount that is degraded by osteoclasts is exactly equivalent to the amount of bone synthesized by the osteoblasts. Thereby bone balance is preserved. If, however, more bone is resorbed than newly built, the balance between bone formation and bone resorption is disturbed and may lead to the development of metabolic bone diseases, such as osteoporosis.

Bone remodeling is regulated by numerous cytokines, growth factors, and hormones. As mentioned above, M-CSF, RANKL, OPG, TNF-α, IL-1, IL-6, and IL-17 are important regulators of bone resorption. Further examples of factors that are critically involved in the regulation of bone remodeling are insulin-like growth factor 1 (IGF-1), transforming growth factor-β (TGF-β), fibroblast growth factors (FGFs), bone morphogenetic proteins (BMPs), and leptin.

3.5 Hormones That Regulate Bone Turnover

Vitamin D_3 is synthesized in the skin under the influence of UV light from 7-dehydrocholesterol or is alternatively taken up from the intestine. In order to be fully active, two hydroxylation steps are required: the first step occurs in the liver and the second in the kidneys (see Chap. 10). The biologically active metabolite $1,25(OH)_2$-VitD_3 increases the resorption of calcium and phosphate in the small intestine and stimulates the tubular reabsorption of calcium and phosphate in the kidney. Moreover $1,25(OH)_2$-VitD_3 stimulates collagen synthesis in the bone and increases bone resorption. Deficiency of VitD_3 thus may severely impair bone metabolism.

Parathyroid hormone (PTH) is a polypeptide that is important for calcium retention by the kidneys and at the same time inhibits phosphate reabsorption. PTH activates bone remodeling and thus mobilizes calcium and phosphate from bone. Moreover, PTH stimulates the synthesis of the active vitamin D metabolite $1,25(OH)_2$-VitD_3 in the kidneys. On a systemic level PTH increases the serum level of calcium but decreases the level of phosphate.

Calcitonin is another polypeptide involved in the regulation of calcium homeostasis. Calcitonin has been postulated as an antagonist of PTH. Thus calcitonin inhibits bone resorption and the tubular reabsorption of calcium. Consequently calcitonin decreases the calcium serum level.

The sexual steroid 17β-estradiol inhibits the generation and activation of osteoclasts. Thus, estrogen deficiency (as is seen in postmenopausal women) may be associated with increased bone resorption and the development of osteoporosis (Chap. 12). Typically, estrogen deficiency is associated with increased bone remodeling: the number of BMUs is increased, and the extent of bone resorption and bone formation is elevated. Nevertheless, bone resorption prevails over bone formation. Therefore bone mass declines and bone structure deteriorates. The aforementioned alteration of bone remodeling has also been referred to as "high bone turnover." Similar to estrogens also androgens are critically involved in the regulation of bone remodeling. Acute testosterone deficiency (e.g., as a consequence of castration) in humans and in rodents has been described to lead to high bone turnover and a loss of bone mass.

Examples of further systemic regulators of bone are glucocorticoids, thyroid hormone, and growth hormone. Augmented production of thyroid hormone leads to high bone turnover. In contrast, glucocorticoid excess is characteristically associated with low turnover osteoporosis. In low turnover osteoporosis, osteoblast activity is significantly diminished. Excessive production of growth hormone (e.g., by a pituitary adenoma) may stimulate bone formation and consequently increase bone mass.

Acknowledgements The authors are grateful to Mrs. Birgit Schwarz for her help with the preparation of the manuscript and to Nadine Ortner, MSc, and Christiane Lamm, BSc, for their assistance with Fig. 3.1.

Further Reading

Bartl R, Bartl C (2004) Osteoporose manual: Diagnostik, Prävention, Therapie. Springer, Berlin, Heidelberg

Bilezikian JP, Raisz LG, Martin TJ (eds) (2008) Principles of bone biology, 3rd edn. Academic, Amsterdam

Guerrouahen BS, Al-Hijii I, Tabrizi AR (2011) Osteoblastic and vascular endothelial niches, their control on normal hematopoietic stem cells, and their consequences on the development of leukemia. Stem Cells Int 2011:375857

König HE, Liebich H-G (eds) (2005) Anatomie der Haussäugetiere. Lehrbuch und Farbatlas für Studium und Praxis. 3. Aufl., Schattauer GmbH, Stuttgart

Pietschmann P (ed) (2012) Principles of osteoimmunology: molecular mechanisms and clinical applications. Springer, Wien, New York

Rauner M, Sipos W, Pietschmann P (2007) Int Arch Allergy Immunol 143:31–48

Wehner R, Gehring W (1995) Zoologie. 23. Aufl., Georg Thieme Verlag, Stuttgart

Zhu J, Garrett R, Jung Y, Zhang Y, Kim N, Wang J, Joe G, Hexner E, Choi Y, Taichman RS, Emerson S (2007) Osteoblasts support B-lymphocyte commitment and differentiation from hematopoietic stem cells. Blood 109:3706–3712

Locomotor Principles: Anatomy and Physiology of Skeletal Muscles

4

Josef Finsterer

Contents

J. Finsterer (✉)
Krankenanstalt Rudolfsstiftung, Vienna, Austria
e-mail: fifigs1@yahoo.de

E. Jensen-Jarolim (ed.), *Comparative Medicine*,
DOI 10.1007/978-3-7091-1559-6_4, © Springer-Verlag Wien 2014

Abstract
The musculature consists of individual muscles which are mainly built up of muscle cells (muscle fibres, myocytes, myotubes), and its main function is to move and support the bones of the skeleton. In human beings, the musculature is the largest organ, consists of 640 single muscles, makes up 50 % of the body weight, moves 200 bones, and has 2,200 points of attachment. Skeletal muscles are under the control of the central nervous system, the myelon, and motor nerves. Histologically, the skeletal muscle belongs to the striated muscle type. In addition to the voluntary motor function, which is the elementary task of the organ, the muscle is also required for involuntary motor control, stabilisation of joints, and heat production, but has also immunological and endocrine functions. In addition to muscle cells, a muscle is built up of a number of other cell types and tissues. The following chapter is designated to describe and discuss basic knowledge about the anatomy and physiology of the human skeletal muscles in comparison with the skeletal muscle of other species.

4.1 Anatomy of Skeletal Muscles

4.1.1 Development of Muscle Tissue

Muscle tissue originates from the mesoderm, which is one of the three germinal layers, and develops between the amnion cave and the chorion cave. It is located in the middle between the two other germinal layers, the ectoderm and the endoderm. With progression of the embryonic development, the mesoderm forms somites along the neural tube. From there precursor muscle cells (myoblasts (developmental progenitor cells)) migrate to the sites of muscle formation in the limbs and body wall, where they give rise to muscle fibres (Partridge 2009). Migration of precursor cells is stimulated by the stimulating factor Pax3. Precursor muscle cells fuse and form a syncytium (single cell deriving from the fusion of several precursor cells), which contains hundreds of nuclei from the precursor cells and expresses the contractile apparatus. Attached to myotubes on the outside of the cell membrane are the so-called satellite cells (stem cells), which can be transformed into myotubes, whenever regeneration is necessary. Number of satellite cells and capacity of stem cells to regenerate to myotubes are limited.

4.1.2 Muscle Fibre

The muscle cell is a long, cylindrical, multinucleated, and tubular structure of up to several centimetres in length, which is surrounded by the cell membrane, called **sarcolemma**. The sarcolemma encloses the cytoplasm, which is called sarcoplasm. The sarcoplasm contains myofibrils, which are built up of the **myofilaments** actin, myosin, troponin, tropomyosin, and some other nonfilamentous proteins. Nuclei are unusually flattened and pressed against the inside of the sarcolemma. Mitochondria (sarcosoma) are located between the myofibrils. Myofibrils are surrounded by the

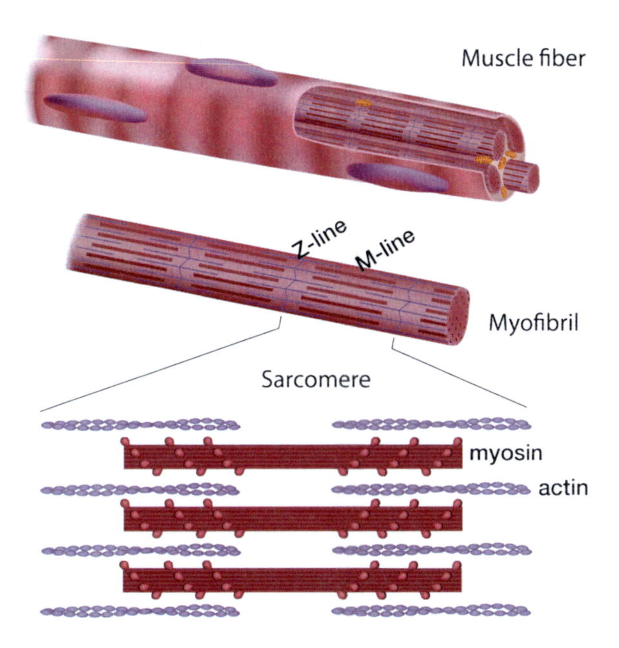

Fig. 4.1 *A muscle fibre* is composed by sarcomeres, which are the smallest contraction units. In the sarcomeres myofibrils (filaments), actin and myosin are arranged in a typical manner. Myosin filaments are connected in the Z-line; in between the M-line is a dominant characteristic. This gives the skeletal muscle a striated appearance. © [Alila Medical Images]—Fotolia.com

endoplasmatic reticulum, called sarcoplasmic reticulum, which stores calcium ions needed for muscle contraction. The sarcoplasmic reticulum ends periodically in dilated sacs, known as terminal cisterns. They cross the entire muscle fibre from one side to the other. Between two terminal cisterns, there is a tubular infolding of the sarcolemma called the transverse tubule (T-tubule). Two terminal cisterns and a transverse tubule form a structure called "triad". The T-tubule is the pathway for the excitation to signal the sarcoplasmic reticulum the release of calcium ions. Each component or compartment of a muscle cell is arranged to ensure optimal function.

4.1.3 Sarcomere

The contractile apparatus is the main morphological and functional unit of a myotube. It is build up of **sarcomeres** (single contraction units) and attached to the sarcolemma by an absorbing system (Fig. 4.1). A sarcomere is built up of the myofibrils (filaments) **actin** and **myosin**, which are arranged in a typical manner. Myosin has a diameter of about 15 nm and actin of about 5 nm. Multiple sarcomeres are arranged in a repeating manner within the myotube, resulting in the typical striated appearance of the myotubes. The term "striated" derives from this typical pattern of a myotube on histological examination. The staining best demonstrating striation of skeletal muscle cells is the phosphotungstic acid haematoxylin stain. Dark colour refers to myosin (A-band) and light colour refers to actin (I-band). All muscles moving bones (skeletal muscle), but also those responsible for mimics or ear movements, are of the striated type. A single myocyte of the biceps brachii muscle contains about 100,000 sarcomeres.

Table 4.1 Characteristics of muscle fibre types

Fibre type	Type-I MF	Type-IIaMF	Type-IIx fibres	Type-IIb MF
Contraction time	Slow	Fast	Fast	Very fast
Motor neuron size	Small	Medium	Large	Very large
Resistance to fatigue	High	Fairly high	Intermediate	Low
Mitochondrial density	Very high	High	Medium	Low
Capillary density	High	Intermediate	Low	Low
Oxidative capacity	High	High	Intermediate	Low
Glycolytic capacity	Low	High	High	High
Major stored fuel	Triglycerides	CP, glycogen	ATP, CP, low glycogen	ATP, CP
Activity	Aerobic	Long-term anaerobic	Short-term anaerobic	Short-term anaerobic
Force production	Low	Medium	High	Very high

CP creatine phosphate

4.1.4 Types of Muscle Fibres

After fusion of myoblasts to myocytes, which takes place already before birth, muscle cells continue to grow and to differentiate according to their twitch capabilities into slow-twitch fibres (type-I muscle fibres) and fast-twitch fibres (type-II muscle fibres). These fibres are differentiated by their content of proteins, content of fuel, and their functions. Concerning the fuel content, there are three sources of high energy phosphates to fill the ATP pool, creatine phosphate, glycogen (undergoes glycolysis and is degraded to lactic acid and phosphate), and glucose (used by the cellular respiration).

4.1.4.1 Slow-Twitch Fibres (Type-I Muscle Fibres, Slow Oxidative Fibres, Red Fibres)

Slow-twitch fibres have a slow speed (velocity) of contraction and a less well-developed glycolytic capacity compared to type-II muscle fibres (Table 4.1). They generate energy in the form of ATP by means of a long-term system of aerobic energy transfer (aerobic system). They are fuelled by **glycogen**, which is split into glucose molecules, as source of their energy production. The activity of the adenosine triphosphate (ATP)ase, the enzyme which splits ATP into phosphate and adenosine diphosphate (ADP), is low and its splitting rate is slow. Slow-twitch fibres contain many and large mitochondria and thus a high concentration of mitochondrial enzymes and large amounts of myoglobin (binds O_2, stores O_2), which is why they are fatigue resistant. Slow-twitch fibres contain myosin-heavy chain type 7 (MYH7). Slow-twitch fibres are supplied by many blood capillaries. Because of the high content of myoglobin, slow-twitch fibres are also termed red muscle fibres. Slow-twitch fibres render 10–30 contractions per second (10–30 Hz) (Scott et al. 2001). Type-I muscle fibres are mainly activated if a weak contraction is needed. Slow-twitch fibres are also needed for long-term exercise and for endurance activities. Slow-twitch fibres are thus particularly found in postural muscles and are predominantly activated by endurance athletes.

4.1.4.2 Fast-Twitch Fibres (Type-II Muscle Fibres)

Fast-twitch fibres, on the contrary, are characterised by fast propagation of action potentials along the sarcolemma, the capability to split ATP very quickly, and the capability to rapidly release or uptake calcium from or to the sarcoplasmic reticulum respectively. Fast-twitch fibres rely on the short-term glycolytic system of energy transfer and contract and develop tension at two to three times the rate of slow-twitch fibres. Fast-twitch fibres render 30–70 contractions per second (30–70 Hz) (Scott et al. 2001). Type-II muscle fibres are further differentiated according to the presence of myosin isoforms. In humans, three subtypes of type-II muscle fibres are differentiated: type-IIa, type-IIx, and type-IIb fibres. Type-IIa fibres contain MYH2, type-IIx fibres MYH1, and type-IIb fibres MYH4.

Type-IIa fibres (red fibres, fast oxidative, fatigue resistant) are characterised by large amounts of myoglobin, many mitochondria, large amounts of glycogen, many blood capillaries, a high capacity to generate ATP by oxidation (reaction of oxygen and electrons to produce energy and water), very rapid rate of ATP splitting, a high contraction velocity, and fatigue resistance lower than in slow oxidative fibres. Type-IIa fibres move five times faster than type-I fibres. Type-IIa fibres are activated if a contraction stronger than provided by a type-I fibre is required and are activated at moderate strain. They are activated to assist type-I muscle fibres (Table 4.1). A similar amount of type-I and type-IIa fibres is activated by middle distance event athletes. Type-IIx fibres (fast-twitch, fast glycolytic, white fibres) are characterised by a fast contraction time, an intermediate resistance to fatigue, low myoglobin content, a medium density of mitochondria, few blood capillaries, a large amount of creatine phosphate, and quick splitting of ATP and very quick fatigability and are needed for sprinting (Table 4.1). Type-IIb fibres are mainly fuelled by creatine phosphate; have a low density of mitochondria, a low oxidative capacity, and a low density of capillaries; are innervated by very large motor neurons; have a very fast contraction time; tire easily; and are used for short-term anaerobic exercise, which lasts <1 min, but produce a high power. Type-IIb fibres are activated for maximal contractions, are always activated at last, and are used for ballistic activities. Type-IIb fibres move ten times faster than type-I fibres. Type-IIb fibres are predominantly activated by sprint athletes. Differences between muscle fibres in muscles are summarised in Table 4.1 (Larsson et al. 1991).

4.1.5 Fibre-Type Composition and Training

Individual muscles are usually composed of all four muscle fibre types (I, IIa, IIx, IIb). Most likely, in humans, there are no sex or age differences concerning fibre distribution, but the composition and proportion of fibre types varies from muscle to muscle and between individuals. Sedentary adults and young children have 55 % type-I fibres and 45 % type-II fibres. Also the number of different skeletal muscle fibres is most probably fixed early in life and does not change in healthy subjects. Generally, the number of muscle cells is regulated by myostatin. Myostatin is a cytokine produced by muscle fibres that inhibits proliferation of muscle fibres. Since it inhibits the same type of cell in which it is produced, it is called a chalone. Whether a certain type of exercise can change the muscle fibre composition within a muscle is under debate, but there are indications that type and training alters fibre-type composition. After high-intensity endurance training, type-IIb fibres enhance the oxidative capacity such that they are capable to perform oxidative metabolism as effectively as slow-twitch fibres of untrained subjects. This would be brought about by an increase in mitochondrial size and number but not in a change in fibre type. Endurance exercise, such as running or swimming, causes gradual transformation of type-IIb fibres into type-IIa fibres. Such transformed type-IIb fibres show a slight increase in diameter, mitochondria, blood capillaries, and strength. Exercises that require great strength for short periods, such as weightlifting, increase size and strength of type-IIb fibres. The increase in size of these fibres is due to increased synthesis of actin and myosin filaments.

4.1.6 Organisation of Fascicles

A fascicle is defined as contraction unit built up of variable numbers of muscle fibres. Within the fascicle, muscle fibres are adjusted in a parallel way. Fascicles, on the contrary, can be organised in different shapes. In parallel muscles, which most of the skeletal muscles are, fascicles run parallel to the direction of the muscle. An example of a parallel muscle is the biceps muscle. In convergent muscles, fascicles fan out from a common point of attachment, allowing more versatile types of movements (Martini et al. 2008). Convergent muscles do not pull as strong as parallel muscles since not all fibres pull in the same direction, but in different directions at opposite ends (Martini et al. 2008). Examples of a convergent muscle are the pectoralis major muscle and the temporal muscle. In pennate muscles, one or more tendons run through the body of the muscle with the fascicles forming an oblique angle to the corresponding tendons, which is why fascicles in pennate muscles pull less strong on their tendons than in parallel muscles. Nevertheless, pennate muscles usually generate greater tension since they are built up of a greater amount of muscle fibres than similarly sized parallel muscles (Martini et al. 2008). Examples of pennate muscles are the rectus femoris muscle and the extensor digitorum communis muscle. The fourth type of fascicle organisation is realised in sphincter muscles. In sphincter muscles, fascicles are arranged concentrically around an opening or around a recessus (Martini et al. 2008). With contraction of a sphincter muscle, the opening gets smaller, which is why this type of muscle is usually found around entrances or exits of internal or external passageways (Martini et al. 2008). Examples of sphincter muscles are the orbicularis oculi and the anal sphincter muscle.

4.1.7 Building Up a Muscle

Though muscle cells make up the vast majority of the muscle, other specialised cells and tissues can be also found in muscles and contribute to proper muscle function. These include connective tissue forming fascia or septa within the muscle, such as the perimysium, epimysium, or endomysium; vessels, such as arteries (transport blood from the heart to other organs), capillaries (enwrap the myotube); veins (transport blood from the organs to the heart) or lymph vessels (transport interstitial fluids from the periphery to the heart), lymph nodes, tendons, and nerves (motor nerve fibres, sensory nerve fibres, vegetative nerve fibres); and receptors, such as the muscle spindles or pain receptors. Nearly every muscle is attached to a bone by bundles of collagen fibres, known as tendons.

4.1.8 Classification of Muscles, Organisation in Functional Units

Muscles are classified according to various criteria. Regarding their location between the skin and the inside of the body, superficial muscles (near to the skin) and deep muscles (inside the body) are differentiated. Skeletal muscles are predominantly located along bones and are most pronounced in size where they intensively move a part of the body. Accordingly, the largest muscles can be found at the thigh, the calves, and the shoulder girdle. Muscles, which move only the skin, like most facial muscles, or the cartilage, like the ear muscles, are thus less prominent. According to their function muscle groups are classified as facial muscles, extra-ocular muscles, chewing muscles, neck flexors, neck extensors, shoulder girdle muscles, inside- or outside rotators of the arm or hip, extensors or flexors of the arm or leg, abductors or adductors of the shoulder or hip, extensors or flexors of the elbow or knee, extensors or flexors of the hands and fingers, finger straddlers, trunk extensors or flexors, extensors or flexors of the foot or toes, or as muscles responsible for inversion or eversion of the foot. Each individual muscle has a distinct origin, a distinct attachment, a distinct movement function, and a specific motor nerve, which innervates each muscle. Each muscle has also a characteristic innervation zone, the area where motor nerves insert into the muscle via its **motor end plates** (Fig. 4.2a). To study insertion, attachment, innervation, and function of the individual muscles, the reader is referred to anatomical textbooks.

4.1.9 Innervation of Muscles

Anatomy of motor nerves, which innervate muscles, is complex and requires profound anatomical knowledge. Basically, motor nerves originate from anterior horn cells inside the **myelon**. They exit the myelon via the anterior roots to form either an individual nerve or, in the region of the cervical and lumbar myelon, trunks, which then split into divisions and thereafter cords to constitute

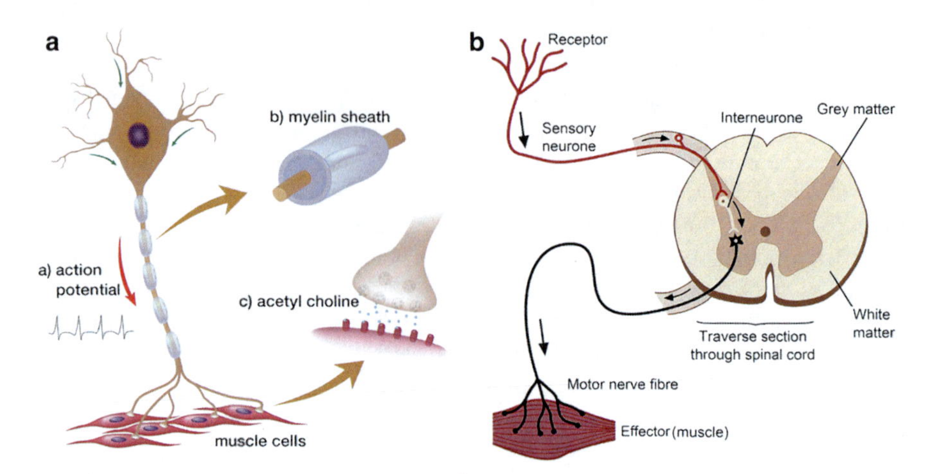

Fig. 4.2 (**a**) *The making of a neuromuscular end plate*. Motor signals to the periphery are propagated via action potentials (a) along myelinated (b) nerves. At the motor end plate, the release of neurotransmitter acetylcholine (c) leads to a depolarisation and impulse to contract muscular fibres. © [Alila Medical Images]—Fotolia.com. (**b**) *A spinal reflex arc*: sensory nerves are via interneurons connected to motor nerves in the grey matter of the myelon. The signal is forwarded thus to muscles and lead to involuntary contraction, i.e. retraction from a danger signal. This can be exploited diagnostically to test for neuronal function. © [Balint Radu]—Fotolia.com

the cervico-brachial or lumbosacral plexus. From these plexuses the main motor nerves of the arm respectively leg origin. Motor nerves originating from the brachial plexus include the median, ulnar, radial, musculocutaneous, axillary, thoracodorsal, subscapular, and lateral pectoral nerves. Motor neurons originating from the lumbar plexus include the iliohypogastric, genitofemoral, femoral, sciatic, and obturator nerves. Branches of these nerves lead to the individual muscles and split intramuscularly into subdivisions to lastly split into terminal endings shortly before arriving at the innervation zone. The innervation zone has a particular architecture and varies between muscles of a single species but also between different species.

4.2 Physiology of Skeletal Muscles

4.2.1 Physiology of Voluntary Muscle Contraction

4.2.1.1 Motor Impulse Generation and Propagation Along the Motor Nerve System

The major task of the skeletal muscle is to contract and thus to move bones. Contraction of the muscle can be voluntary or involuntary (reflectory). For voluntary contraction an impulse originating from the motor cortex of the cerebrum is propagating along the pyramidal tract to the end of the first motor neuron. There, the electrical impulse is transferred into a chemical message in form of the

neurotransmitter glutamate. The chemical message is recognised at the anterior horn cells, located in the anterior horn of the myelon, by glutamatergic receptors. Stimulation of these receptors depolarises the anterior horn cell. From there the impulse propagates along the myelinated peripheral nerves to the **neuromuscular junction** (**motor end plate**), where the electrical impulse in form of an action potential is transferred into chemical information in form of the transmitter **acetylcholine**, which then induces an action potential at the postsynaptic membrane (Fig. 4.2a). This current propagates along the muscle membrane to the T-tubule. Here, the impulse induces the release of calcium ions from the sarcoplasmatic reticulum to induce contraction of the sarcomeres.

To protect the body from damage, involuntary reflexes are important. Sensations such as pain are translated into a motor neuron reaction leading to muscle contraction and retraction from the site of danger. In this setting, the switch is made in the next spinal segment before propagating the sensation to the brain (Fig. 4.2b).

4.2.1.2 Generation and Propagation Along the Sarcolemma

Each cell membrane is charged but the type and degree of charging is determined by the cell type. Non-excitable cells establish only a membrane potential, which is defined as difference in electrical polarity between the interior and the exterior of a cell. The membrane potential is typically −60 to −80 mV in amplitude. Cells that are excitable, like neurons, muscle cells, or secretory cells, generate not only a membrane potential but, after appropriate stimulation, also an **action potential**. Membrane and action potentials are mainly determined by ion concentrations inside and outside the sarcolemma. These concentrations are mainly dependent on the selective permeability of ion channels, ion transporters, or exchangers of these ions located inside the cell membrane. Ions contributing most to the membrane and action potentials are sodium, potassium, chloride, and calcium. The resting membrane potential is stable among all species. An action potential is a short depolarisation during which the membrane potential rises and falls, following a consistent trajectory (see Fig. 4.2a; (a)). In muscle cells the action potential is a key step in the chain of events leading to muscle contraction. The action potential is generated by specific types of voltage-gated ion channels embedded in the sarcolemma (Barnett and Larkman 2007). They are shut when the membrane potential is near the resting potential, but they rapidly open in case of an increase in the membrane potential to the threshold value. After opening, channels allow an inward flow of sodium ions, which produces a further increase in the membrane potential. This causes other sodium channels to explosively open until all are open, resulting in a further increase of the membrane potential. The sodium influx causes the polarity of the sarcolemma to reverse, which in turn inactivates the ion channels. Sodium ions now have to be actively transported outwardly. Potassium channels are then activated, resulting in an outward current of potassium and in returning the electrochemical gradient to the resting state. After the action potential, a negative shift known as after-hyperpolarisation or refractory period due to additional potassium currents occurs. Sodium-based action potentials last for <1 ms (Bullock et al. 1977).

4.2.1.3 Motor Unit

A motor unit is a functional innervation unit defined as a single anterior horn cell of the myelon plus all muscle fibres innervated by this anterior horn cell (illustrated in (b) of Fig. 4.2b). The number of muscle fibres innervated by a single anterior horn cell varies greatly between different muscles. Motor units in muscles responsible for highly precise movements have only a few muscle fibres per anterior horn cell. Motor units of the muscles controlling the larynx have 2–3 fibres per motor neuron, and those controlling the extra-ocular eye muscles have ~10 fibres. Motor units, which need to produce high strength, like the lower leg muscles, have 2,000 muscle fibres per motor unit. A single motor unit contains only a single type of muscle fibre. The electrophysiological correlate of a motor unit is the motor unit action potential, which is the sum of all muscle fibre currents of a motor unit within the uptake area of a recording electrode evoked during voluntary or involuntary contraction. Activity of individual motor units follows a distinct pattern (size principle). At the beginning of a contraction, smaller motor units are activated first, but with increasing strength and duration of exercise, increasingly larger motor units are activated. The pattern of activation of individual motor units is determined by the brain and spinal cord.

4.2.1.4 Electrochemical Coupling

At the site of the T-tubule, the sarcoplasmatic reticulum is in close contact with the sarcolemma, which is why their function is closely related. After arrival of the electrical excitation at the T-tubule, **calcium ions** are released from the sarcoplasmatic reticulum, which is the storage reservoir of calcium ions within the myotube. Calcium ions are essential for muscle contraction, since it accomplishes the cyclic binding between the major **contractile elements** actin and myosin. After release from the sarcoplasmatic reticulum, calcium ions interact with the regulatory protein troponin located on the actin fibres. Calcium-bound troponin undergoes a conformational change that leads to the movement of tropomyosin, subsequently exposing the myosin-binding sites on the actin filament. This allows for myosin and actin ATP-dependent cross-bridge cycling and shortening of the sarcomere. The interaction between actin and myosin on the molecular level starts with movement of a myosin head (myosin subfragment S1) towards a neighbouring action subunit by hydrolysis of ATP. As soon as actin and myosin bind, ADP is released and the myosin head undergoes a conformational change, such that it turns down and drags the rest of the myosin filament behind. Now, another ATP is bound to myosin and the binding with actin is released. ATP hydrolysis let the myosin head return to its previous conformation and another cycle begins. According to this pattern, each myosin molecule slides along the actin filament so that the two filaments increasingly overlap. Each myosin filament carries about 500 myosin heads, which runs through the cycle five times per second.

4.2.1.5 Muscle Tone

A fundamental characteristic of each muscle is the muscle tone. Two types of muscle tone are differentiated, resting tone and dynamic tone. Resting tone (tonic stretch reflex) is required for the muscle to be prepared to immediately respond with a contraction in case of acute voluntary innervation. Resting tone results from activation of a few motor units at all times even at rest. As one set of motor units relaxes, another set takes over. Dynamic tone (dynamic stretch reflex) is required to adapt contraction forces during movements. The dynamic tone greatly influences the degree of monosynaptic tendon reflexes. Muscle tone is regulated by the gamma motor system, which innervates the muscle spindles, which are the core of the regulatory system for the muscle tone. A muscle spindle is either built up of gamma-1 spindle fibres ("Kernsackfasern") or gamma-2 spindle fibres ("Kernkettenfasern"). Gamma-1 spindle fibres are innervated by gamma-1 motoneurons and record the tone of the fibres via so-called annulospiral endings located in the middle of the cells. Gamma-2 spindle fibres are innervated by gamma-2 motoneurons and record the tone of the fibres via so-called flower-spray endings located in the periphery of the cells. Information of the annulospiral receptors is transmitted to the myelon via 1a axons. The information of the flower-spray receptors is transmitted to the myelon via 1b axons. Muscle tone can be normal, increased, or decreased. Reduced muscle tone, also reflected as muscle hypotension, occurs in the acute stage of ischemic stroke or in peripheral nerve or myopathic lesions. Muscle tone may be increased in Parkinson's disease (rigour) or patients with lesions of the pyramidal tract (spasticity) (Bähr and Frotscher 2005).

4.2.1.6 Types of Skeletal Muscle Contraction

The skeletal muscle usually links two bones across its connecting joint. When these muscles contract or shorten, the bone moves. There are, however, also muscles which do not cross a joint but support or move structures like the shoulder, the larynx, and the diaphragm or those which have lost the ability to move a structure at all. Only few striated muscles move the skin or the cartilage, like the facial muscles, which are responsible for the mimics, or the muscles of the ear, which formerly moved the ear. Generally, a muscle may contract in a shortening, isometric, or lengthening manner (contraction with or without change in muscle length). Contraction, however, may be also classified as isokinetic or non-isokinetic (constant or variable angle velocity), static or dynamic (opposing muscles contract against each other without or with changing muscle length), concentric or eccentric, (shortening or lengthening of muscle during contraction), or incremental or constant. A contraction is called isotonic (same tension) if the muscle is allowed and not prevented to shorten. Muscle force is proportional to the physiologic cross-sectional area, and muscle velocity is proportional to muscle fibre length (Quoted from National Skeletal Muscle Research Center; UCSD, Muscle Physiology Home Page—Skeletal Muscle Architecture, Effect of Muscle Architecture on Muscle Function). The strength of a contraction, however, is determined by biomechanical parameters, such as the distance between muscle insertions and pivot points and muscle size.

4.2.1.7 Coordination of Muscle Activity

Though any of the 640 muscles can act individually, muscles usually work together with other muscles, organised in functional units. The pattern of activity and the design of a movement are determined in the CNS. According to the movement patterns, individual muscles are activated in a concerted manner. To accomplish these patterns, muscles must act in functional muscle groups in a synergistic way, as agonists or as antagonists. Muscles are generally arranged in opposition so that as one group of muscles contracts, another group relaxes or lengthens. Voluntary and involuntary innervation of involved muscles must be graded appropriately to reach the goal of an intended movement. Since voluntary or involuntary movements change the position of the body, the head, the limbs, or the trunk, nearly all muscles are affected in case of any muscle activity. Action in one muscle evokes reaction in many others. The more extensive a movement, the more extensive is the reaction of other not primarily involved muscles.

4.2.1.8 Muscle Fatigue

A main characteristic of a skeletal muscle is that it fatigues with exercise. Depending on the type of exercise, muscle fatigue develops earlier or later after onset of the exercise. Fatigue in response to exercise (exercise (-induced) fatigue) can be enhanced by mental disorders, by organic CNS abnormalities (central fatigue), or by peripheral nervous system (PNS) dysfunction or skeletal muscle disease (peripheral, muscle, contractile, or mechanical fatigue, contractile impairment, loss of force generating capacity) (Boyas and Guével 2011). Non-neurological causes of feeling tired in response to exercise include cardiac, pulmonary, haematological, renal, metabolic, neoplastic disease, chronic obstructive pulmonary disease, overtraining syndrome, or chronic fatigue syndrome (Spruit et al. 2005). The cause of muscle fatigue to exercise is largely unknown, but it can be speculated that energy production becomes insufficient or that oxidative stress occurs and disrupts cell functions. The phenomenon of muscle fatigue can be also regarded as reaction of the tissue to prevent it from damage by overuse.

4.2.2 Other Types of Muscle

The skeletal muscle is one of the three muscle types, the two others are the smooth muscle and the cardiac muscle (Fig. 4.3a). Though most processes are similar, the three muscle types act differentially.

4.2.3 Cardiac Muscle

Contrary to skeletal muscle cells, cardiac muscle cells are self-contracting, autonomically regulated, and continue to contract in a rhythmic fashion throughout life. Cardiac muscle cells are Y shaped and shorter and wider than skeletal muscle cells (Fig. 4.3a). Cardiac muscle cells are predominantly mononucleated. Contrary to the

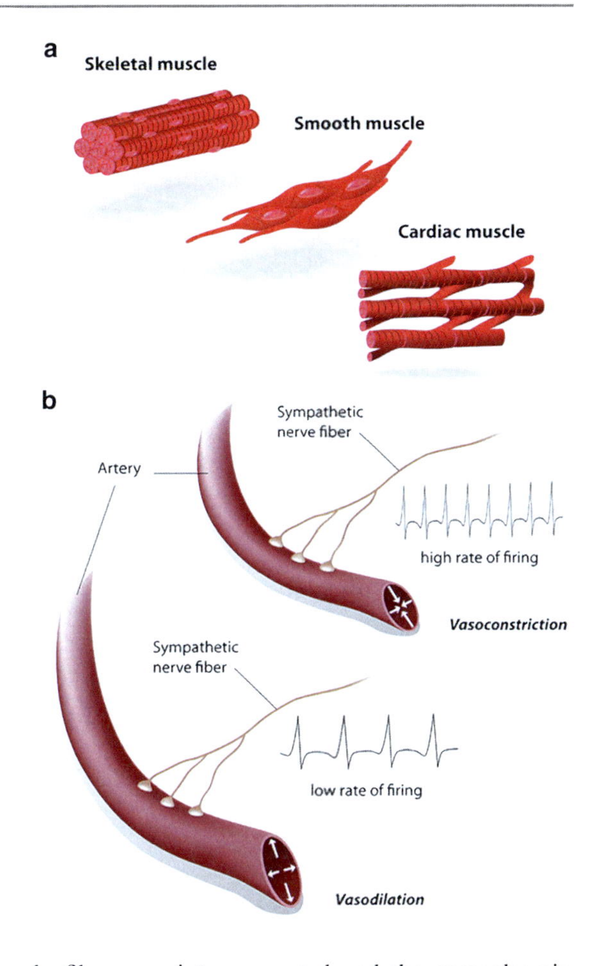

Fig. 4.3 (**a**) *The three muscle types*: skeletal muscle, smooth muscle, and cardiac muscle. © [Designua]—Fotolia.com. (**b**) *Smooth muscle function* is directed by the sympathetic or parasympathetic vegetative nervous system. Stress and activity lead to enhanced tonus of most smooth muscles, whereas parasympathetic stimuli lead to relaxation. This principle is illustrated for arteria. © [Alila Medical Images]—Fotolia.com

skeletal muscle, cardiac muscle fibres are interconnected and the sarcoplasmic reticulum is less well developed. Additionally, contraction of cardiac muscle fibres is actin regulated, meaning that calcium ions for the contraction derive not only from the sarcoplasmic reticulum but also from the extracellular space, as in smooth muscle cells. The arrangement of actin and myosin is similar to myofibres. Cardiac muscle cells are auto-rhythmic, which means that they contract without innervation by a nerve (pacemaker cells). Cardiac muscle fibres are separated by intercalated discs, which contain gap junctions to provide communication channels between the cells and to facilitate waves of depolarisation to sweep across the membranes, allowing synchronised muscle contraction. Depolarisation of cardiac muscle cells is similar to that of a myocyte, but repolarisation takes much longer to occur, why cardiac muscle cells cannot be stimulated at high frequencies and are prevented from titanic contractions. More details about anatomy and physiology of the cardiac muscle will be provided in the chapter about the heart.

4.2.4 Smooth Muscle

Smooth muscles of internal organs or vasculature cannot be stimulated voluntarily, but are controlled by the **sympathetic** and **parasympathetic** system (Fig. 4.3a, b). Contrary to myotubes, smooth muscle cells are spindle shaped, much smaller than striated muscle cells, and without striation or sarcomeres. Instead of a sarcomeric organisation of myofibrils, thin and thick filaments, corresponding to actin and myosin filaments of the striated muscle, are **organised in bundles**. Troponin and tropomyosin are absent in smooth muscle cells. Intermediate filaments, which are interlaced through the cell like a fishnet, anchor the thin filaments and correspond to Z-discs of the striated muscle. As in striated muscle, contraction of smooth muscle cells involves the formation of crossbridges between thick and thin filaments, such that thin filaments slide past the thick filaments. Contrary to striated muscle, however, shortening occurs in all directions whereupon intermediate filaments draw the cell up, like closing a drawstring purse. Calcium ions from the extracellular space bind to the calmodulin-myosin light chain, an enzyme, which then breaks up ATP and transfers phosphorus directly to myosin to activate it and to form crossbridges with actin. Both cardiac and smooth muscle can contract without being stimulated by the nervous system.

4.2.5 Comparison of Muscle Anatomy and Physiology Between Different Species

Though skeletal muscles largely function similarly among different species, there are a number of differences that need to be highlighted. Differences concern the evolution of muscles, the number of muscles, muscle size, muscle distribution, proportion of muscle to body mass, muscle types, sarcomere structure, and muscle proteins. The most obvious difference between species is the number of skeletal muscles. Whereas humans have 640 muscles, the **elephant** is equipped with 40,394 muscles. Insects are equipped with a few hundred muscles and other species with a few thousand (Triplehorn et al. 2005). Whereas the nose of the human cannot be moved at all anymore, the highly flexible trunk of the elephant is moved by about 40,000 individual muscles. Concerning the architecture and composition of muscle, there can be significant differences between vertebrates and other species. In male **salmons** or tuna fish, the amount of muscle in relation to total body weight can reach up to 70 %. Another difference between species concerns the proportion of the muscle mass to the total body weight. Whereas the proportion of muscle mass is 50 % in humans, it is much lower in bees or other insects. The proportion of muscle mass to body mass is also larger in **birds** compared to **bats** (Norberg and Norberg 2012). Differences between species refer also to the presence of other muscle types. Whereas vertebrates have both striated and smooth muscles, insects have only striated muscles. Muscle strength is also different between species, whereas an ant can lift up to 50 times its own body weight; a weightlifter lifts only three to four times his own body weight. An **ant** has a strength advantage because of the ratio surface area to volume. The strength of a muscle is proportional to the surface area of its cross section, which is proportional to the square of its length. Volume on the other hand is three-dimensional and thus proportional to the cube of its length. Larger animals have a greater disparity between mass and strength. To lift an object, muscles of a large animal must also move a greater volume or mass of its own body. There is also a difference between species concerning the composition of contractile filaments. Whereas humans have only myosin types MYH1, 2, 4, or 7, animals have a much wider variability of myosins. Furthermore, increase in muscle mass in mice is not only achieved by an increase of the thickness of individual fibres, like in humans, but by attracting more myoblasts to fuse to a myotube. In animals not only voltage-gated sodium channels but also **voltage-gated calcium channels** contribute to the development of the action potential. In humans, action potentials last <1 ms, whereas in animals action potentials may last **up to 100 ms**.

4.2.6 Synopsis

This chapter should have helped the reader to understand basic development of muscle, anatomy and histology of skeletal muscles, and basic prerequisites of muscle function. Though statements, descriptions, and explanations were reduced to comprehensible messages, it was intended to provide an overview about all aspects of muscle anatomy and physiology using the human as example. Furthermore, it was attempted to identify differences concerning muscle organisation, architecture, function, and performance between different species. Despite these intentions, the reader is advised to collect more detailed knowledge about various aspects of the topic, if more profound insights into the world of muscles in human beings and other species are desired.

Acknowledgement Thanks to Prof. Erika Jensen-Jarolim, MD, and crossip communications, Vienna, Austria, for support in preparation of the figures, for the arrangement of illustrations.

References

Bähr M, Frotscher M (2005) Neurologisch-topische Diagnostik, 9th edn. Thieme, Stuttgart

Barnett MW, Larkman PM (2007) The action potential. Pract Neurol 7:192–197

Boyas S, Guével A (2011) Neuromuscular fatigue in healthy muscle: underlying factors and adaptation mechanisms. Ann Phys Rehabil Med 54:88–108

Bullock TH, Orkand R, Grinell A (1977) Neurobiology, neurons, neurophysiology, physiology. Freeman WH, San Francisco

Larsson L, Edström L, Lindegren B, Gorza L, Schiaffino S (1991) MHC composition and enzyme-histochemical and physiological properties of a novel fast-twitch motor unit type. Am J Physiol 261:C93–C101

Martini FH, Timmons MJ, Tallitsch RB (2008) Human anatomy, 6th edn. Benjamin Cummings, San Francisco, CA, pp 251–252

Norberg UM, Norberg RÅ (2012) Scaling of wingbeat frequency with body mass in bats and limits to maximum bat size. J Exp Biol 215:711–722
Partridge T (2009) Developmental biology: skeletal muscle comes of age. Nature 460:584–585
Scott W, Stevens J, Binder-Macleod SA (2001) Human skeletal muscle fiber type classifications. Phys Ther 81:1810–1816
Spruit MA, Thomeer MJ, Gosselink R, Troosters T, Kasran A, Debrock AJ, Demedts MG, Decramer M (2005) Skeletal muscle weakness in patients with sarcoidosis and its relationship with exercise intolerance and reduced health status. Thorax 60:32–38
Triplehorn CA, Johnson NF, Borror, DeLong (2005) Introduction to the study of insects, 7th edn. Brooks/Thomson Cole, USA

Lifeblood Flow: The Circulatory Systems

5

Claudia Stöllberger

Contents

Abstract

The function of the circulatory system (CS) is **transport** for supply of nutrients like amino acids, vitamins or electrolytes, gases, hormones, and blood cells; **removal** of toxic agents and metabolic products; and **regulation** of body temperature and volume. In some animals like flatworm or jellyfish, the CS is absent. In these animals every cell is able to obtain nutrients/water and oxygen without the need of the transport system.

C. Stöllberger (✉)
Krankenanstalt Rudolfstiftung, Vienna, Austria
e-mail: claudia.stoellberger@wienkav.at

E. Jensen-Jarolim (ed.), *Comparative Medicine*,
DOI 10.1007/978-3-7091-1559-6_5,

5.1 Closed and Open Circulatory Systems

In vertebrates and humans the CS is a **closed system** (Fig. 5.1a, b). A closed system is characterized by the basic principle that the blood never leaves the network of arteries, veins, and capillaries. A closed CS is composed of three components:

(a) The **blood vessels** are the "roads" for transportation. Arteries transport the blood from the heart to the organs. The capillaries are the places where exchange occurs. Nutrients are delivered and metabolic products are removed. The veins transport the blood from the organs to the heart. The lymphatic vessels transport fluid from tissue and organs to the heart.

(b) **Blood and lymphatic fluid** are the transported materials.

(c) The **heart** is a muscular pump that maintains the circulation.

In some invertebrates the CS is an **open system** (Fig. 5.2). In these animals a fluid—"hemolymph"—supplies the organs directly with oxygen and nutrients (Wasserthal 2007). Muscular movement facilitates the movement of the hemolymph like a "mixer." When the heart relaxes, hemolymph is drawn back toward the heart through open-ended pores. Hemolymph surrounds all cells and is composed of water, electrolytes, and proteins (Cammarato et al. 2011).

5.2 Circulatory System of the Fish

The circulatory system of the fish consists of the **heart with two chambers**, one atrium and one ventricle. Between the chambers are **valves**. The vascular system of the fish consists of veins, capillaries of the **gills**, arteries, and capillaries in the organs. Oxygen-deprived blood is pumped from the ventricle of the heart to the gills (Fig. 5.1c). Within the capillaries of the gills, the gas exchange takes place and oxygen-enriched arterial blood is transported to the capillaries of the organs. Oxygen-deprived venous blood returns to the heart. The musculature of the gills enhances the pumping function of the heart. The anatomy of the ventricles of a fish heart is characterized by extensive trabeculations and by the lack of coronary arteries. The transparent zebra fish embryo is used for research about myocardial growth and development (Verkerk and Remme 2012). In contrast to vertebrates the zebra fish heart has been shown to have the capacity of regeneration if parts of the heart are removed (Williams 2010).

5.3 Circulatory System in Endotherms (Including Mammals and Humans)

5.3.1 Systemic and Pulmonary Circulation

The circulatory system in endothermic species consists of the heart with four chambers: the right atrium and ventricle and the left atrium and ventricle (Figs. 5.1b and 5.3). Four valves are found at the entrance and exit of the ventricles. The

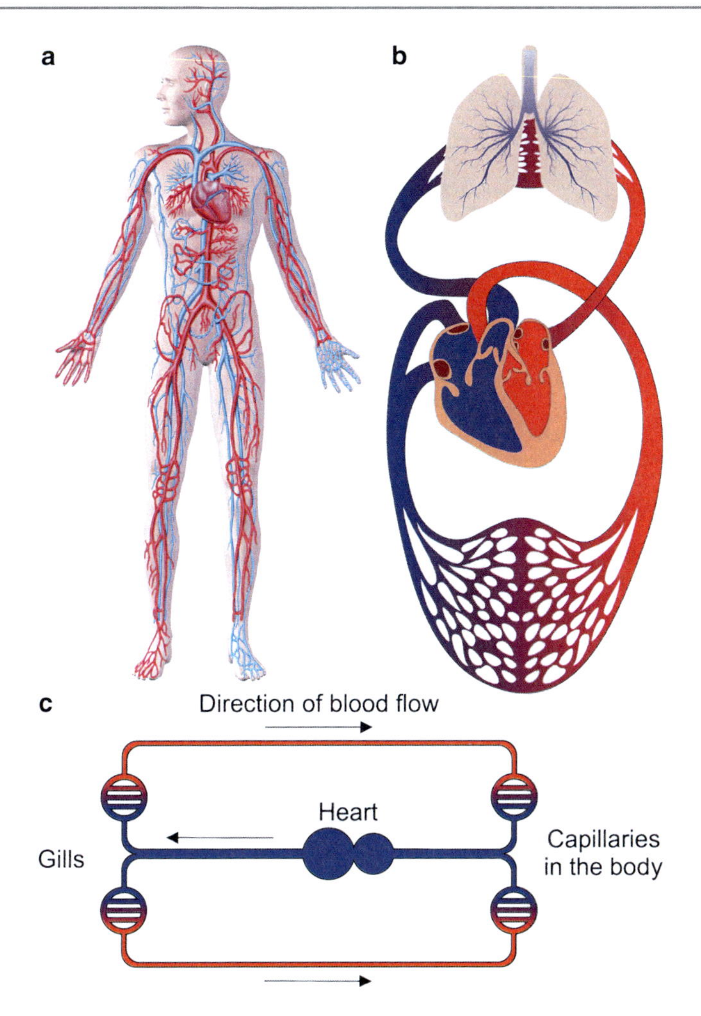

Fig. 5.1 (**a**) *The closed circulatory system in humans:* arteries are depicted in *red* and veins are depicted in *blue*. © [pixelcaos]—Fotolia.com. (**b**) *Circulatory system in endotherms* including mammals and humans: arteries are depicted in *red* and veins are depicted in *blue*. The heart as a central pump organ collects peripheral venous blood in the right heart side and forwards it to the lung for gas exchange (see Chap. 11). Oxygenated blood then enters the left heart side and is pumped to the periphery again. Valves between the atria and ventricles and at the exit points to the pulmonary artery and aorta play an important role in directing the bloodstream and preventing back stream. By definition, arteries are vessels leaving the heart, whereas veins represent vessels entering the heart, independent whether they carry oxygenated blood or not. © [p6m5]—Fotolia.com. (**c**) *Schematic drawing of the circulatory system of the fish*: arteries are depicted in *red* and veins are depicted in *blue*. In fish the gills take in the role of oxygen load to erythrocytes

vascular system consists of the **systemic circulation** with arteries transporting oxygen-rich blood from the left ventricle to the organs and veins transporting oxygen-deprived blood to the right atrium. Additionally the vascular system

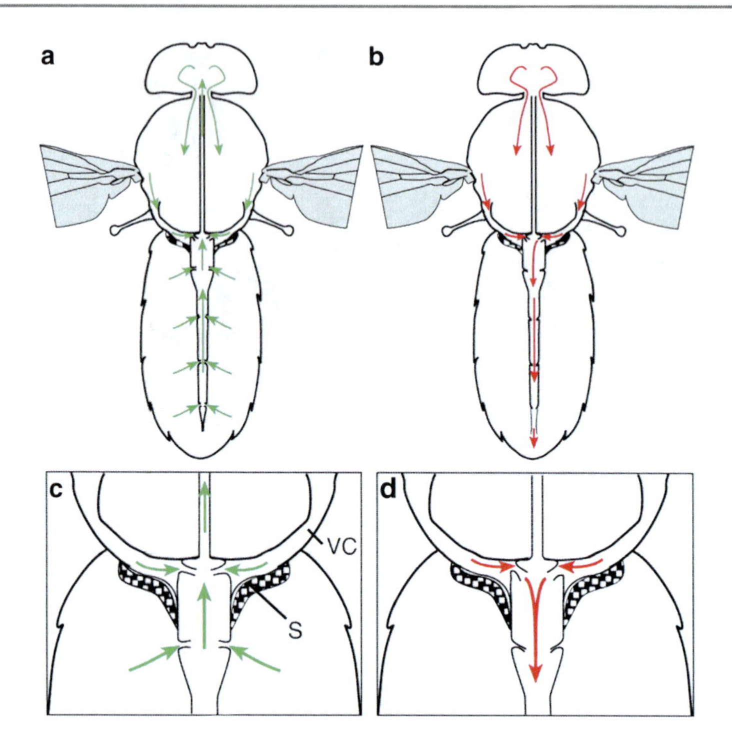

Fig. 5.2 *Graph summarizing hemolymph circulation* through the forward (**a**, **b**) and backward beating heart (**c**, **d**). (**a**, **c**) Overview; (**b**, **d**) detail, showing the central location of the anterior heart and its strategically favorable connection to the venous channels (VC). The lateral thoracic hemolymph flows through these channels directly to the first ostia. During diastole of the backward beating heart, thoracic hemolymph enters these anterior ostia exclusively. During forward beating, hemolymph is aspired from the heart through all ostia, including the first ones. Thus a lateral circulation in the thorax is maintained independently of heartbeat direction. S. pericardial septum. Reproduced with permission, from Wasserthal et al (Wasserthal 2007), J Exp Biol 2007;210:3707–3719

consists of the **pulmonary circulation** where arteries transport oxygen-deprived blood from the right ventricle to the lungs and veins transport oxygen-rich blood from the lungs to the left atrium.

5.3.2 Blood Vessels

The blood vessels are composed of different layers. The most inner layer of arteries, capillaries, and veins is the endothelium. It is a smooth surface of a cellular layer. Outside the endothelium is a layer of fibrous tissue being surrounded by a muscular layer. The outermost layer of the vessel wall is a layer of cells, which connect the vessel with the surrounding tissue. Depending on the size and function of the vessels, the layers vary in their thickness. The thinnest vessel walls are found in the capillaries.

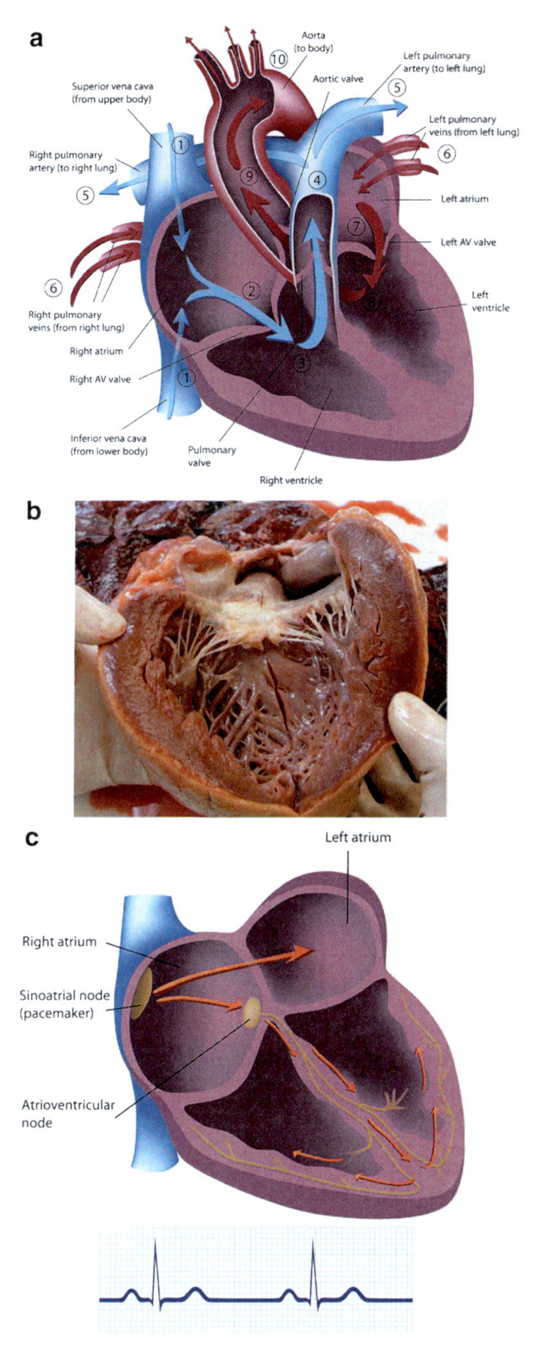

Fig. 5.3 (**a**) *The pathway of blood flow through the heart:* venous blood from the upper and lower parts of the body is collected by the superior and inferior vena cava (1) and enters the right atrium (2). After passing the tricuspid (right atrioventricular) valve, venous blood enters the right

The volume capacity of the blood vessels is much larger than the volume of the blood. Thus, maintenance of the muscular tone of the vessel wall is essential for metabolism, volume, and thermoregulation and is regulated by the sympathetic and parasympathetic nervous system (see Fig. 5.3b). Capillaries are able to distend from 6 to 30 μm. There are more veins than arteries. More than 50 % of the blood stays in the veins. This phenomenon is termed "venous pooling."

The vascular tone is regulated by:

(a) **Nerves** around the blood vessels: sympathetic (adrenergic) activation leads to vasoconstriction (narrowing of the vessel), and parasympathetic (cholinergic) activation leads to vasodilatation (widening of the vessel).
(b) Various **hormones** which are released into the blood in different parts of the brain as a consequence of various stimuli. Further hormones, like natriuretic peptides or renin, are released in the cardiac cavities and in the kidneys (see Chap. 10).
(c) **Receptors** in the vascular wall: pressoreceptors measure changes of the volume status and chemoreceptors measure changes of the oxygen saturation and CO_2 content of the blood.

5.3.3 Blood

Blood consists of cellular and liquid components. Blood accounts for 6–8 % of body weight. Inside the vessels, the blood in physiological conditions is always liquid. Clot formation occurs either if the vessel wall is damaged or if blood leaves the vessel. Only in pathologic conditions clot formation may occur inside the vessel. A clot that is formed inside the vessel is termed "thrombus."

Cellular components of the blood:

(a) Red cells, also termed erythrocytes—they are instruments for oxygen transportation besides other metabolic transport functions.

Fig. 5.3 (continued) ventricle (3). Venous blood leaves the right ventricle via the pulmonary valve and enters the pulmonary artery (4). From there the venous blood is led either to the right or to the left lung by the right or left pulmonary artery (5). Within the lungs, gas exchange takes place and oxygen-enriched arterial blood is transported by the right and left pulmonary veins (6) to the left atrium (7). After passing the mitral (left atrioventricular) valve, the blood enters the left ventricle (8). The blood leaves the left ventricle via the aortic valve and enters the aorta (9). From there, the blood either enters the vessels to the head and upper limbs or follows the aorta into to the lower parts of the body (10). © [Alila Medical Images]—Fotolia.com. (**b**) *Autoptic specimen of a human heart*. It is opened in the long-axis direction and shows the left atrium in the upper portion, the mitral valves and papillary muscles, and the left ventricle. (**c**) *The electrocardiogram* (ECG, EKG) records electrical activity within the cardiac conduction system (*orange lines* and *yellow spots* in the figure). In regular sinus rhythm, the electrical activity starts within the atrium and is conducted into the ventricles. In physiological sinus rhythm, the electrocardiogram shows a small wave, indicating activity of the atrium, followed by a large wave, indicating electrical activity within the ventricles. © [Alila Medical Images]—Fotolia.com

(b) White cells, also termed leukocytes—they are instruments for defense against endogenous and exogenous harms, like bacteria. Depending on their chemical properties, they are classified into granulocytes, lymphocytes, and monocytes. The ratio of the red to white cells varies among the species and is between 500 and 1,500:1.
(c) Platelets, also termed thrombocytes—they are important for clot formation. The number of platelets is relatively constant among vertebrates between 200.000 and 800.000/μL.

Fluid components of the blood:

(a) Plasma consists of proteins, water, and electrolytes. Proteins are needed for nutrition. Additionally proteins may act as hormones, enzymes, and as soluble factors of innate and adaptive immunity, such as antibodies which are needed for the specific defense against pathogens. Further proteins comprise coagulation factors needed for clot formation and fibrinolytic proteins needed for clot resolution.
(b) Lipids are transported from the gut to the organs where they are metabolized or stored.
(c) Inorganic substances like electrolytes and buffers are important for muscular activation and CO_2 elimination.

5.4 Heart

The heart of endotherms including birds, mammals, and humans is a conic muscular organ with four cavities (Fig. 5.3a, b). The heart is located in a sack (pericardium) within the thorax between both lungs. The right ventricle pumps the oxygen-deprived blood into the lungs, and the left ventricle pumps the oxygen-enriched blood into the systemic circulation. Cardiac valves, located between the atria and ventricles and in the outflow region of the ventricles, ensure that the blood flow is directed into the appropriate direction. The lymphatic vessels drain into veins or into the right atrium. The relative heart weight of the different species is influenced by several factors like metabolic range (= the amount of energy expended/day) and training state (Table 5.1). The function of the muscle cells including heart muscle cells is described in Chap. 4.

5.4.1 Coronary Arteries and Veins

Two coronary arteries with many branches supply both ventricles and atria. Many cardiac veins transport metabolic products and oxygen-deprived blood via coronary sinus into the right atrium.

Table 5.1 Relative heart weight of several species (Habermehl et al. 2005)

Species	Heart weight as percentage of body weight (%)
Fish	0.06–0.20
Rabbit	0.25
Pig, rat	0.30
Human	0.40–0.50
Cattle	0.50
Cat	0.55
Mouse	0.60
Horse	0.60–1.04
Chicken	0.50–1.42

5.4.2 The Skeleton of the Heart

The skeleton of the heart stabilizes the cardiac structure and function. It is located close to the valves and close to the insertion of the interventricular septum (= border wall between the right and left ventricle). The skeleton of the heart consists of fibrous tissue in most species. Additionally, cartilage is found in horses and bone in cattle.

5.4.3 The Cardiac Cycle

Contraction of the heart is termed **systole.** Relaxation of the heart is termed **diastole.** Contraction and relaxation constitute the heart beat ("pulse"). The heart rate is counted in beats per minute. During the cardiac cycle, blood flows through the cardiac chambers in a specific manner and direction. The cardiac valves prevent backward flow. The heart rate varies among species and is influenced by several factors like metabolic range (= the amount of energy expended/day), physical activity, and temperature of the environment (Table 5.2).

5.4.4 Cardiac Impulse Generation and Conduction

Hearts of several species continue beating even when they are removed from the body. This observation suggests that the cardiac rhythm evolves within the heart. This phenomenon is termed **cardiac autonomy.**

In all species, even in invertebrates, the impulse is generated in a distinct location of the heart, termed **pacemaker** (Sláma 2012). The impulse spreads from there either via muscular cells or via the cardiac conduction system. Nerves and hormones are additional extracardiac regulators of the cardiac rhythm. The impulse generation and conduction can be visualized by the surface or intracardiac electrocardiogram (Fig. 5.3c).

Table 5.2 Resting heart rate in several species (Penzlin 2005)

Species	Beats/min
Whale	15–16
Elephant	25–30
Cat	110–140
Cattle	55–80
Rat	350–450
Mouse	550–650
Bat	up to 972
Turkey	93
Crow	342
Canary bird	800–1,000
Eel	46–68
Frog	35–40
Turtle	11–37
Zebra fish	130
Octopus	33–40
Daphnia	250–450
Asellus	180–200
Drosophila melanogaster	360
Human	60–80

Table 5.3 Systolic and diastolic blood pressure in different species (Penzlin 2005)

Species	Blood pressure—normal values (mmHg)
Giraffe	300/230
Horse	114/90
Human	120/80
Cat	125/75
Mouse	147/106
Cock (hen)	191/154 (162/133)
Starling	180/130
Sparrow	180/140
Frog	27
Eel	35
Dogfish	31/16

5.4.5 Blood Pressure

The systolic blood pressure (SP) is defined as the maximal pressure at systole, and the diastolic blood pressure (DP) as the maximal pressure during diastole. The mean arterial pressure (MAP) is the average arterial pressure during the cardiac cycle and can be estimated by the following formula:

$$\text{MAP} = \text{DP} + 1/3\ (\text{SP} - \text{DP})$$

The blood pressure differs among species, is higher in birds than in mammals, has no relation to the body size, but is influenced by the relative height distance between the heart and the different organs (Table 5.3).

Acknowledgement Thanks to Prof. Erika Jensen-Jarolim, MD, and crossip communications, Vienna, Austria, for support in the arrangement of illustrations.

References

Cammarato A, Ahrens CH, Alayari NN et al (2011) A mighty small heart: the cardiac proteome of adult Drosophila melanogaster. PLoS One 6:e18497. doi:10.1371

Habermehl KH (ed), Nickel R, Schummer A, Seiferle E (2005) Lehrbuch der Anatomie der Haustiere, Band III Kreislaufsystem, Haut und Hautorgane. Parey, Hamburg

Penzlin H (2005) Lehrbuch der Tierphysiologie. Spektrum, Heidelberg

Sláma K (2012) A new look at the comparative physiology of insect and human hearts. J Insect Physiol 58:1072–1081

Verkerk AO, Remme CA (2012) Zebrafish: a novel research tool for cardiac (patho)electrophysiology and ion channel disorders. Front Physiol 3:255. doi:10.3389/fphys.2012.00255

Wasserthal LT (2007) Drosophila flies combine periodic heartbeat reversal with a circulation in the anterior body mediated by a newly discovered anterior pair of ostial valves and 'venous' channels. J Exp Biol 210:3707–3719

Williams R (2010) Thanks be to zebrafish. Circ Res 107:570–572

Steering and Communication: Nervous System and Sensory Organs

6

Josef Finsterer
Hanna Schöpper and Sabine Breit

Contents

J. Finsterer (✉)
Krankenanstalt Rudolfstiftung, Vienna, Austria
e-mail: fifigs1@yahoo.de

H. Schöpper (✉) • S. Breit
Institute of Anatomy, Histology and Embryology, University of Veterinary Medicine, Vienna, Austria
e-mail: Hanna.Schoepper@vetmeduni.ac.at

E. Jensen-Jarolim (ed.), *Comparative Medicine*,
DOI 10.1007/978-3-7091-1559-6_6, © Springer-Verlag Wien 2014

6.1 The Human Central and Peripheral Nervous System

Josef Finsterer

6.1.1 Abstract

The mammalian, specifically the human nervous system is an organ which affects nearly every other system and exerts controlling, regulatory, monitoring, creative, and executive functions. It is divided into the central nervous system (CNS), which is composed of the **brain** (cerebrum) and the **myelon** (spinal cord), and the **peripheral nervous system** (PNS), composed of the peripheral nerves and the musculature (skeletal muscles) (Trepel 2008). Peripheral nerves are divided into **cranial nerves** (originate from the brain) and the **spinal nerves** (originate from the spinal cord). Both, the CNS and the PNS, have a motor, sensory, and autonomic division (Platzer 2005). The motor division and sensory division include a somatic and autonomic fraction. The autonomic fraction comprises the sympathetic and the parasympathetic division (see below Fig. 6.4). This chapter aims to give a short overview and basics about the developmental, anatomic, histological, and physiological aspects of a highly developed nervous system taking the human as a prototype of mammalians.

6.1.2 Introduction: Development of the Nervous System

After conception the zygote multiply divides to give way to the blastocyst, a vesicle bounded by a single layer of stem cells (Reus 2010). After invagination (gastrulation) of this vesicle, two caves evolve, the amniotic cave and the adjacent chorionic cave. At the contact point of these two caves three germinal layers develop, the ectoderm, the mesoderm, and the endoderm. From the endoderm originate the inner organs, from the mesoderm the muscles, connective tissue, and the bones, and from the **ectoderm** the nervous system. First, the ectoderm differentiates into the **neural plate**, followed by the neural groove and folds. Neural groove and folds form the neural tube and the neural crest. After closure of the frontal pore of the neural tube, three vesicles develop from the frontal part of the neural tube (Reus 2010). They are known as **prosencephalon** (forebrain), the **mesencephalon** (midbrain), and the **rhombencephalon** (hindbrain). From the forebrain originate the telencephalon and the diencephalon, from the mesencephalon the midbrain, and from the rhombencephalon the metencephalon and the myelencephalon.

6.1.3 Anatomy of the Brain

Anatomically, the entire mass of the CNS is built up of two **hemispheres**, with areas predominantly containing **grey matter** (contains cell bodies of neurons (nerve cells)) and areas containing predominantly **white matter** (contains myelinated or unmyelinated axons (extensions of neurons)). The amount of grey and white matter varies between different CNS areas and is characteristic for a CNS region. The anatomy of the brain is depicted in Fig. 6.1a–c.

6.1.3.1 Telencephalon

The telencephalon is built up of the cortex (grey matter), the white matter (white matter), the basal ganglia (grey matter), the corpus callosum (white matter), the hippocampus (grey and white matter), and the rhinencephalon (grey and white matter) (Feneis 1974).

6.1.3.2 Cortex

The cortex builds up the outer surface of the telencephalon and is densely folded, giving rise to **sulci** (grooves) and **gyri** (elevations) (Fig. 6.1a). From outside only one third of the cortex is visible. The other two thirds are hidden inside the sulci and gyri. Due to their high individual variability, only a few sulci can be constantly found in all human individuals: the central sulcus (Rolandi), the Calcarina sulcus, the lateral sulcus, also known as Sylvian groove, the gyrus cinguli, and the sulcus temporalis superior. The main sulci divide the cortex of the human telecephalon into four lobes, the frontal lobe, the parietal lobe, the occipital lobe, and the temporal lobe (Feneis 1974).

6.1.3.3 White Matter

The white matter makes up the major portion of the telencephalon and is built up of **myelinated axons**, **glia**, and vessels. It is limited by the **cortex**, the **lateral**

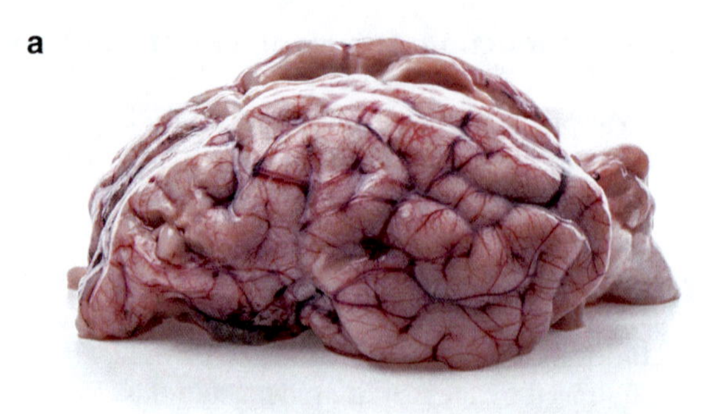

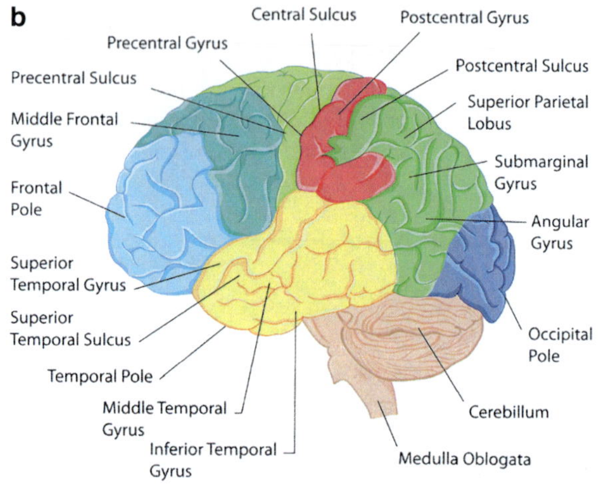

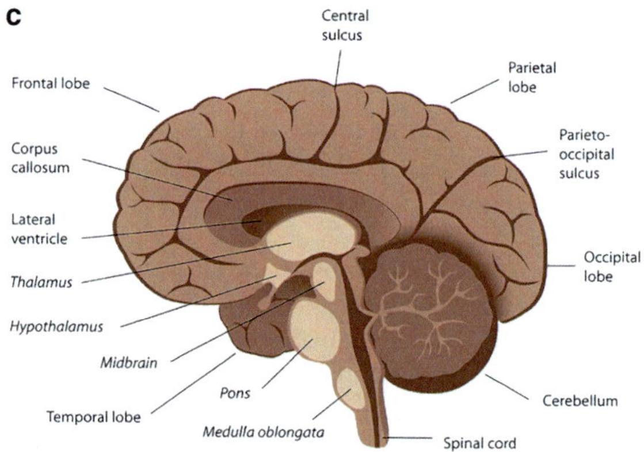

Fig. 6.1 (a) **Brain of a sheep** from left (temporal) side. Note sulci and gyri. ©[msk.nina]—Fotolia.com. (**b**) **Side view on human brain** from left side. ©[snapgalleria]—Fotolia.com. (**c**) **Sagittal cut through human brain**, from left side. ©[Alila Medical Images]—Fotolia.com

ventricle, and the **basal ganglia**. Axons of the white matter predominantly contain projection fibers which are either corticofugal (they descend from the cortex and are made up of cortico-nuclear, cortico-pontine, cortico-spinal, cortico-striatal, cortico-thalamic, cortico-nigratal, cortico-reticular, cortico-rubral, and cortico-tectal portions) or corticopetal fibers (Feneis 1974).

6.1.3.4 Basal Ganglia

Basal ganglia (striatum) are composed of the **caudate nucleus** and the (lens-like) **lentiform nucleus**. The lentiform nucleus houses the **putamen** and the **globus pallidus**. Caudate nucleus and putamen form the neostriatum and the globus pallidus the paleostriatum. Some anatomical books also include the claustrum and the amygdala as part of the basal ganglia (Feneis 1974). The main task of the basal ganglia is to **control movement**. Dysfunction of the basal ganglia results in the development of movement disorders, such as Parkinson's disease, Huntington's disease, hemiballism, or dystonia (Platzer 2005).

6.1.3.5 Corpus Callosum

The corpus callosum is the largest of the **interhemispheric connections** and contains approximately 300 million axons. It is divided into splenium, truncus, genu, and rostrum corporis callosum. Despite functional differences between right and left hemisphere, interhemispheric information transfer is carried out via this structure (Feneis 1974).

6.1.3.6 Hippocampus

The hippocampus is one of the evolutionary oldest parts of the brain and located at the mesiotemporal portion of the cerebrum and visible either from the bottom of the brain or after dissection of part of the temporal lobe. It plays a major part in the storage and processing of **short- and long-term memory** contents. In other mammals the hippocampus plays a major role in spatial orientation (Platzer 2005; Feneis 1974).

6.1.3.7 Rhinencephalon

The rhinencephalon includes the olfactory bulb, the olfactory tract, the anterior olfactory nucleus, the anterior perforated substance, the medial olfactory stria, parts of the amygdala, and the prepyriform area. The main task of the rhinencephalon is the conduction and **processing of olfactory input** from the olfactory mucosa at the upper nasal concha (Trepel 2008; Platzer 2005; Feneis 1974).

6.1.3.8 Diencephalon

The diencephalon is composed of the retina (the ectodermal parts of the eye are part of the brain), the thalamus, the hypothalamus, the subthalamic nucleus, the metathalamus consisting of the corpus geniculatum mediale and laterale, the epithalamus, comprising the epiphysis and the habenulae, and the pituitary gland (Feneis 1974).

Retina

The retina of the **eye** is responsible for transforming light into bioelectrical information, which then is conducted along the **optic nerve**, the chiasm, the optic tract, the corpus geniculatum laterale, and the visual radiation of Gratiolet and Meyer's loop to the primary and secondary **visual cortex**. There, processing of the visual input results in realization of the optic reality in relation to previous, memorized input.

Thalamus

The thalamus, located in close relation to the brainstem and the basal ganglia (Fig. 6.1c), plays a central role in the **processing of sensory input** via spinal and cranial nerves and receives its afferences via the spinothalamic tract, the medial lemniscus, the pallidum, the putamen, the cortex, the reticular formation, and the dentate nucleus. Vice versa its efferences reach the cortex, the putamen, and the pallidum. The term thalamus originates from the old Egyptian word "thalem," which means gate to consciousness (Platzer 2005; Feneis 1974).

Hypothalamus

The hypothalamus plays a central role in the control and regulation of vegetative functions and is regarded as the **origin of the central sympathetic tract** (Fig. 6.4b). Provided with numerous afferences from all kinds of receptors, it regulates the minute volume, blood pressure, heart rate, tidal volume, basal metabolic rate, diameter of precapillary sphincters, sweat and saliva production, contraction states of detrusors, the release of adrenalin and other hormones, peristalsis, and the pupillary reaction. The hypothalamus also governs the pituitary gland (Platzer 2005; Feneis 1974).

Subthalamic Nucleus

The subthalamic nucleus is a small lens-shaped nucleus ventral and below the thalamus, dorsal to the substantia nigra, and medial to the internal capsule. The precise function of the nucleus is unknown but it is regarded to be involved in the basal ganglia control system, which is regarded to perform **action selection**. The nucleus seems to suppress impulsivity in individuals with two equally rewarding stimuli (Feneis 1974).

Metathalamus

The metathalamus consists of the corpus geniculatum mediale and laterale and is regarded as a relay station for **visual and auditory input** and modifies these stimuli.

Epithalamus

The epiphysis plays a central role in the **regulation of circadian and seasonal rhythms**, which is why it is regarded as part of the photoendocrine system. The habenulae are part of the olfactory system. They function as relay stations, connecting the rhinencephalon with autonomic centers in the brainstem (Trepel 2008).

Pituitary Gland

The pituitary gland is an endocrine organ functionally connected to the hypothalamus (see Chap. 8). It secretes ten hormones that regulate homeostasis. The gland is divided into an adenohypophysis and a neurohypophysis. The adenohypophysis produces the growth hormone, thyroid stimulating hormone, the adrenocorticotropic hormone, beta-endorphine, prolactine, luteinizing hormone, follicle-stimulating hormone, and the melanocyte-stimulating hormone. The neurohypophysis produces oxytocin and the antidiuretic hormone (Platzer 2005).

6.1.3.9 Midbrain (Mesencephalon)

The midbrain or mesencephalon is located between the prosencephalon and the rhombencephalon and is built up of the crura cerebri, the tegmentum, and the tectum. It functions as a relay between frontal and posterior parts of the cerebrum and houses important tracts, which connect centers of the forebrain with those of the hindbrain and the myelon. The most important of these are the corticobulbar tract, the **cortico-spinal tract** (**pyramidal tract**), the medial and lateral **lemniscus**, and the lateral and medial **spinothalamic tracts**. Important nuclei located in the midbrain include the substantia nigra, the nucleus ruber, the medial and lateral geniculate body, the nuclei of cranial nerves III, IV, and the mesencephalic tract of cranial nerve V, and the superior and inferior colliculus (Trepel 2008; Platzer 2005; Feneis 1974).

6.1.3.10 Metencephalon

The metencephalon is built up of the **pons**, the **pedunculi cerebelli**, and the **cerebellum** (Fig. 6.1c). The pons contains the pyramidal tract, and motor and sensory nuclei of cranial nerves V, VI, and VII. The cerebellum is divided into the spinocerebellum, the cerebrocerebellum, and the vestibulocerebellum. The spinocerebellum is involved in the regulation of muscle tone and coordination of skilled voluntary movements. The cerebrocerebellum is involved in the planning and modulation of voluntary activity, and the location for storage of procedural memory. The vestibulocerebellum is involved in the maintenance of **balance** and control of **eye movements**. The inferior cerebellar peduncle contains fibers originating from the inferior olives, and proprioceptive input from the spinocerebellar tract. The middle cerebellar peduncle contains fibers connecting the pons and the cerebellum. The superior cerebellar peduncle contains efferent fibers heading for the thalamus and the ruber nucleus (Trepel 2008; Feneis 1974).

6.1.3.11 Myelencephalon

The myelencephalon, also known as **medulla oblongata** (Fig. 6.1c), contains a number of important afferent and efferent tracts. It also houses the cross-over of the pyramidal tracts, the olives, and motor, sensory, and autonomic nuclei of cranial nerves V (descending tract), VIII, IX, X, XI, and XII. Most of the lower (bulbar) cranial nerves originate from the medulla. The medulla oblongata together with the pons and the midbrain form the **brainstem**. The brainstem contains motor, sensory, and autonomic tracts, and motor, sensory, and autonomic nuclei of cranial nerves. The pyramidal tract runs through the brainstem, which is accompanied by the extrapyramidal system, which modulates motor functions, and is built up of the neostriatum, the globus pallidus, the substantia nigra, the red nucleus, and the subthalamic nucleus (Feneis 1974).

6.1.3.12 Cranial Nerves

Except for cranial nerve I, **cranial nerves originate from the brainstem** (Bähr and Frotscher 2003) (Fig. 6.2a). **Cranial nerve I** (**olfactory nerve**) is purely sensory and responsible for the conduction and distribution of olfactory input. **Cranial nerve II** (**optic nerve**) is not a real nerve but an extension of the cerebrum and pure sensory. It carries input from the retina to the centers of visual processing in the occipital lobe.

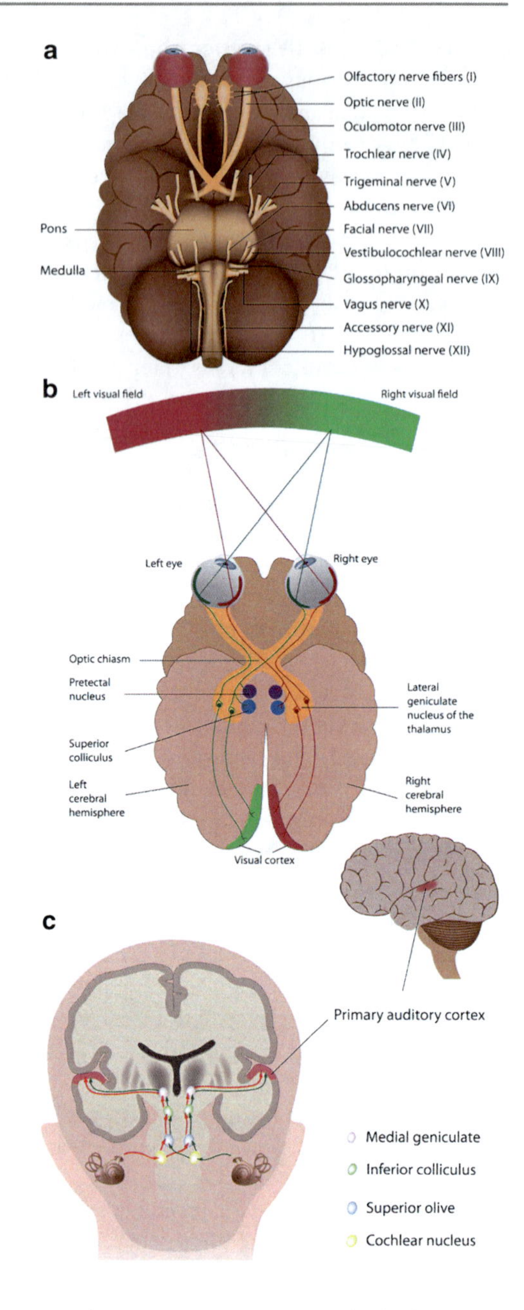

Fig. 6.2 (a) **Scheme showing cranial nerves I–XII**. ©[Alila Medical Images]—Fotolia.com. (b) **The visual projection pathway**. ©[Alila Medical Images]—Fotolia.com. (c) **The auditory pathway**. ©[Alila Medical Images]—Fotolia.com

Cranial nerve III (**oculomotor nerve**) carries motor and parasympathetic fibers and is involved in the execution of eye movements, pupillary width, and tear production. **Cranial nerve IV** (**trochlear nerve**) is pure motor and together with **cranial nerve III and IV** involved in bulb movements. **Cranial nerve V** (**trigeminal nerve**) carries motor and sensible functions. It is involved in chewing and the sensibility of the face and tongue. **Cranial nerve VI** (abducens nerve) is pure motor and involved in bulb movements. **Cranial nerve VII** (**facial nerve**) carries motor, sensory, sensible, and autonomic fibers, innervates the facial muscles, and is additionally involved in tasting, sensibility of the tongue, and production of saliva. **Cranial nerve VIII (statoacustic nerve)** is pure sensory and involved in the conduction of input from the ear and the balance organ. **Cranial nerve IX (glossopharyngeal nerve)** carries motor, sensory, sensible, and autonomic fibers and is involved in swallowing, tasting, sensibility of the tongue and the mouth, and production of saliva. Cranial nerve **X (vagal nerve)** also carries motor, sensory, sensible, and autonomic (parasympathic) fibers (Fig. 6.4b) and is involved in speaking, peristalsis, tasting, sensibility of the tongue, saliva production, and control of heart rate, breathing rate, and bronchial function. **Cranial nerve XI (accessorius nerve)** is pure motor and involved in head anteflexion, head rotation, and shoulder elevation. **Cranial nerve XII (hypoglossus nerve)** is pure motor and responsible for tongue movements (Trepel 2008; Platzer 2005; Feneis 1974; Bähr and Frotscher 2003).

6.1.3.13 Cerebral Blood Circulation

The cerebrum receives its blood supply from the heart via the aorta, the **carotid arteries** (anterior circulation), and the **vertebral arteries**, which unify to the basilary artery (posterior circulation). From the internal carotid artery originate the medial cerebral artery and the anterior cerebral artery. The basilary artery divides into the posterior cerebral arteries. Anterior and posterior circulation are connected via the posterior communicans arteries. The right and left anterior cerebral arteries are connected via the anterior communicans artery (Bähr and Frotscher 2003). Blood transport from the brain to the heart is carried out via the venous system, which is built up of the so-called **sinuses**, which all lastly flow into the **jugular vein** (Bähr and Frotscher 2003).

6.1.3.14 Protection of the Brain

Since the cerebrum is one of the most important organs, it became well protected during evolution. Structures involved in brain protection include the meninges, the cerebrospinal fluid (CSF), the blood–brain barrier (BBB), the calvaria, and the cerebral immune system. The **meninges** are composed of the **pia mater**, which is directly attached to the cerebral tissue, the arachnoid layer, which spans the subarachnoid space where the CSF is circulating and where the cerebral arteries are located before they enter the cerebrum, and the **dura mater**, which immediately touches the skull and the arachnoid layer and provides space for the **venous sinuses**. The ventricular system is composed of four main caves (ventricles) within the cerebrum. CSF is produced by the choroid plexus of ventricles 1, 2, and 4 (about 320 ml/day). CSF circulates through the ventricular system and the subarachnoid space. From the subarachnoid space the CSF is reabsorbed into the sinuses via the Pacchioni granulations (Platzer 2005).

6.1.4 Histology of the Brain

6.1.4.1 Neurons

The dominant cell type within the CNS is the neuron (Fig. 6.3a). It is a unique cell type with lots of branches (**axons**, **dendrites**), which connect them to other neurons,

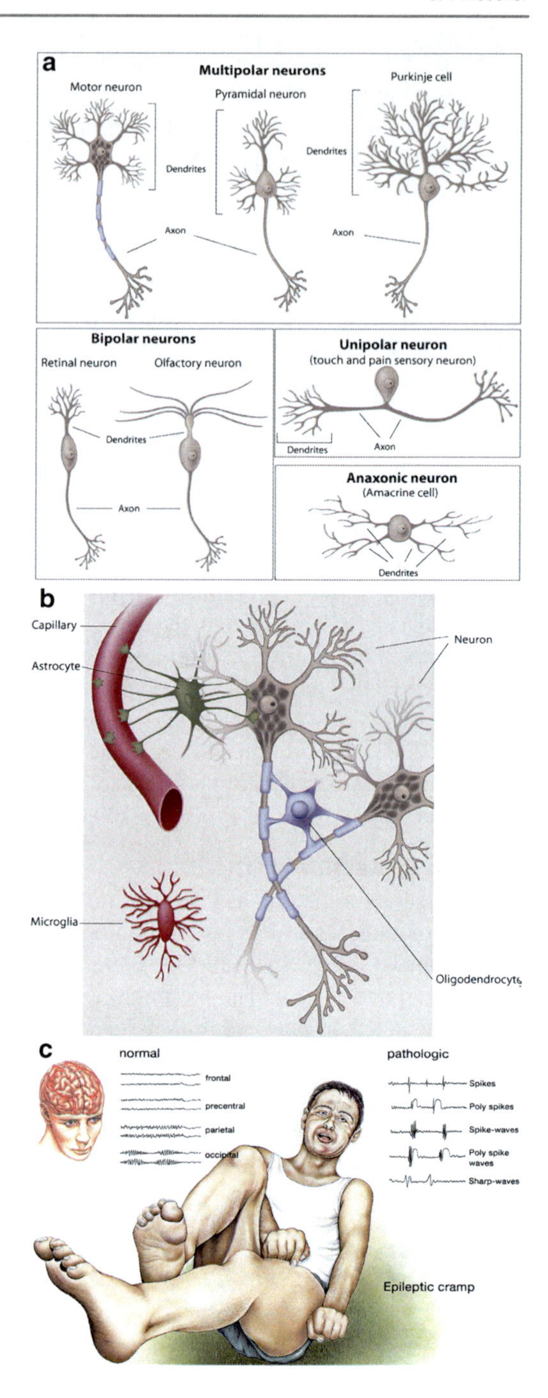

Fig. 6.3 (**a**) Different types of neurons. (**b**) **The cells of the brain** are constituted by neurons, macroglia (astrocytes and oligodendrocytes), and microglia each composed of different cell types with different functions. There is a certain plasticity between the cells; e.g., astrocytes, being pluripotent cells, may turn into neurons. ©[Alila Medical Images]—Fotolia.com. (**c**) **The electroencephalogram** (**EEG**) records simultaneously from left and right brain hemisphere, and from frontal, precentral, parietal, or occipital area. Therefore, two almost identical curves can be deduced from the two hemispheres of a brain in healthy condition (*left* side: normal). In pathologic conditions (*right* side) aberrant curves with various excitation types (spikes, waves) can be deduced, allowing diagnosis. © [Henrie]—Fotolia.com

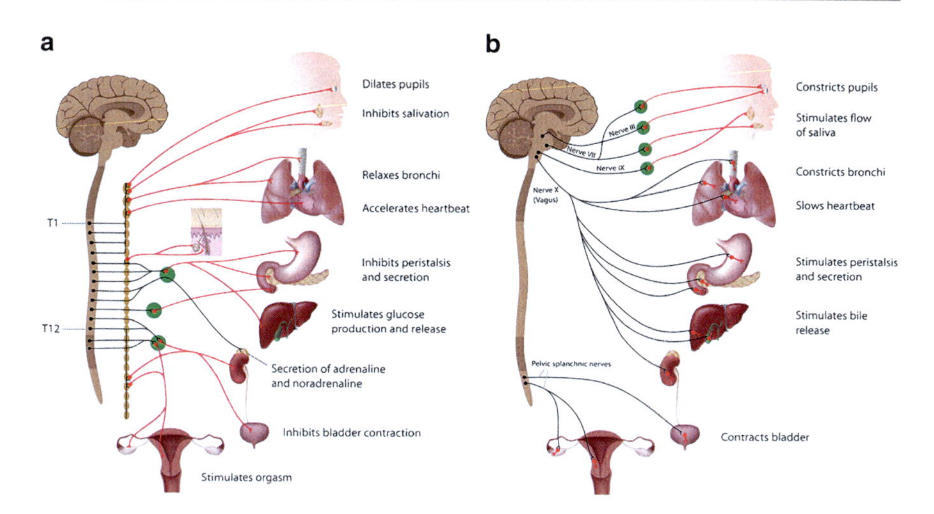

Fig. 6.4 (a) **The sympathetic nervous system** derives from central nuclei and the myelon and is responsible for stress reactions and the tonus of smooth muscles. The sympathetic nerve fibers are converted into postsynaptic fibers in peripheral ganglia, such as thoracic ganglia (T1, T12). © [Alila Medical Images]—Fotolia.com. (**b**) **The parasympathetic nervous system** is predominantly organized by the Nervus vagus (cranial nerve no. 10) and converted into postsynaptic fibers in the peripheral organs only. The parasympathetic autonomous nervous system is responsible for relaxation of the system and organ functions in this condition. ©[Alila Medical Images]—Fotolia.com

muscle cells, or receptors via synapses. Neurons are excitable and generate action potentials. Neurons interact with other neurons, glial cells, or other cell types. Neurons are classified according to their polarity into monopolar, bipolar, or multipolar neurons. According to the direction of the conduction, they are classified as **afferent or efferent neurons**, or as **interneurons** which conduct into both directions. According to their morphology several different cell types (Betz, Purkinje, anterior horn, granular, etc.) are differentiated. Neurons are also classified according to the **transmitter** which they use for communication (e.g., acetylcholine, serotonin, GABA). According to their discharge pattern tonic (regular spiking), phasic (bursting), and fast spiking neurons are differentiated (Sobotta and Histologie 2010).

6.1.4.2 Glial Cells

Glial cells are the second most frequent cell type within the CNS. Three types of glial cells are known, **astrocytes**, **oligodendrocytes**, and **microcytes** (Fig. 6.3b). Astrocytes synapse with neurons and capillaries. Oligodendrocytes form sheaths around axons, similar to myelin sheaths in the PNS. Glial cells are regarded as the glue of the CNS. They support neurons in the formation of the BBB. They are involved in the clearing of neurotransmitters after release, participate in the synaptic transmission in the hippocampus and cerebellum, and regulate the synaptic plasticity (synaptogenesis). They are capable of mitosis and crucial for the development of the CNS (Sobotta and Histologie 2010).

6.1.4.3 Blood–Brain Barrier

The BBB is a mechanical, functional, and immunological barrier between the capillaries and the cerebrum and between the epithelium of the plexus and the CSF. The BBB is composed of the **endothelial cells**, the **basal membrane** of the capillary, the **pericytes**, and the **astrocytes**. Astrocytes form part of the outer layer of the capillary tube. The BBB is permeable for oxygen, CO_2, water, electrolytes, and monocytes (Sobotta and Histologie 2010).

6.1.5 Physiology of Brain

6.1.5.1 Membrane Potential

A membrane potential originates from differences in polarity between extracellular and intracellular compartment. This polarity difference is provided by the flux of ions through pores within the cell membrane. Sodium and chloride ions are found in high concentrations at the extracellular side, whereas high concentrations of potassium are found on the intracellular side. Due to the distribution of these ions the outside of the membrane is positively charged and the inside is negatively charged. This difference in chargement is called **membrane potential**, which amounts to −70 to −90 mV (Reus 2010; Huppelsberg and Walter 2009).

6.1.5.2 Action Potential

An action potential is a triggered, typical sudden change in polarity, between the intra- and extracellular side of the membrane. As soon as the resting potential is decreased to the trigger level, depolarization of the membrane to a value of +50 mV is initiated. The overshoot is followed by repolarization down to an undershoot below the resting potential and return of the potential to the resting stage. Depolarization is mainly carried out by a sudden influx of sodium into the cell, whereas repolarization is carried out by an outflux of potassium ions. Also the undershoot below the resting potential is accomplished by potassium outflux. A characteristic of an action potential is its tendency to induce opening also of neighboring sodium and potassium channels and thus generation of neighboring action potentials resulting in spreading of the depolarization along the membrane. The propagation of the action potential may be carried out either in a continuous manner in the absence of myelination of the axon or in a saltatory (jumping) way if the axon is myelinated. The propagation velocity is higher in myelinated axons since only those parts of the axon are depolarized which locate within the constrictions of the myelin sheath (nodes of Ranvier) (Reus 2010; Huppelsberg and Walter 2009).

6.1.5.3 Synapses

Synapses are a crucial structure and functional unit of the CNS and PNS where they serve as **connections** for the information transfer between neurons, between neurons and muscle cells, or between receptors and neurons. According to their connections axonosomatic, axonodendritic, and axonoaxonic synapses are differentiated. Synapses may be also differentiated according to the transmitter released (Reus 2010).

6.1.5.4 Neurotransmitters

Neurotransmitters used by neurons may be categorized as biogenic amines, which include acetylcholine, adrenalin, noradrenalin, serotonin, and histamine, peptides such as endorphins, encephalines, substance P, etc. as inhibitory aminoacids (GABA, glycine, alanine, taurine, etc.) or as excitatory aminoacids (glutamate, cysteine, etc.). Any of the neurotransmitters may bind to a specific **receptor** site at the **post- or presynaptic membrane** where it induces opening or closure of an ion channel. Excitatory transmitters induce ion flux to trigger a postsynaptic action potential and thus further propagation of the depolarization. Inhibitory transmitters prevent depolarization of the postsynaptic membrane (Reus 2010).

6.1.5.5 Circuitry of Neurons

Distribution of neuronal input follows various patterns of neuronal circuitry (motherboard). Input may be distributed within the same pathway, may be distributed to multiple pathways, may result in convergence from multiple sources or a single source, may be distributed via a reverberating circuit, or via a parallel after-discharge circuit (Huppelsberg and Walter 2009).

6.1.5.6 Electroencephalogram

The electroencephalogram (EEG) is a powerful tool to record cortical electrical activity noninvasively or invasively. Various rhythms or waves can be recorded, which include the alpha-waves (8–12 Hz), the beta-waves (>12 Hz), theta-waves (4–7 Hz), or delta-waves (0.1–3 Hz). Exclusively over the motor cortex also the mu-rhythm (8–13 Hz) can be recorded. In pathological conditions cortical activity can change dramatically and specific or nonspecific paroxysmal activity may be recorded. This includes spikes, sharp-waves, spike-waves, spike-slow-waves, sharp-slow-waves, polyspike-waves, or myocloni. Paroxysmal activity may go along with or without clinical manifestations (seizures). Paroxysmal activity with clinical manifestations may be continuous, discontinuous, or intermittent. Paroxysmal activity may be also focal or generalized. Epilepsy is present if there is one non-provoked seizure and paroxysmal activity on electroencephalography or abnormalities on cerebral MRI (Fig. 6.3c).

6.1.5.7 Sleep

Sleep is essential for recreation of the body. According to the EEG-activity four stages of sleep are differentiated. Stage 1 is characterized by the change from alpha-waves to theta-waves. Stage 2 is characterized by the presence of theta-waves and the occurrence of sleep spindles and K-complexes. Stage 3 is characterized by the occurrence of delta-waves in a frequency of 20–50 %. Stage 4 is characterized by delta-waves in a frequency of >50 %. A fifth stage is not characterized by the EEG-activity but by involuntary eye movement. This pattern of sleep is the so-called rapid-eye-movement (REM) sleep. For sufficient and effective sleep at least 4–5 periods of stage 4 are required per night (Reus 2010; Huppelsberg and Walter 2009).

6.1.6 Sensory Organs of a Human

6.1.6.1 Vision

The visual system is built up of complex structures, including the eye, the optic nerve, the chiasma, the optic tract, the corpus geniculatum laterale, the colliculus superior, the visual tract, and the visual cortex for processing of the visual input (Fig. 6.2b). Four neurons are involved in the transmission of the visual input to the processing area. The mechanism of light perception is based on photoreceptors, being a highly conserved biological principle (see Sect. 6.2). Three of the neurons in humans are localized within the retina and the fourth after the corpus geniculatum. Proper vision, however, requires more than the visual pathways. Further essentials for proper vision include the corneal reflex, a projection device against injuries to the eyes, the pupillary reflex, the convergence reaction, and the coordination of the bulb movements. Coordination of bulb movements is provided by the medial longitudinal fasciculus, which carries information about the direction that the eyes should move. This system also gets input from the vestibular system, the cerebellum, and the cortex (Reus 2010; Schmidt and Schaible 2005).

6.1.6.2 Hearing

The auditory system is built up of the outer ear, the middle ear, and the inner ear, the vestibulocochlear nerve, the cochlear nuclei on the floor of the fourth ventricle, the projections to the superior olivary nucleus, to the inferior colliculus, the medial geniculate body, and the superior temporal gyrus (Heschel'sche Querwindung) (Fig. 6.2c). However, not only ascending projection for the propagation of the input to the cortex are available but also descending projections to the cochlea via the same system. Sound from outside the body is converted into oscillations of liquid columns within the cochlea where these oscillations are transformed into electrical energy within Corti's organ (Reus 2010; Schmidt and Schaible 2005).

6.1.6.3 Smelling

Smelling is mediated by sensory cells of the nasal cavity where they are assembled in structures known as glomeruli. These cells carry olfactory receptors to which odorant molecules bind at specific sites. Contrary to many other species, only a single olfactory system exists in humans. Chemoreceptors of the glomerula detect volatile chemicals (main olfactory system in species other than humans) and fluid-phase chemicals (accessory olfactory system in species other than humans) simultaneously. The glomerula transmit the olfactory signals to the olfactory bulb, which forms the most frontal part of the olfactory nerve. From there mitral cells synapse to five major cerebral regions, the anterior olfactory nucleus, the olfactory tubercle, the amygdala, the piriform cortex, and the entorhinal cortex. The entorhinal cortex is involved in emotional and autonomic responses to odor and in motivation and memory via projections to the hippocampus. Odor is stored in long-term memory and is strongly related to emotional memory (Reus 2010; Schmidt and Schaible 2005).

6.1.6.4 Tasting

Tasting or gustatory perception (gustation) requires a sensory system that starts with the taste buds (gustatory calyculi) of the tongue, mouth, and pharynx. The highest concentration of taste buds (up to 5,000 in humans) can be found at the tip of the tongue. Taste buds represent chemoreceptors being stimulated by a chemical reaction between molecules of the mouth and the receptor. Taste buds are built up of taste receptor cells and basal cells and of nerve endings. Afferences originating from taste buds project their input via three cranial nerves (VII, IX, X) to the brain stem. From there the input is conducted to the thalamus and from there to the gustatory cortex at the lower parietal lobe. Five basic taste qualities are differentiated, sweet, salty, sour, bitter, and umami. Sweetness indicates energy-rich food whereas bitterness serves as a warning of poisons. Other sensations like "coolness" (e.g., menthol) or "hotness" (pungency) are perceived via chemestesis (chemical sensibility of skin and mucosa). Taste together with smell and trigeminal input (touch (mechanoreceptors), temperature (thermoreceptors), pain (pain receptors)) determine flavors (sensory impression of food) (Reus 2010; Schmidt and Schaible 2005).

6.1.6.5 Balance, Equilibrium

Balance means the ability to maintain the vertical line of the center of gravity of a body within the base of support with minimal sway (Shumway-Cook et al. 1988). Sway is the horizontal movement of the gravity center even if a person stands still (Shumway-Cook et al. 1988). Maintaining balance requires the coordinated input from the vestibular, somatosensory, and visual system. The vestibular system regulates the equilibrium by processing directional information of relative head positions during gravitational, linear, or angular acceleration. In addition to balance the vestibular system contributes to spatial orientation. Together with the cochlea, the vestibular system constitutes the labyrinth of the inner ear. In line with movements that consist of rotations and translations, the vestibular system comprises two components, the semicircular canal system, which recognizes rotational movements, and the otoliths, which recognize linear accelerations. The vestibular system sends information to the longitudinal medial fascicle (controls eye movements, vestibulo-ocular reflex) and the muscles to keep the body upright. From the brainstem projections are led to the thalamus and parietal cortex, where imbalances become conscious (Reus 2010; Schmidt and Schaible 2005).

6.1.6.6 Somatovisceral Sensation

See below Sect. 6.2

Acknowledgement

Thanks to Prof. Erika Jensen-Jarolim, MD, and crossip communications, Vienna, Austria, for support in the arrangement of illustrations.

6.2 Comparative Aspects of the Nervous System and Sensory Organs

Hanna Schöpper and Sabine Breit

6.2.1 Abstract

A characteristic of life is the ability to react in response to a stimulus. Therefore, even primitive organisms need to be able to sense the stimulus—mostly by some kind of local excitation—transfer the information within the body, and arouse a reaction to it. Eventually this reaction will increase the chance of survival and potential reproduction and hence is highly selected for. Already single-celled protozoa are able to react to external stimuli like temperature or light. They either show a topic (approach the stimulus) or phobic reaction (flee). The movement is enabled by drift of the cytoplasm and pseudopodia. In multicellular metazoans, the site of stimulus is not necessarily the site of reaction and, therefore, conduction of information is needed.

6.2.2 Introduction

The nervous system and associated structures compose a complex communication network that enables an organism to adjust and maintain homeostasis. It has **sensory components** that are able to convert the perceived stimulus, a **pathway to conduct and to process** this information, and **effector components** that induce an adequate reaction. The underlying mechanisms of function are similar in all species. Nevertheless, morphology and degree of complexity have evolved according to the respective requirements of species. Some of the major differences will be described in the following.

As outlined in Sect. 6.1 of this chapter, the smallest functional unit of the nervous system, the nerve cell, also called **neuron**, is composed of a cell body containing the nucleus and two types of cytoplasmic processes: dendrites conducting information towards (cellulipetally) and an axon conducting away from the cell body (cellulifugally). If a neuron possesses one dendrite it is termed unipolar, if dendrite and axon emerge from the same part of the cell body pseudounipolar and bipolar if from opposing parts. Multipolar neurons own more than just one dendrite. Neurons are generally built similar and are present in all living animals except some sponges and basic invertebrates. Also the function of all neurons is based on the **action potential** as a single principle, an electrochemical impulse that always follows the same pattern. The selective permeable membrane of the nerve cell allows exchange of charged ions building up an electrical gradient: the **resting membrane potential**. A stimulus may then open voltage-gated ion channels and therewith rapidly change the ion flow and cause a depolarization of the neuron membrane. The subsequent re- and hyperpolarization lead to the initial situation of resting membrane potential. This sequence of change in electrical membrane potential is called action potential and may occur at each point along the neuron membrane, generally starting at the region of the cell body close to the axon (axon hillock) and running along the axon.

Despite the uniformity of action potentials, the efficiency of conduction is differing greatly between specialized neurons and species. Temperature influences velocity of conduction and, therefore, ectotherms may be at disadvantage in cold surroundings. While in arthropods, annelids, and molluscs the axon diameter is enlarged to increase speed of conduction (giant axons), vertebrates may as well increase speed of information transmission by a multitude using saltatory conduction via myelinated neurons. In myelinated axons the action potential only depolarizes the membrane at intervals (nodes of Ranvier) between two myelin sheaths that otherwise prevent depolarization. Therefore, the depolarization leaps the myelinated parts and is much faster. At the termination of the axon the action potential is transferred to the next cell by either electrical or chemical synapses. **Electrical synapses** are formed by gap junctions between cell membranes that can either be closed or open to enable exchange of ions and rapidly forwarding an action potential to the next cell. Electrical synapses show no time lag and most of the time they work bidirectionally. They are mainly seen in arthropods, annelids, and molluscs but are also present as gap junctions in vertebrate cardiac muscle cells, for example. In higher animals **chemical synapses** are predominant and transfer the excitation via chemical substances called neurotransmitters to the postsynaptic membrane of the next cell. The stimulus is passed to an effector organ like a myocyte via the motor end plate (see Chap. 4, Fig. 4.2a), adenocyte, or another neuron.

The organization of the nervous system is highly influenced by the complexity of the organism and, therefore, considerable differences are evident between species. Cnidaria, for example, have a simple **diffuse nerve net** with interconnected neurons that cover the entire body. If a stimulus excites a neuron the action potential runs in all directions with the same speed, thereby allowing no specific transmission. The subsequent development of **ganglia** can be seen as a major evolutionary step, because this co-localization of nerve cell bodies enables increased interaction of different neurons. Ganglia are present in almost all animals and allow the introduction of the term **centralization**. By centralization differentiation between a central nervous system (CNS; ganglia) and a peripheral nervous system (PNS; peripheral nerves) is possible and in consequence a target-oriented transfer of information. In segmental animals like annelids and arthropods two ganglia per segment are interconnected to the next pair and thus represent a higher level of processing information than the peripheral nerves. This form of nervous system is called **ventral cord system** as the row of ganglia pairs runs through the ventral parts of the body. If the first pair of ganglia is enlarged and/or fused, a further hierarchy is apparent: **cephalization** and formation of a brain. The localization of the brain at the head end is reasonable, as most sensory organs are located here, food intake takes place, and movement of the body is oriented to this direction.

Vertebrates show this form of cephalization and therewith several functions are connected: high information storage capacity, enhanced velocity, and complexity of the brain. With those prerequisites, vertebrates are able to display behavior and even form associations. In addition to the brain, the spinal cord contributes to the **CNS** as one of the anatomical divisions making up the nervous system. The spinal cord, embedded in liquor for mechanical protection, reveals a central hollow canal and is located within the vertebral canal. Hereby, two characteristics clearly distinguish the vertebrate nervous system from the simpler version of ventral cord system described above that is compact in structure and situated below the alimentary canal. Next to the CNS, the **PNS** consisting of the neurons carrying information to and from the periphery completes the vertebrate nervous system. Like outlined for humans in Sect. 6.1, in vertebrates there are different types of neurons classified by their function: afferent (sensory) neurons that convey signals from sensory receptors and carry information to, while efferent (motor and autonomic) neurons carry information from the CNS to an effector organ like a muscle or gland. In addition, interneurons that are the vast majority of neurons of the CNS are interconnecting different neurons and integrate the information.

Some cell bodies of neurons have migrated from the brain and spinal cord and formed ganglia that are now part of the PNS, mostly with vegetative function. Outside the CNS, a bundle of parallel nerve fibers is generally termed nerve. Remember, those originating within the brain are called cranial nerves (I–XII) (see Fig. 6.2a), whereas those originating within the spinal cord are called spinal nerves. Information can be processed and integrated in the ganglia, spinal cord, or brain.

6.2.3 Central Nervous System of Vertebrates

A characteristic of the CNS of vertebrates is the aggregation of neurons that integrate the information received. Interneurons following a sensory afferent neuron are organized in functional groups and will integrate and modulate the received information and pass the stimulus by their many processes to several efferent motor neurons. Thereby, the number of possible answers to a stimulus is increased and the number of sensations leading to a response, too. The brain and also the spinal cord have a key role as a feedback control system and by increased levels of integration led especially the brain evolve to a center of association enabling sensation, memory, complex behavior, and awareness. In both, brain and spinal cord, grey and white matter can be distinguished macroscopically. The grey matter accommodates neuronal cell bodies and the white matter is composed of axons that are myelinated in most instances. Axons running parallel are called tract within the spinal cord.

6.2.3.1 Brain (Encephalon), Intelligence, and Behavior

By developing many complex centers, the brain takes over more and more the duties of a control and modulatory center over the whole body and its functions. But also automatisms, like spontaneous arousal without a preceding external stimulus, are possible and may lead to some kind of activity. Nevertheless, organs may as well be viable without the brain as can be seen in explants of heart or intestine.

Through animal phyla the brain size varies considerably and can give us a rough idea of how complex the animal's behavior is. To account for differences in size, a ratio of brain to body size is given. Fishes and also amphibians and reptiles have a comparably small brain size, which is getting larger in birds and mammals. But even within mammals, relative brain size is variable, as the numbers of different domestic mammals show. In addition, domestication reduces brain weight up to 30 %. Another possibility to estimate intelligence of certain species is to compare the weight of brain and spinal cord. In fishes and amphibians it is evenly distributed, while the brain of a human is 55 times heavier than the spinal cord. By actually counting neurons within the brain it became clear that a large brain size does not implicate a large number of neurons—with the interneuronal space becoming an interesting potential for building up connections between different neurons and therewith increasing complexity.

Nevertheless, it is noticeable that different regions of the brain are developed in different degrees, suggesting different capacities of, for example, sensations. Generally, the brain is divided into different regions according to the phylo- and ontogenetic development. **Prosencephalon** (forebrain) and **rhombencephalon** (hindbrain) are to be distinguished quite early during development whereas the vertebrate **mesencephalon** (midbrain) is characteristically seen as a third primary brain vesicle during development only in birds. Subsequent organization leads to five (secondary) brain vesicles, of which the telencephalon and diencephalon arise from the prosencephalon, and the rhombencephalon divides further into metencephalon and myelencephalon. In different phyla and species these portions of the brain are diverse in shape and relative size. As the structure is correlated to its function and use one can see brain morphology in association with the animal's environment. Sensation or other function that is mostly needed will lead to a dominance of the respective part of the brain. Morphologically, the brain may be divided into the cerebrum (dorsal part of the telencephalon), the cerebellum (dorsal part of the metencephalon), and the brainstem.

The **telencephalon** controls voluntary motor functions, perceives conscious sensations—especially smell—and integrates and processes this data in association centers. One can distinguish a pallium and subpallium or basal nuclei. The pallium can be divided into phylogenetically "older" areas like the paleopallium and archipallium, while the neopallium is a quite recent innovation of mammals. Fishes and amphibians feature aspects of paleo- and archipallium, which are highly developed and are responsible mainly for olfaction and early emotions. Both areas may also be summarized as the rhinencephalon. In reptiles a primordium of neopallium is represented in addition as a multilayered area, while still older regions predominate. Still in mammals smell is considered to be one of the "oldest" sensory functions and the lobe of paleopallium responsible called the olfactory lobe. Even within mammals considerable differences in smelling capacity and development of the olfactory lobe are evident. [For more detailed information see Sect. 6.2.8.]

Another important aspect of the phylogenetic old parts of the telencephalon—especially the archipallium—is the limbic system including the hippocampus with substantial regulation of vital and also species-preserving functions. The hippocampus is involved in learning and memory, and is thought to be the anatomic site of instinct. In contrast, the neocortex (grey matter of the neopallium) is a relatively "young" acquisition of vertebrates and is even in mammals highly diverse in development. While the surface of the telencephalon is smooth in rodents and birds, for example (called lissencephalic brain), other mammals show an increase in neocortex area by folding and forming grooves and wrinkles (called gyrencephalic brain) (see Fig. 6.1a). In the cortex association areas are present and conscious perception is facilitated. Association allows problem-solving and abstract creative thinking.

Due to different microscopic structure of the avian telecephalon, birds were considered to lack a neopallium and only exhibit reflexive behavior based on increased basal nuclei. However, many avian species like crows or parrots show great memory capability, some sort of language, and tool-using strategies—believed to be associated with a more sophisticated brain structure. Nowadays, the telencephalon of birds is considered to also consist of pallium and subpallium and many homologue structures within the avian telencephalon have been identified that may differ in appearance but serve similar functions compared to the mammalian neopallium (Reiner et al. 2005).

Within the **diencephalon** especially thalamus and hypothalamus illustrate main functions. While the thalamus integrates sensory perception to a conscious sensation at the cortex, the hypothalamus is a major control center of autonomic functions integrating the control over homeostasis and temperature regulation and it holds high endocrine activity. Emotions of rage and aggression seem to originate in the hypothalamus, too.

In the **mesencephalon**, nuclei (i.e., accumulations of neuron cell bodies with similar functions) and the tectum serve as centers for visual and auditory stimuli. This part of the brain is already highly developed in fishes and amphibians, as visual and auditory information is passed here and the mesencephalon acts as a center for integration and mediation of sensory stimuli. In higher animals integration of information has mostly been transferred to the prosencephalon, whereas the visual and auditory reflexes are still mediated in the mesencephalon.

Main components of the **metencephalon** are cerebellum and pons, with the cerebellum controlling and involuntary coordination of equilibrium and posture are associated. Species that move close to the ground like amphibians and reptiles have a poorly developed cerebellum reflecting the reduced necessity for complex locomotion. Bony fishes, however, need to be oriented in a three-dimensional space and consequently have a larger cerebellum despite lower hierarchy in phylum. The most complex cerebellum is seen in birds and mammals that need precise motor coordination and equilibrium. Here the cerebellum is not only larger but also the surface is increased and subsequently lies in folds (folia). The pons then links the information of the cerebellum and acts as a control area between prosencephalon and spinal cord.

The **myelencephalon**, termed medulla oblongata because of the impression to serve as a continuation of the spinal cord, regulates basal vital functions like respiration, heart rate, and swallowing. This part of the brain is phylogenetically very old and shows least differences between species.

6.2.3.2 Spinal Cord

The caudal continuation of the myelencephalon situated in the vertebral canal is the spinal cord. It consists of grey and white matter that is localized around a central canal connected to the inner fluid-filled spaces of the brain. The **reflex arch** is the functional principle of the spinal cord with its simplest form of receiving information via afferent fibers from a receptor organ (see Chap. 4, Fig. 4.2b). Within the spinal cord these are changed over to an efferent neuron forwarding the message to the effector organ. Usually, association neurons are interposed between afferent and efferent fibers and, therefore, information is integrated and coordinated independently of the brain.

In invertebrates like branchiostoma the spinal cord is not yet closed to form a central canal. Therefore, the cross section looks triangular in shape. In addition, white and grey matter cannot be distinguished grossly due to lack of myelination of fibers. In fishes and amphibians this discrimination can be clearly made and the spinal cord is fully fused with a central canal being prominent. The cross section has a nearly round shape. At the level of limbs the cross section of the spinal cord is enlarged reflecting the increased need for nervous integration. The same can be seen in amniota with the cervical swelling being especially prominent in species that focus on forelimb movement (e.g., flying birds), while in species with bipedal locomotion (e.g., ostrich, human) the lumbar enlargement is pronounced. The cross section is more or less horizontally oval. Dorsal and ventral roots of spinal nerves originate at the same segment and merge already within the vertebral column, whereas in lower animals they first leave the vertebral canal to join later.

6.2.3.3 Meninges of the Central Nervous System

The CNS needs to be well protected due to its vital functions and because of its reduced capability to heal in case of injury. Therefore, the brain and spinal cord are located within bony structures of skull and vertebral column and a variable number of protective membranes, termed meninges. The CNS of fishes is just surrounded by a sheath of connective tissue called primitive meninx. Amphibians and reptiles already have a double-layer system of outer dura mater and an inner structure comparable to a combined pia-arachnoid membrane. In birds pia mater and arachnoid membrane start to be distinguishable, while mammals show three distinct meninges with **dura mater** being the outermost layer. The dura mater is adherent to the skull, whereas within the vertebral column the dura mater is separated by fat from the wall of the vertebral canal. The **arachnoid membrane** builds up a net-like structure connecting dura and **pia mater**—the innermost layer of meninges that is tightly bonded to the nervous tissue. In between the pia mater and arachnoid membrane **cerebrospinal fluid** (CSF, *liquor*) is apparent, bathing the CNS. This cavity is connected to the inner ventricles of the brain and to the central canal of the spinal cord. All vertebrates have CSF in- and outside to the CNS—even though meningeal structures may be slightly different. Functions of CSF include mechanical and immunological protection.

6.2.4 Peripheral Nervous System of Vertebrates

Both spinal and cranial nerves are part of the PNS as described above. The cell bodies of many neurons of the cranial and spinal nerves are located within the CNS, while some others have left it to form the ganglia in the PNS. A possibility to group the PNS is to divide it into functional sections. (1) The **motor system** carries signals to skeletal muscles, mainly in response to external stimuli. The control of skeletal muscle can be voluntary or involuntary as a reflex (see Chap. 4). (2) The **autonomic nervous system** mainly regulates internal environment by controlling smooth and cardiac muscles and organs and may again be divided into **parasympathetic and sympathetic divisions** (see Fig. 6.4). (3) The **sensory system** provides the motor and autonomic nervous system with external and internal information. In general the PNS is comparable between vertebrates and the differences known are too specific to be described in detail here.

6.2.5 Sensory Organs

Next to the function of transferring, integrating, and processing information, the perception of environmental stimuli is essential for reactions of the organism. Sensations result from stimuli that initiate afferent impulses, which in mammals eventually reach a conscious level in the cerebral cortex. Even though theoretically any cell should be reactive, highly specialized sense organs for this task have evolved that are all based on sensory receptors. In general, sensory receptors can be categorized into various groups: according to localization (general vs. local), depth (somatic vs. visceral), origin of stimuli (external vs. internal), type of stimulus (pain vs. temperature vs. mechanical vs. chemical vs. electromagnetic), and probably many more of which we not yet know of.

The functional principle of sensory receptors is the conversion of specific stimuli that the receptor is specialized on into an electrical signal. The conversion is also called **sensory transduction** and relies on the change of membrane (or in these cases receptor) potential. Thereby, a stimulus triggers a receptor leading to a signal transduction pathway. This pathway includes changing ion concentrations due to opening ion channels and a subsequent receptor potential. The receptor potential leads to neurotransmitters crossing the synaptic cleft to the subsequent neuron. The neuron will transmit action potentials, the common language of the nervous system, to the CNS. The intensity of the sensation will be reflected in the rate of action potentials and lead to adequate responses of the effector organs.

6.2.5.1 Somatovisceral Sensation

Somatovisceral sensation comprises perception of different stimuli like touch, pressure, pain, temperature, and tension. Of those stimuli **touch** is considered to be one of the traditional "five senses" (together with vision, hearing, olfaction, and taste). Unlike the more specialized senses, somatovisceral sensation is seldom confined to special sensory organs. Receptors sensitive to touch are found all over the body surface but are especially concentrated in areas, which are important for exploring the environment like the face in general, the lips or the beak in birds, the tongue, or the acra of the limbs.

The simplest form of a somatovisceral receptor is a **free nerve ending** like the nociceptor (pain receptor). Nociceptors may be localized in many tissues but are found in highest densities in the skin of mammals. They are relatively unspecific and respond to different stimuli like pressure, heat, or chemical irritation and are intended to make the animal leave the area of potential danger. But also nonmammalian species have been found to possess nociceptors; for example, larval stages of drosophila could exhibit multidendric neurons underneath the epidermis, which react to heat and mechanical stimuli (Tracey et al. 2003). Some fish species like the trout (Sneddon et al. 2003) have been detected to react to noxious treatment, too.

Besides the free nerve ending there exist more specialized receptors detecting stimuli of touch, pressure, temperature, and tension—not only in the skin but also from within the animals' body (e.g., internal organs or locomotory system). These **chemo-**, **thermo-**, and **mechanoreceptors** act according to the basic functional principles of receptors being prevalent in other sensory organs, too. Generally, there are different kinds of mechanoreceptors present in all tetrapods, all relying on the principle of influencing the membrane of the receptor cell, thereby leading to a change of permeability transduced into a subsequent receptor potential. Examples of occurrence of mechanoreceptors of the somatovisceral system are around vibrissae, within muscle spindles, tendons, joints, or blood vessels as well as being part of the **lateral line system** in fishes and some amphibians. This organ is located at the side of the body and on the head and consists of the so-called neuromasts, groups of mechanoreceptors that consist of hair cells covered by a gelatinous cupula. Water movement along the animal leads to deflection of the cupula and therewith deviation of the embedded stereocilia of the sensory cells—followed by the transduction pathway described above. The frequency of action potential of the afferent nerve then depicts a broad picture of both spatial and temporal information. The lateral line system is best developed in active fish species especially those having low visual perception and is a distant touch receptor system detecting wave vibrations. It serves a variety of functions like navigation (obstacle avoidance, predator avoidance, prey capture) and communication.

Many migratory species like birds, fishes, and reptiles but also bees use the geomagnetic field for navigation and even though there have been numerous years of studying this phenomenon no specific **magnetic receptor** has been detected yet (Beason 2005). Not only magnetic, but electrical fields may be detected by many species, too. **Electrical fields** may be created by muscular movement of prey animals. Mainly aquatic animals are capable of using detection of electrical fields probably due to better conduction of the stimulus. Thereby, perception is provided by derivates of the lateral line system. However, only recently it could be demonstrated (Clarke et al. 2013) that even bumblebees may use electrical charge differences between flowers and the environment for orientation and decision-making.

Cold and warm is detected by **thermoreceptors**, either consciously within the skin or unconsciously from within the body. The hypothalamus regulates body temperature of mammals and birds to stay in a constant and quite small range. While perception of visible light will be described in Sect. 6.2.5.2, infrared light is detected by specialized **pit organs** in certain snakes. The pit organ allows sensing of infrared radiation and thereby body heat of prey facilitating precise hunting. Detection of infrared radiation was demonstrated to be associated with a thermal signal transduction rather than photochemical reaction and may therefore be rather considered a thermoreceptor (Gracheva et al. 2010).

6.2.5.2 Vision

Detection of light is essential in the animal world and even though many different organs have evolved, the underlying mechanism is quite conserved. The sensory receptor is called **photoreceptor** and contains pigment that absorbs the incident light. The most primitive form of detecting light is via **sensitive cells on the body surface** of unicellular species. But even some of those already feature-specialized organs like the **ocelli** in flatworms that are composed of a multitude of photoreceptors and allow detection of light and simple movements. In invertebrates two different principles of highly developed eyes have evolved: the compound eye of arthropods and a lens-containing eye of squids. The **compound eye** is composed of a multitude of ommatidia—little eyes, each consisting of a lens and several photoreceptors. Thereby, the field of vision is built up of numerous single pictures fusing to a visual impression. This type of eye is especially aimed to detect motion and, therefore, fits the fast movements of flying insects. The **single-lens eye of squids**, however, is built of a single lens only and facilitates focused and high-resolution pictures by a variable distance between lens and photoreceptor containing retina. Even though the **vertebrate-eye** also contains a single lens—both systems are believed to have evolved independently and exhibit some distinct differences. Two major distinctions are: in vertebrates the focus is achieved by a modifiable shape of the lens itself and photoreceptors are oriented inversely and absorb the light reflected from the pigmented layer behind the receptors.

In general, the vertebrate eyeball is built up of three different layers: the outer fibrous layer (support), which is composed of the transparent cornea and the white sclera; the middle layer (vascular), which contains the iris, the ciliary body, and the choroid; and the inner nervous layer (sensory), assembling the **retina**. Two types of photoreceptors can be distinguished: rods and cones (see drawing of human eye in Fig. 6.5a). While rods are associated with black and white sight and because of advanced light sensitivity may be used in low-level light, the cones are associated with color vision but require higher light levels. Physiological

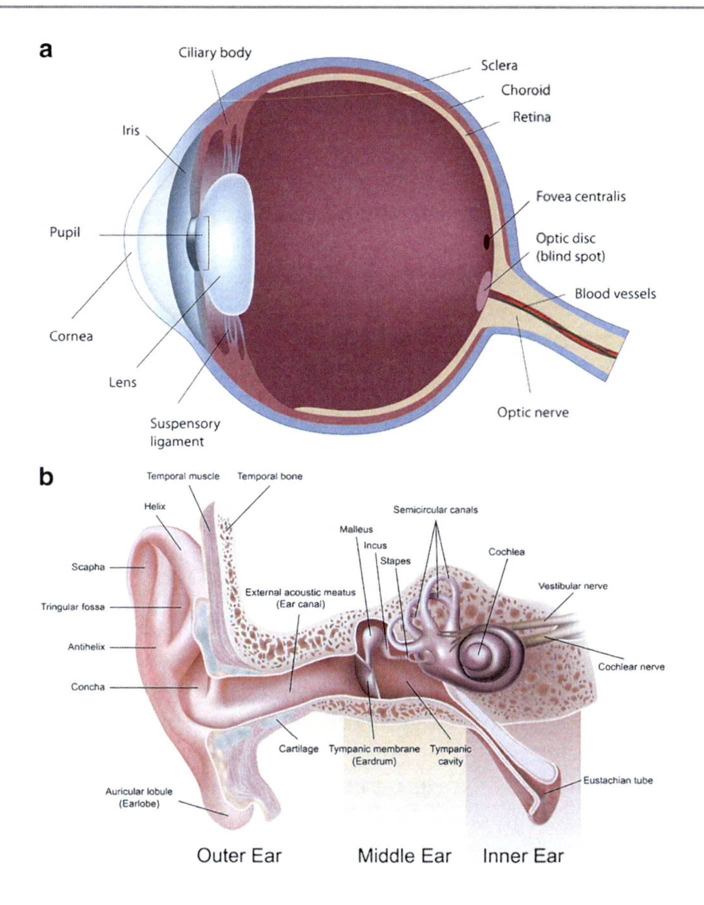

Fig. 6.5 (a) **The building of a human eyeball** by three layers: the *outer* fibrous layer, composed of the transparent cornea and the white sclera; the *middle* vascular layer which contains the iris, the ciliary body, and the choroid; and the *inner* nervous layer, assembling the retina. In the area where the optic nerve is bundled, no light receptors are present (blind spot). The site of the highest receptor density in the Fovea centralis is called Macula densa. Two types of photoreceptors can be distinguished: rods and cones. ©Alila Medical Images—Fotolia.com. (**b**) In **the inner ear of a vertebrate organism** (shown here for a human) both the auditory and equilibrium receptors are located side by side. The auditory signals are translated mechanically via tympanic membrane, malleus, incus, and staples until they reach the auditory epithelium in the cochlea, where the signal is transferred to the cochlear nerve. For the sensation of (im)balance, three semicircular canals/loops in the *inner* ear along the x, y, and z axis of the three-dimensional room are responsible which are connected to the vestibulum and vestibular nerve. Movements of fluid in these rooms by physical movements along one or the other axis elicit irritation of balance receptors indicating a change of the body's position. ©peterjunaidy—Fotolia.com

measurements showed that humans have three different types of cones: S, M, and L. Some women do have a fourth type of cones and have increased **color vision**. The majority of mammals are capable of bichromatic color vision, while most marine mammals are only monochromatic in sight. But also tetrachromatic vision is apparent in reptiles, amphibians, and birds. Some of them even have five cone types and pigeons are believed to even have more than that. Retinas of domestic animals mostly contain rods that are extremely sensible to light and vision in low-light situations. This light-sensitive sight is further supported by the **tapetum lucidum**. This layer is located behind the pigmented layer (which does not contain pigments in this area) of the retina and reflects incoming light to allow more light to be absorbed by the photoreceptors. The tapetum lucidum is seen especially in nocturnal species like the cat or deer and those living in deep water as, for example, sharks. Color and location of the tapetum lucidum vary between species. To protect the eye from excess light either the pupil (formed by the iris) is reduced in diameter or like in some fishes a pupillary operculum can be expanded to reduce incoming light.

The visible spectrum in humans ranges between 380 and 700 nm. Electromagnetic radiation with longer wavelengths is termed infrared light and can be used by some fish species. Light with shorter wavelengths than visible light (as low as 10 nm) is termed ultraviolet light and many vertebrates like birds, fishes, some mammals and reptiles are able to perceive it (Jacobs 1992). For example, predator birds like the kestrel were found to be able to see ultraviolet reflection of rodent urine, marking common pathways of prey species (Viitala et al. 1995).

The field of vision is different between species and allows insight into life styles. Predatory animals tend to have a higher proportion of binocular vision to increase acuity of the picture and estimate distances. In contrast, monocular sight is predominant in prey species, consistent with a larger field of vision. Insects have a field of vision of almost 360° and may therefore detect predators rapidly.

Concerning species, some characteristics can be described. In fishes, the sclera is stiffened by cartilage or bone and accommodation is performed by protrusion of the spherical lens. Cones are present in some bony fish species, others lack them. Most fishes have no upper and lower eyelid comparable to those of mammals and therefore cannot close their eyes. Amphibians are the first to develop eyelids and tear glands after metamorphosis and all terrestrial vertebrates possess some kind of those moisturizing facilities. Small lens muscles move the lens in order to accommodate similar to fishes. Reptiles and birds have a cartilage support of sclera in addition accompanied by a bony scleral ring to prevent distortion, when ciliary muscles are active. In contrast to fish and amphibians the lens of amniota is soft and may change shape if tension is applied. The ciliary muscle is built of striated musculature, facilitating very fast accommodation. In mammals, the sclera is lacking cartilage or bone and the ciliary muscle is smooth.

The **nictitating membrane** is a translucent third eyelid well developed even in some fishes (that otherwise lack eyelids), amphibians, reptiles, birds, and also some mammals, but it is highly reduced in humans and some other primates. Unlike the upper and lower eyelid this membrane serves to reduce excessive illumination (in fishes), and to mechanically protect and moisten the eye while maintaining visibility (other vertebrates).

To monitor ambient light a **third or parietal eye** is located between the lateral eyes and is functional in some fishes, amphibians, and many reptiles. The parietal eye is considered to be part of the diencephalon and develops either as an isolated organ or as the light-sensitive part of the epiphysis (pineal gland). In higher animals the pineal gland has lost its light-sensing capacity, whereas still maintaining important endocrine functions in vertebrates mainly related to the circadian rhythm. The third eye consists of photoreceptors and sometimes even of a lens and resembles lateral eyes in many cases. It could be shown in lizards that the photoreceptors may act inversely and hyperpolarize if light is too bright, therefore giving information of very dim light in dusk and dawn (Solessio and Engbretson 1993).

6.2.5.3 Hearing/Equilibrium

The ear reassembles two senses of an animal: hearing and equilibrium. Both are based on the same principle of mechanoreceptor activation and may be traced back to the lateral line system of fish that is also discussed in connection with some basic acoustic perception. During ontogeny one placode is sinking below the superficial layers and forms a hollow vesicle that is becoming increasingly complex and contains nervous compounds, too. The middle aspect of the placode forms the primitive **membranous labyrinth**, which is surrounded by the **otic capsule**; the other (pre- and postotic) aspects develop into the lateral line system. The otic capsule forms the **bony labyrinth** in higher animals. The membranous labyrinth represents a sac filled with **endolymphe** and is embedded in the cavity of the **bony labyrinth** that is filled with **perilymphe** in higher animals. Bony and membranous labyrinth form the **inner ear** that can be subdivided into **vestibular and cochlear systems** according to their function. While the vestibular system is much older and mainly devoted to equilibrium, the cochlear system is primarily designed for hearing and modifies as animals become terrestrial. Both senses rely on the same basic principle of mechanical motion of hair cells that is transduced to nerve impulses.

In all jawed vertebrates, the vestibular system consists of three membranous **semicircular ducts** aligned at right angles to each other (one for each axis of rotation) allowing perception of rotational acceleration and two membranous chambers called **saccule** and **utricle** perceiving linear acceleration and gravity. Saccule and utricle are also termed the **otolith organ**. In the wall of the membranous labyrinth groups of mechanoreceptors are located at specific areas of the semicircular ducts (ampullary crests) and the otolith organ (macula utriculi, macula sacculi). Each of those mechanoreceptors is built up of several hair cells embedded in a gelatinous mass (cupula). In the otolith organ the gelatinous mass contains crystals (otolithic membrane) in some species, whereas it is represented by **otoliths** in others. During movement, changes in gravity make the cupula and the otolithic membrane (the otolith respectively) change position on top of the hair cells causing shearing forces to the kinocilium and evoking action potentials in the respective nerve. The membranous labyrinth is only partly enclosed by bone in bony fishes and in those animals the otoliths qualify for species determination due to specific shape. Despite differences in shape and dimensions the general principle of the vestibular system is very old and conserved in all vertebrates, apart from some primitive jawless species like the hagfish or lamprey, in which only one or two semicircular ducts are present. Even invertebrates have a similar system of a hollow structure called statocyst, in which a mineralized statolith is placed on sensory hairs that transmit nerve impulses in case of motion.

The cochlear system is devoted to hearing, with sound waves being transmitted to fluid generating waves of the perilymphe and subsequently also to the endolymphe as both are only separated by the wall of the membranous labyrinth. At areas where the membranous labyrinth of the cochlear system contains mechanoreceptors, the movement of peri- and endolymphe is transduced to a nerve signal by hair cells again. Fishes may perceive sound waves via receptors in the saccule but do not seem to have a very well developed hearing capacity, whereas amphibians already have sound detecting mechanoreceptors arranged as the papilla amphibiorum within the wall of the saccule. A distinct cochlear system is formed in terrestrial animals initially represented by an elongation of the saccule, which is first developed in reptiles as a short **lagena** containing an area of mechanoreceptors—the **basilar papilla**. In some crocodiles and in birds the lagena is further enlarged to form the **cochlear duct**, which in mammals (except monotremes) is curled and housed in the bony **cochlea**. The basilar papilla is also comparably enlarged and becomes the main acoustic organ or **organ of Corti**.

Hearing is based on acoustic waves hitting the animal and being transduced to first mechanical and then nerve impulses. Vibrations under water will be recognized by the lateral line system of fishes when produced nearby. However, waves produced further away are too delicate to be detected and need further amplification. Amplification of mechanical waves hitting the body in some fish species is achieved by a gas-filled chamber/resonator otherwise used to remain in buoyancy—the swimbladder. Vibration of gas is then transduced to affect the mechanoreceptors in the inner ear. In terrestrial animals, acoustic signals are mostly reflected by the body surface and sound transmission is performed by the **middle ear**. The middle ear consists of the **tympanic membrane** perceiving vibrations from the environment, an air-filled middle ear cavity or **tympanic cavity**, and one (termed stapes or columella) or more **auditory ossicles** (like malleus, incus, and stapes in mammals) transferring vibrations from the tympanic membrane to the perilymphe of the inner ear. In some species like mammals excessive vibrations may even be damped by musculature to prevent the delicate middle and inner ear structures from injury. Amphibians and some reptiles respond especially to low-frequency sounds. Many of them still lack a tympanic cavity, with the stapes representing the only ossicle being attached to skin or bone and, therefore, these species especially may conduct vibrations from the ground.

Mammalian ears are characteristic in some features: (1) the cochlea, (2) three auditory ossicles (**malleus**, **incus**, **and stapes**), and (3) a fully developed **outer ear** (see human example in Fig. 6.5b). The outer ear includes the **pinna** (auricle), which is present only in terrestrial mammals and the **ear canal** (external auditory meatus) first developed in some reptiles and present in all other higher vertebrates. Both are used as an acoustic horn to better detect sound. The pinna helps to catch incoming sound.

Not all sound frequencies are normally audible to all animals. Many animals use sound to communicate with each other, but frequencies capable of being heard range widely between animals. Elephants are believed to use infrasound for communication, whereas dogs are able to hear ultrasound (the principle of a "silent" dog whistle). Some bats and aquatic species like toothed whales use ultrasound for echolocation assisting navigation through poor lighting conditions and identification of prey or obstacles.

6.2.5.4 Olfaction

The oldest reactions to environmental stimuli are considered to answer chemical sensation. Even single-celled organisms react to oxygen content or foodstuff nearby with movement towards/away, called chemotaxis. In multicellular organisms this sensory competence is maintained and organized in senses like olfaction and taste. As most sensory functions concerned with environmental stimuli they are located at the head end of the organism, ensuring control of surroundings, food intake, as well as mate choice and avoidance behavior. Olfactory signals are received by chemoreceptors and are vital for foraging, recognizing danger, and social interactions. One can distinguish animals, whose olfactory senses are well developed (macrosmatic, e.g., dogs) and those with low abilities for olfactory perception (macrosmatic, e.g., dogs).

Main parts of the forebrain are devoted to olfaction being one of the oldest senses. To activate olfactory neurons, chemicals have to be in solution. In water-living animals this is always the case and therefore **olfactory epithelium** may be localized around the outer nose parts (**nasal pits**) or in the **olfactory sacs**, while in terrestrial animals the nasal cavity contains mucus and serous glands to dissolve chemicals close to the olfactory epithelium with its filaments extending into the mucous layer. Filaments have different microstructure between species and this fact is believed to be responsible for species-specific sensitivity to certain smells. Only some birds are able to smell (vultures for example) but almost all mammals show this sensory function. Mammals that live in aquatic environment are somehow disadvantaged as they can only smell when breathing air, thus not during diving.

An additional organ associated with olfaction is the **vomeronasal organ**, a blind-ending duct opening into the nasal and/or oral cavity. This organ may be homologue to the olfactory sacs in fishes and contains olfactory epithelium connected to a different brain region than the olfactory epithelium of the nasal cavity. Amphibians, reptiles, and most mammals have it. In humans, turtles, and birds the vomeronasal organ, that is believed to be associated with pheromone olfaction in particular, is rudimentary or absent. Reptiles like the snake and lizards use it by transferring scent molecules to the organ with the tips of the forked tongue.

Other chemoreceptors are located within vessels and the brain and monitor oxygen content of the sensitive tissue and have been demonstrated in a variety of vertebrates (Milsom and Burleson 2007).

6.2.5.5 Tasting

Similar to olfaction, chemicals dissolved in fluid may activate chemoreceptors aggregated at the so-called **taste buds**, mainly localized on the tongue. Each taste bud is composed of supportive cells and taste cells, found in groups on gustatory papillae with surrounding serous glands. Cells of taste buds are exposed and constantly renewed. There are five types of taste that can be distinguished (salty, sour, sweet, bitter, and umami) with not all animals being comparably sensitive to them. In fishes, taste receptors are localized not only in the oral cavity and pharynx but also in the gill cavity, on the outer body surface and on barbels used to detect food. Other water-living animals like amphibians show a comparable localization, while terrestrial animals need to better protect gustatory receptors from drying out and taste buds are therefore restricted to the oral cavity. Birds and reptiles that even have very keratinized and dry tongues move their taste buds back to the pharyngeal region.

References

Bähr M, Frotscher M (2003) Duus'Neurologisch-topische Diagnostik, 8th edn. Thieme, Stuttgart

Beason RC (2005) Mechanisms of magnetic orientation in birds. Integr Comp Biol 45:565–573

Clarke D, Whitney H, Sutton G, Robert D (2013) Detection and learning of floral electric fields by bumblebees. Science 340:66–69

Feneis H (1974) Anatomisches Bildwörterbuch, 4th edn. Thieme, Stuttgart

Gracheva EO, Ingolia NT, Kelly YM, Cordero-Morales JF, Hollopeter G, Chesler AT, Sánchez EE, Perez JC, Weissmann JS, Julius D (2010) Molecular basis of infrared detection by snakes. Nature 464:1006–1011

Huppelsberg J, Walter K (2009) Kurzlehrbuch Physiologie, 3rd edn. Thieme, Stuttgart

Jacobs GH (1992) Ultraviolet vision in vertebrates. Am Zool 32:544–554

Milsom WK, Burleson ML (2007) Peripheral arterial chemoreceptors and the evolution of the carotid body. Respir Physiol Neurobiol 157:4–11

Platzer W (2005) Taschenatlas Anatomie. Nervensystem, 10th edn. Thieme, Stuttgart

Reiner A, Yamamoto K, Karten HJ (2005) Organization and evolution of the avian forebrain. Anat Rec A Discov Mol Cell Evol Biol 287A:1080–1102

Reus B (2010) Onto- und Phylogenese des Nervensystems. Lehrbuch Vorklinik. Teil B. Anatomie, Biochemie, und Physiologie des Nervensystems, der Sinnesorgane und des Bewegungsapparates. Dt. Ärzteverlag

Schmidt RF, Schaible H-G (2005) Neuro- und Sinnesphysiolgie (Springer Lehrbuch), 5th edn. Springer, Berlin

Shumway-Cook A, Anson D, Haller S (1988) Postural sway biofeedback: its effect on reestablishing stance stability in hemiplegic patients. Arch Phys Med Rehabil 69:395–400

Sneddon U, Braithwaite VA, Gentle MJ (2003) Do fishes have nociceptors? Evidence for the evolution of a vertebrate sensory system. Proc R Soc Lond B 270:1115–1121

Solessio E, Engbretson GA (1993) Antagonistic chromatic mechanism in photoreceptors of the parietal eye of lizards. Nature 364:442–445

Tracey WD, Wilson RI, Laurent G, Benzer S (2003) Painless, a Drosophila gene essential for nociception. Cell 113:261–273

Trepel M (2008) Neuroanatomie. Struktur und Funktion, vol 4. Urban & Fischer, München

Viitala J, Korpimäki E, Palokangas P, Koivula M (1995) Attraction of kestrels to vole scent marks visible in ultraviolet light. Nature 373:425–427

Welsch U (2010) Sobotta Lehrbuch Histologie, 3rd edn. Urban & Fischer, München

Further Readings

Aspinall V, Capello M (2009) Introduction to veterinary anatomy and physiology, textbook, vol 2. Butterworth-Heinemann, Oxford

Hickman CP Jr, Roberts LS, Keen SL, Eisenhour DJ, Larson A, L'Anson H (2011) Integrated principles of zoology, 15th edn. McGraw-Hill, New York

Hildebrand M (1974) Analysis of vertebrate structure, 1st edn. Wiley, New York

Nickel R, Schummer A, Seiferle E (2004a) Lehrbuch der Anatomie der Haustiere—Band IV Nevensystem, Sinnesorgane, Endokrine Drüsen, 4th edn. Parey, Stuttgart

Nickel R, Schummer A, Seiferle E (2004b) Lehrbuch der Anatomie der Haustiere—Band V Anatomie der Vögel, 3rd edn. Parey, Stuttgart

Reece WO (2009) Functional anatomy and physiology of domestic animals, 4th edn. Wiley-Blackwell, Aimes

Reece JB, Taylor MR, Simon EJ, Dickey JL (2012) Campbell biology—concepts & connections, 7th edn. Pearson, San Francisco

Starck D (1982) Vergleichende Anatomie der Wirbeltiere auf evolutionsbiologischer Grundlage, 1st edn. Springer, Berlin

Stoffel MH (2011) Funktionelle Neuroanatomie für die Tiermedizin, 1st edn. Enke, Stuttgart

Von Engelhardt W, Breves G (2010) Physiologie der Haustiere, 3rd edn. Enke, Stuttgart

Surface, Barrier, and Interface Zone: Comparative Aspects of the Skin

7

Lucia Panakova and Krisztina Szalai

Contents

L. Panakova (✉)
Department of Dermatology, Internal Clinic for Small Animals, University of Veterinary Medicine, Vienna, Austria
e-mail: Lucia.Panakova@vetmeduni.ac.at

K. Szalai
Comparative Medicine, Messerli Research Institute, University of Veterinary Medicine Vienna, Medical University Vienna, University Vienna, Vienna, Austria
e-mail: Krisztina.Szalai@vetmeduni.ac.at

E. Jensen-Jarolim (ed.), *Comparative Medicine*,
DOI 10.1007/978-3-7091-1559-6_7, © Springer-Verlag Wien 2014

Abstract
The skin is the main body barrier of vertebrates covering the entire body. Via the evolution, this outer layer underwent substantial changes; moreover, different structures and derivatives were developed as a consequence of adaption to the lifestyles of different animal species. In general, the skin is the largest organ and the first barrier, therefore being important to interfacing with the environment and protecting the body against heat, dehydration, light, injury, and infections. For these functions also several substances are secreted, like salts or different organic molecules.

7.1 Structural Features of the Skin Among selected Animal Species and Human

Dependent on their lifestyle, nutrition, sex, and reproductive status, the thickness of the skin and its cellular composition vary widely among animal species (Fig. 7.1).

In this chapter, structural and functional differences of the animal skins will be outlined in fishes, frogs, reptiles, birds, nonhuman mammals, and separately in humans.

7.1.1 Fish

The skin of fishes is composed of cuticle, epidermis, dermis, and hypodermis.

The most superficial layer is the **cuticle**. It is only 1 μm thick, consisting of mucus and cellular debris to prevent abrasion and reduces friction of the skin. In this layer, IgM antibodies and enzymes with antimicrobial properties are also deposited.

The **epidermis** is composed of two layers. The upper *stratum basale* consists of columnar or cylindrical cells, while the lower layer called *stratum germinativum* consists of oval or round cells, which become increasingly flatened on their way to the surface. Epidermal cells are capable of mitosis in all layers but particularly when being close to the basal membrane. The epidermis is mainly composed of keratinocytes, but also mucus-producing "Goblet cells" in the superficial part of the epidermis, alarm substances secreting "Club cells," and specialized sensory cells can be found. In this layer also pigment cells are localized, being responsible for coloration and camouflage as well as for the communication. They are at the dermo-epidermal interface as well as at the boundaries between the dermis and hypodermis. Pigment cells can be differentiated based on their produced pigment, like melanophores being responsible for black and brown color, xanthophores for yellow, erythrophores for orange and red, leucophores for white, and iridophores for reflective coloration. Changes of color are caused either due to the neuroendocrine effects (e.g., adrenergic, cholinergic, or melatonin stimulation) or due to water quality, temperature, or salinity.

As in other species, the **basement membrane** separates the *epidermis* and the *dermis*.

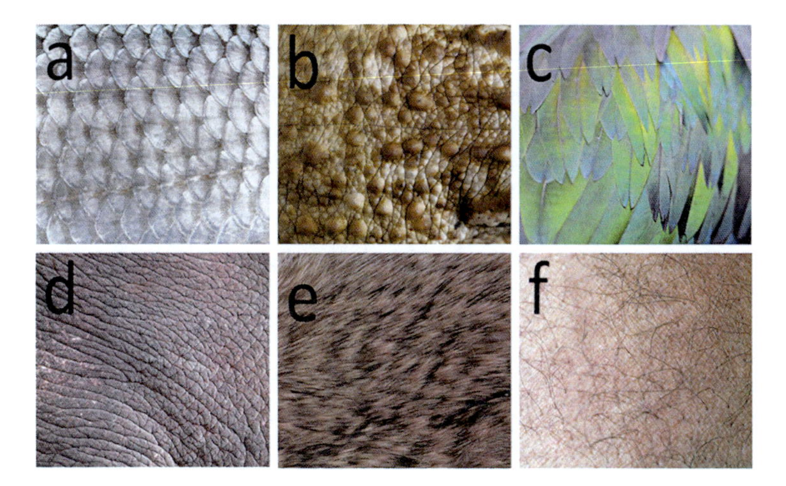

Fig. 7.1 The skin differentiates diversely in the species according to environment and demands for barrier function, camouflage, and mating: (**a**) fish © [Kondor83]—Fotolia.com, (**b**) frog © [Eric Isselée]—Fotolia.com, (**c**) bird © [panuruangjan]—Fotolia.com, (**d**) elephant © [jakgree]—Fotolia.com, (**e**) rabbit © [moonrise]—Fotolia.com, and (**f**) human © [Santhosh Kumar]—Fotolia.com

The dermis in the fish has two layers: the upper *stratum spongiosum* and lower *stratum compactum*. *Stratum spongiosum* is a connective tissue, built up by collagen and reticulin fibers. Due to the dense collagen in *stratum compactum*, the dermis keeps its structural strength.

In fishes, according to their size, shape, and structure, different types of scales are recognized, such as cycloid, ctenoid, ganoid, placoid, and cosmoid. Cycloid and ctenoid scales are found in the teleosts. Cycloid scales build a translucent disc with smooth margins, and ctenoid scales have additional tiny teeth (ctenii), so they have a rough texture and can grow in size.

The lowest, deepest layer of the fish skin is called **hypodermis**, composed of loose fatty tissue, and connects the skin to the underlying structures, like muscle and bones.

7.1.2 Frogs

In the frog, the skin structure comprises of the **epidermis** and the **dermis**. The outermost part *of the epidermis* (*stratum corneum*) consists of 4–5 layers of epithelial cells, while the innermost part (*stratum malpighii*) is located on the **basal membrane**.

The **dermis** is a thick, connective tissue, containing muscle and nerve fibers, blood vessels, capillaries, pigment cells, and cutaneous glands. Two layers of the dermis can be distinguished: *stratum spongiosum* and *stratum compactum*.

7.1.3 Reptiles

In reptiles, skin condition is enormously influenced by the health and nutritional status as well as by environmental factors of the animals. Additionally, the skin supports exchange of water, gas, and temperature. In chameleons and geckos color changes provide camouflage of the host in their surroundings. Skin will also protect reptiles from physical influences, like dehydration, abrasion, and UV radiation. It is modified into scales and is composed of two layers **epidermis** and **dermis**.

The epidermis is divided into *stratum corneum* (outermost and heavily keratinized layer), *intermediate zone*, and *stratum germinativum*.

Stratum corneum is the outer layer of the skin and is further subdivided into "*Oberhautchen layer*," "*β-Keratin layer*" (hard and brittle, creating surface of the scales), and "*α-Keratin layer*" (soft and pliable, making connections between the scales). The *intermediate zone* is comprised of keratinocytes in various stages of development, while the *stratum germinativum* is the deepest layer of the epidermis, from where the cells originate.

The dermis being a connective tissue with interlaced collagen fibers, blood and lymphatic vessels, as well as smooth muscle fibers, nerves, chromatophores, and bony structures forms the dermal skeleton. In some chelonians and lizards, scales are underlain by osteoderms and osteoscutes (bony plates). In the turtle, there are two parts of the shell: the upper portion is called "carapace" and the bottom half is called "plastron" and both having blood and nerve supplies. Osteoderms fuse with ribs and vertebrae dorsally and with the sternum ventrally. Additionally, in reptiles there are two groups of cells, being responsible for coloration. In the basal layer of the epidermis, melanocytes are localized; in the outer part of the dermis, chromatophores are layered, which are pigment-containing and light-reflecting organelles.

Ecdysis is a process of skin molting of reptiles, where the old skin is replaced and renewed. During ecdysis, cells of the intermediate layer start to replicate so that three new layers of epidermis are formed. After this procedure is finalized, lymph diffuses into the area. The cleavage and separation of the old skin is caused by enzymes, contained in the lymph.

Between the phases of ecdysis, reptiles live normal lives. The so-called resting stage (period between two ecdysis) takes weeks to months. Shedding frequency and length of shedding is influenced by the age (young animals shed every 5–6 weeks, adults 3–4×/year), temperature (higher temperature increases the metabolism and so it increases the frequency), nutrition of the animal, as well as environmental humidity. Disease, scarring, and hormonal imbalance also influence the duration and frequency of skin renewing. For example, thyroid gland hormones are well known for their influence on ecdysis in different reptiles. There are different shedding patterns between the reptile groups, but shedding continues throughout the lifetime of all reptiles. Lizards and chelonians shed small pieces of skin, while terrestrial chelonians shed small flakes, and aquatic chelonians shed entire scutes. Snakes and some lizards shed their skin as a single piece including their spectacles, and during this time, the snake will become anorectic and will avoid contact.

Handling snakes during this period can be hazardous for the animal. Approximately 14 days prior to the shed, the snake will have a dull (grayish) appearance. The spectacles become bluish approximately 7–10 days prior to the shed and then clear 2–3 days later. The dull color is associated with the lymphatics and enzymes that fill the space between the old and new epidermal layers. The snake should be provided appropriate cage furniture (e.g., rock), so that they can rub their rostrum against a hard surface and facilitate the shedding process.

7.1.4 Birds

Avian integument is thin, elastic, and loosely attached to the body, giving birds the freedom of movement needed for flight. Its epidermis is both keratinized and lipogenic and acts as a sebaceous secretory organ. The skin is covered by feathers over most of the body. Many birds show colored naked skin or cutaneous outgrowths on the head and neck. Cornified epidermis covers the beak, claws, spurs, and the scales on the legs and feet. Most of these structures contain β-keratin. Birds have sebaceous secretory glands at the base of the tail (uropygial gland) and in the ear canals. Feathers are the most numerous and diverse of avian integumentary derivatives.

Histologically the skin of birds comprises **epidermis** and **dermis**.

Avian epidermis is simpler as in reptiles. In general it is thin in areas covered by feathers and thick in bare areas, its germinative layer is like in reptiles, but the cornified layer is much thinner than in reptiles. In birds, the feathers that provide mechanical protection and the reduction of body weight are advantages for flight. In birds, as in mammals and in the soft parts of reptilian skin, the cells become filled with α-keratin. Avian skin lacks sweat glands and sebaceous glands; the epidermis itself, in a variety of species, produces neutral fats and phospholipids. Avian epidermal cells include both lipogenesis and keratinization in their differentiation. The sebum serves as a moisture barrier and also probably helps to maintain the pliability of the keratinized epidermis.

As in other species a *basal membrane* (BM) is the interface between epidermis and dermis.

The following layers of the **epidermis** are recognized:

- ***Stratum germinativum*** is responsible for the production of cells, which later mature and form ***stratum corneum***. Stratum germinativum can be further subdivided into:
  - ***Stratum basale***, which is just above the BM.
  - ***Stratum intermedium*** is composed of larger cells of polygonal shape, which are connected by desmosomes.
  - ***Stratum transitivum*** is composed of well-developed cells with signs of keratinization.
- ***Stratum corneum*** is the uppermost part of the skin—localized on the top of *stratum germinativum* and consists of vacuolated and flattened cells.

The dermis is thicker than the epidermis and contains blood vessels, fat deposits, nerves and free nerve endings, several types of neuroreceptor organs, and a complex set of smooth muscles that move the feathers and exert tension on the skin. In birds, the dermis can be also subdivided into:

- ***Superficial dermis***.
- ***Deep dermis***, which can be subdivided into ***compact deep dermis*** and ***loose deep dermis***. Loose dermis is the deepest portion and contains fat and apterial muscle.

For the multiple and very variable coloration of birds, different pigment cells are responsible. **Melanocytes** produce black, brown, and yellow melanin and **carotenoids** and **xanthophyls** yellow and red pigments, which the birds receive via their nutrition. These pigments are deposited in feather follicles and the preen gland. For the final coloration **porphyrins and schemochromes** are responsible, while **uropygial gland pigments** affect in light reflection.

7.1.5 Mammals

Basic structure and function of the skin among the nonhuman mammals (dog, cat, small exotic mammals, bovine, swine) and humans is very similar, though some notable differences among the species exist.

7.1.6 Humans

The human skin is the largest organ of the body, contributing to 16 % of the total body weight, with an average area of 1.5–2.0 m^2. In this very thin layer, 1 square area contains 60,000 melanocytes, 1,000 nerve endings, and 650 glands in human. It can be divided into three major layers: epidermis, dermis, and subcutis (Fig. 7.2a).

The **epidermis** is the outermost layer of the skin, representing stratified squamous epithelium, with variable thickening all over the body, like 0.05 mm on the eyelids and a very thick layer on the soles and palms with 0.8–1.5 mm. The main cells in the epidermis are keratinocytes generated in the deepest epidermis and undergoing maturation within the epidermis. Structurally, the epidermis can be subdivided into five layers (Fig. 7.2b):

- *Stratum basale*
- *Stratum spinosum*
- *Stratum granulosum*
- *Stratum lucidum*
- *Stratum corneum*

The deepest layer of the epidermis is the ***stratum basale*** made up by basal keratinocytes, which are stem cells and constantly renew the epithelial cells in the above layers. Half of these basal cells differentiate and move up to the next layers in the epidermis to continue their maturation. The other half of it stay at the basal layer and divide over and over to replenish the basal cells in this layer. Within the basal

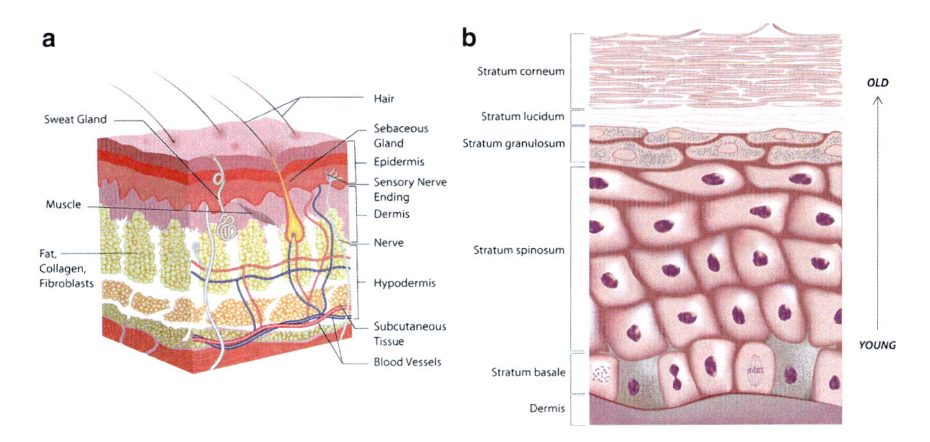

Fig. 7.2 The building plan of mammalian skin. (**a**) The three main layers of the skin are hypodermis (subcutis), dermis, and epidermis, either being differentiated to stratified squamous epithelia as shown here © [snapgalleria]—Fotolia.com. (**b**) The epidermal layers between stratum basale (basal layer) and stratum corneum, the exfoliation zone. During the differentiation process of squamous epithelia, the epithelial cells are flattened, finally die, and build up callused skin as mechanical protection © [Alila Medical Images]—Fotolia.com. Along the mucosal barriers the epidermis is non-callused or built up by a single-layered cylindrical epithelia (not shown)

keratinocytes other cells like melanocytes, Langerhans cells, and Merkel cells (touch receptors) are distributed.

Melanocytes are melanin-producing pigment cells, responsible for skin coloration and protection of the skin from UV light. They give 8 % of all epidermal cells and differentiate slowly and are less mobile. Other cells in the basal layer are the Langerhans cells, which are related to the dendritic cells. They are important antigen-presenting immune cells and therefore act as the first line of defense against microbial antigens (see Chapter 13). Via their morphology, they can be easily identified by their dendritic nature and by Birbeck granules.

The next layer above the basal layer is the ***stratum spinosum***, also termed "prickle cell layer," since desmosomal connections within the adjacent cells give the particular appearance of these cells. This layer is composed of polyhedral keratinocytes, with large pale-staining nuclei, therefore being very easily recognizable. Keratinocytes are the most predominant cells in the epidermis, representing 95 % of all epidermal cells. Keratinocytes are important immune cells, participating in the immune response, inflammation, wound healing, and vitamin D production. The two deepest layers of the epidermis *stratum basale* and *stratum corneum* together are called *stratum germinativum*, since they are responsible for the cell generation in the skin.

The middle layer is called ***stratum granulosum***, where the cells coming from the lower layer lose their nuclei and organelles, becoming nonviable, and continue to flattened form. This layer makes the barrier between living, active cells of the lower layer and dead cells from the outer layers.

The fourth, translucent layer is "shiny" and called ***stratum lucidum***. It has 3–5 layers of dead flattened keratinocytes, filled with eleidin, an intermediate form of keratin.

The most outer layer of the epidermis and the whole skin is the ***stratum corneum*** or horned layer, with 10–30 layers of dead cells, called corneocytes being hexagonal-shaped cells, without cytoplasmic organelles. Corneocytes are terminally differentiated keratinocytes via keratinization or also called cornification. In this process, keratinocytes move from the *stratum spinosum* to the *stratum corneum* and form the epidermal barrier. Via cornification, keratinocytes undergo several cellular changes that can be observed:

- Production of keratin, which is a fibrous protein, and finally incorporation into longer keratin intermediate filaments. It is involved in the tissue to hold water and keep moisture in the skin.
- Formation of cornified cell envelope (loricrin, involucrin).
- Loss of nuclei and organelles of keratinocytes. Therefore, at the final stage, corneocytes are almost fully filled with keratin.

The **dermis** is a tightly connected layer to the epidermis via the basement membrane, between the epidermis and the deepest skin layer hypodermis. It can be distinguished into two layers: the superficial layer adjacent to the epidermis is called papillary dermis, while the deeper and thicker connective tissue layer is called reticular dermis.

- Papillary dermis is an areolar, loose connective tissue and can be identified via fingerlike projections toward the epidermis. This layer consists of a terminal network of blood capillaries and Meissner's corpuscles, rapidly adaptive receptors, being sensitive to light touch and mostly concentrated on the fingers and lips.
- Reticular dermis is a deeper and thicker connective tissue, containing a high concentration of collagenous, elastic, and reticular fibers, supporting the strength, extensibility, and elasticity of this layer. Within this region, roots of hair, nails, sebaceous and sweat glands, and blood vessels are distributed.

The hypodermis or subcutis is the deepest, innermost layer of the human skin. It is mainly composed of fat, being important for the homeostasis and thermoregulation. In this tissue, fibrous bands, blood and lymphatic vessels, roots of hair, free nerve endings, and Pacinian corpuscles (mechanoreceptor for vibration and pressure) are found.

7.1.7 Nonhuman Mammals

The structure of **the epidermis** in nonhuman mammals is composed of the same layers (and cell types) as in humans, namely, *stratum basale*, *stratum spinosum*, *stratum granulosum*, *stratum lucidum*, and *stratum corneum*; however species-specific differences are developed. For example, in dogs and cats the living epidermis is only 3–4 layers thick, and the epidermis and its *stratum corneum* is thicker in non-haired areas (e.g., footpads, planum nasale), while the *stratum* *lucidum* is visible only in the footpads. In pigs, like in humans, BM forms prominent rete ridges (folds of dermo-epidermal junctions). In ruminants, equine, and carnivores, these are present only in the non-haired areas.

The dermis has very similar composition as in humans; however, papillary and reticular dermis are not present in carnivores. Their dermis comprises insoluble fibers (collagens, elastin) and soluble polymers (proteoglycans and hyaluronan). Epidermal appendages, e.g., arrector pili muscles, blood and lymph vessels, and nerves, are localized in the dermis.

Within nonhuman mammals, the thickest skin is found on the forehead, dorsal neck, or thorax. The thinnest skin is localized ventrally and on the lateral surface of the pinna. Skin thickness in the cat comprises 0.4–2 mm; in the dog, 0.5–5 mm; in the horse, 3.8 mm (1.7–7.7 mm); in the cow, approximately 6.0 mm; and in the pig, 2.2 mm. pH of the skin in the dog and cat is 5.5–7.5 and in the horse, 7.0–7.4 and increases up to 7.9 when sweating.

7.2 Derivatives and Appendages of the Skin

In the skin of vertebrates during evolutionary development, a variety of derivatives and appendages, like glands, scales, claws, hair, and nails, are developed, adapting the different species to their environmental conditions.

7.2.1 Glands

Glands are special groups of cells, distributed in the dermis of the skin. They secrete and release different substances on the skin surface, like sweat or oil, thereby taking part in the regulation of body temperature and protection of the skin against fungal and bacterial infections.

7.2.1.1 Frogs

Frogs have two types of cutaneous glands—mucous and granular glands (also called poison glands). These derivatives are located in the dermis. **The mucous glands** are small flask-shaped glands, uniformly distributed over the entire surface of the body, and secrete a colorless watery fluid. This keeps the body surface moist and sticky.

The poison (granular) **glands** are fewer in number and produce milky white poisonous substance, which helps to protect the animal from enemies and infections.

Amphibian skin is a remarkable organ that serves the multiple roles of fluid balance, respiration, and transport of the essential ions. Frog's skin also provides protection to the host by color changes and participates in respiration by direct gas exchange, in absorption of water, as well as in removal of wastes by exfoliation of keratinocytes. Mucous glands keep the skin moist and sticky.

7.2.1.2 Reptiles

Reptiles have generally a very low number of skin glands in comparison to frogs; however, different specialized glands, like in turtles, have evolved. **Musk glands** are paired glands near the bridge and connect the carapace to the plastron in turtles; **mental gland** is localized in the lower jaw of male turtles. **Secretory precloacal pores** exist in geckos and iguanas. Temporal glands situated in lateral commissures of the mouth in chameleons serve defensive behavior. Paired scent glands are found at the end of the tail in snakes and some lizards. Sea turtles possess special salt glands in their head behind each eye, which allow them to drink seawater. Femoral glands important for sexual behavior are described in some lizards. Keratinous structures like the jaw and claws are composed mainly of β-keratin.

7.2.1.3 Birds

There are no **sweat glands** in birds, which predisposes them to hyperthermia.

The uropygial gland (preen gland) is a bilobed gland, localized dorsally near the base of the tail and divided into branched alveolar glands; however, it is absent in some birds (e.g., ostriches). It has a holocrine type of secretion and produces waxy sebum, which is distributed over the body by preening. This provides waterproofing and antimicrobial properties.

7.2.1.4 Nonhuman Mammals

The epitrichial (apocrine) sweat glands are coiled, saccular, or tubular, localized throughout the hairy skin, below the sebaceous gland. The epitrichial duct merges into the pilary canal above the sebaceous gland. They produce sweat, pheromones, and immunoglobulins. These glands are innervated probably only in the horse.

The atrichial (eccrine) sweat glands are localized in footpads of dogs and cats. They produce sweat and have a coiled structure.

7.2.1.5 Humans

In the human skin are two types of glands, **sweat** and **sebaceous**, having different composition and function.

Sweat glands produce sweat and, based on their secretion, can be subclassified into **eccrine** and **apocrine** sweat glands:

- **The eccrine sweat glands** are coiled tubular glands, distributed all over the body, with varying density within the different body areas, however, being more concentrated on the soles, palms, and scalps. These glands produce a clear, odorless substance, consisting of water and NaCl, and secrete directly onto the skin surface. Via this release, the eccrine glands are important for thermoregulation.
- **The apocrine sweat glands** are larger, coiled tubular glands compared to the eccrine ones, and its distribution is limited to the axilla and perianals in human. They discharge their secretion into the canal of a hair follicle, which is cloudy and odorous, also containing pheromones, being important for the chemical communication. They start to secrete at puberty; their

In the human skin, the second type of glands, the **sebaceous glands**, secret an oily, waxy substance, called sebum. These glands are mostly found in hair-covered areas and connected to hair follicles, with greatest abundance on the face and scalps, but they are missing on the palms and soles. Via its chemical properties, the produced sebum maintains the integrity of skin barrier and hydrates the outer layer, the stratum corneum. The sebum also delivers antioxidants and antimicrobial lipids, thereby protecting the body from infections. It also provides vitamin E, and the secreted sebaceous lipids may harbour proinflammatory or anti-inflammatory properties. Interestingly, its defense function is remarkable already in the embryonic stage, protecting the embryonic skin from the amniotic fluid.

7.2.2 Outgrowths of the Skin

7.2.2.1 Birds

Cutaneous outgrowths in birds are modifications almost always expressed on the head or neck, where they are most visible. Outgrowths are commonly larger, or brighter in adult males of a species. Their coloring, which often contrasts with the adjacent plumage, is due to either intrinsic pigments or structural preconditions of the epidermis or to blood in the superficial capillary network.

One of the most impressive and visible outgrowths of birds is the **comb**, which is anatomically divided into base, body, points, and blade, and its tissue is very well supplied by blood vessels. Combs cannot be found in all bird species, but they are always found in chicken.

Wattle can be found under the jaw in many species of domestic poultry and is characterized by thick epidermis and dermis rich in blood vessels.

Frontal process (**snood**) is a structure dorsal to the nasal region and is typical in turkeys. As in other structures, color and size are determined by the density of the blood vessels.

Caruncles are multiple, mostly head protuberancies, sometimes on the upper neck in some species of domestic poultry.

Cere is a keratinized structure at the base of the upper beak in, e.g., budgerigars and some other species. Its color is helpful in determination of the sex of the bird.

Brood patch is an alopecic area in the breast region in some species with a thickened and vascularized dermis. This patch supplies heat for breeding eggs and can be found in both male and female.

Feet have almost no feathers in most species, so they have thickened epidermis, which is transformed into scales. **Scutes** are large scales, localized on the dorsal side of the toes and on the anterior side of the metatarsus in chickens. **Scutella** are small scales on the ventral side of the toes and on the posterior side of the metatarsus in chickens. **Reticulae** are the smallest distinct scales, and **cancella** are tiny scales between reticulae.

Feathers are arranged in pterylae and the areas of these feathers are called apteriae. The feather is composed basically of a shaft, the vane constituted by barb and rachis, and an after-feather (Fig. 7.3a). Pulp is the mesodermal component of

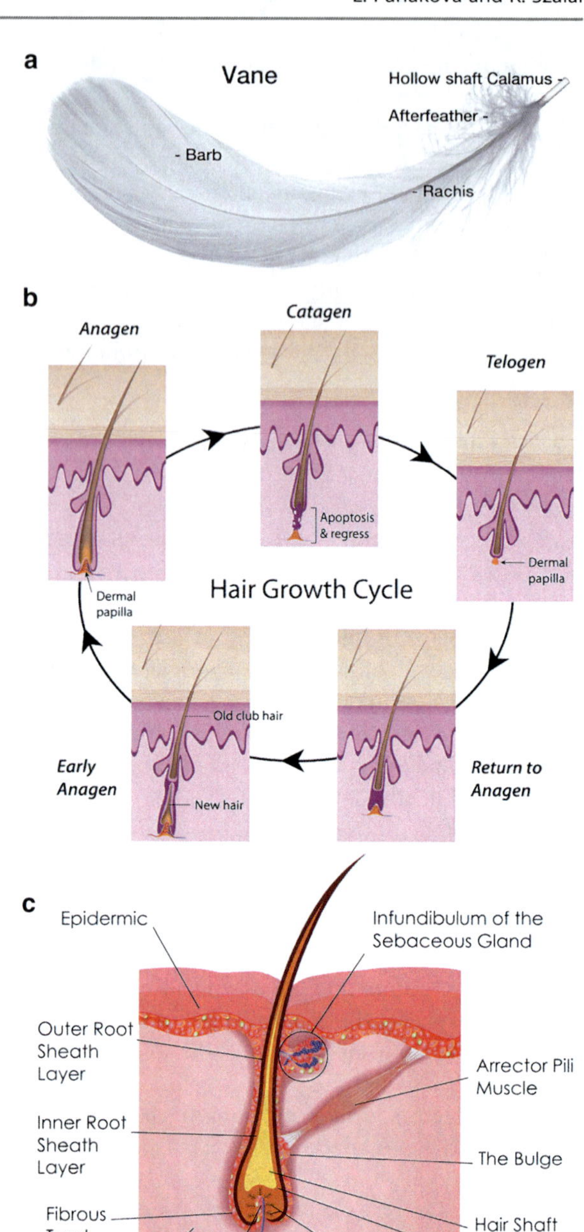

Fig. 7.3 Examples of appendices of the skin. (**a**) Bird feather © [arthurdent]—Fotolia.com. (**b**) The hair growth cycle © [Alila Medical Images]—Fotolia.com. (**c**) The microanatomy of a typical mammalian hair © [snapgalleria]—Fotolia.com

the growing feather consisting of vascular connective tissue and regresses as the feather grows. It is absent in the mature feather. There are different feather types which can be distinguished as (1) natal down feathers, the initial feather covering; (2) juvenile feathers of normal feather appearance, but smaller; and finally (3) feather sheath, also called pin feathers, consisting of cover feathers.

Adult feathers appear at the third molt and are of different types with different functions. Contour feathers are the main feathers and are located on the wings and on the body; flight feathers on the wings are called remiges and on the tail are called rectrices and covert cover remiges and rectrices.

There are two basic components of the feather, the outer epidermis and the inner pulp. The epidermis differentiates into barbed ridges, rachis, and hyporachis as it grows longitudinally. The pulp grows up with the epidermis and distally is covered by the epidermis (pulp cap). As the feather grows longitudinally, the pulp regresses by degeneration and resorption of the connective tissue. The color of the skin and of the feathers is due to pigment, structural conditions, and a combinations of these.

With respect to the structure, there are marked similarities between hair follicle and feather follicle. Feathers grow from the follicles in the dermis. The dermal papilla is the point of attachment between feather and its follicle, at this point the follicle is supplied with the blood and nerves. Herbst's corpuscles are found at the base of the feather follicle and detect vibrations.

7.2.2.2 Nonhuman Mammals

In general, **the hair** is a filamentous derivative of the skin, mainly composed of keratin. It is growing from the skin and it is one of the most characteristic features of mammals. It is also responsible for insulation, thermoregulation, sensory perception, and photoprotection. Further, hair has also a barrier function against chemical, physical, and microbial agents. In different species and breeds, hair is of different length, thickness, density, and medullation.

In dogs, cats, and horses, primary and secondary hairs can be differentiated. *Primary hair* is also called outer or guard hair and is of a greater diameter than the secondary hair. The medulla is always larger than the cortex and full of glycogen or pigment granules in the primary hair. *Secondary hair* builds an undercoat and is of smaller diameter and often wavy. In the secondary hair, the medulla is always thinner than the cortex and is filled with air bubbles, which serve as insulation. In goats, a special, nonmedullated hair exists called *lanugo hair*.

Basically the structure of hair in nonhuman animals is very similar to humans and is built up of three parts: (1) external cuticle, (2) rigid keratin cortex, and (3) vacuolized medulla in the center, filled with glycogen or air.

Similarly as in humans, in mammalian animals, the hair follicle cycle has three sequential stages: anagen, catagen, and telogen (Fig. 7.3b).

Follicular cycle is influenced by seasonal changes, changes of photoperiod, hormonal influences (melatonin, thyroxin, sexual hormones, corticosteroids), temperature of the environment, nutrition, general state of health, and genetics.

Footpad in dogs and cats is a specialized area of the integument. The footpad has different characteristics in comparison to hairy skin, since it is exposed to extreme environmental conditions and mechanical forces during movement. Thick epidermis and large fat deposits protect against mechanical trauma. Atrichial sweat glands are localized only in the footpads and work against traction and are used for scent markings by dogs and cats.

Hair follicles in cows and horses have simple arrangement; in dogs and cats hair follicles are compound. The sebaceous and epitrichial (apocrine) sweat gland and arrector pili muscle accompany hair follicle. Primary hairs are accompanied by sebaceous gland, apocrine gland, and muscle. Secondary hairs are accompanied by sebaceous glands only. Hair follicle layers are identical to the ones in humans.

7.2.2.3 Humans

In humans, **the hair** is very characteristic for most of the people and makes humans individually distinguishable from each other. An average adult person has five million hair follicles all over the body, except the palms of hands and soles of feet. Its thickness varies from 40 to 120 μm between the different areas. In many social cultures, there is a traditional headdress, often connected to social status.; however, in general women wear a longer and men a shorter cut of hair. One of the differences in hair is their color, varying from light blond to brown and black, and the shades of color can be scaled by the Fischer-Saller scale. Hair color is determined by the intensity of two pigment melanins, eumelanin and phaeomelanin, which are produced by melanocytes in the hair bulb. Eumelanin gives the hair a darker color, like brown or black, and is inherited as a dominant allele. In contrast, phaeomelanin makes the hair color lighter and is combined as a recessive allele.

The hair follicle is localized in the dermis of the skin and produces the hair itself. Structurally the hair is composed of three sections (Fig. 7.3c):

- *The hair bulb* is the base of the hair follicle, composed of the dermal papilla and matrix. The dermal papilla is connective tissue, fed by the bloodstream, bringing in nutrition for new hair growth. The matrix in the bulb is formed by epithelial cells. Here cell division takes place forming the new structure of the hair. In this tissue, melanocytes produce pigment materials, providing the color to the hair shaft.
- *The inner root sheath.*
- *The hair shaft* is composed of the inner medulla, cortex, and cuticle. The medulla is the innermost layer, partially keratinized. The layer above is called cortex; it is the greatest portion of the hair shaft, containing keratin. Keratin is a protein chain, with a high concentration of cysteine, enabling the making of disulphide bonds, and providing strength and elasticity to the hair. The outer cuticle of the hair shaft is a thin, colorless layer and protects the hair.

The hair is growing over the time and three sequentially phases can be distinguished:

- *Anagen phase* is the active growth phase of hair growth. A healthy hair shaft grows 1 cm in 28 days; however, scalp hairs stay in this phase even for 2–7 days.
- *Catagen phase* is a short phase, following the anagen phase. It takes 1–2 weeks, where the hair is off from the blood capillary supply and from the basal cell from the hair bulb.

- *Telogen phase* is the last phase of the hair growth. When the dermal papilla separates from the follicle, the hair starts to fall out.

After the telogen phase, the whole hair growth starts form the beginning with the anagen phase (Fig. 7.3b).

In humans, another important outgrowth is the **nails**, which are visible, hornlike hard plates, covering the dorsal, final ends of the fingers and toes of humans and mammals. It is composed of the following parts:

- *The matrix* is a keratogenous membrane, protecting the nail and containing nerves, lymph, and blood vessels. In this matrix, cells are produced, which become the nail plate. It continues to grow as long as it receives nutrition.
- *The nail bed* is located beneath the nail plate and is made up of two types of tissues of the skin: epidermis and dermis.
- *The nail plate* is called as the "body of the nail," made of translucent keratin, which is composed of several layers of compressed and dead cells.
- *Hyponychium* is located under the nail plate, a thin epithelial layer, connecting the nail plate and fingertip, protecting the nail bed.
- *The nail wall* is a cutaneous fold covering the sides and the proximal end of the nail.

7.3 Synopsis

The outer body barrier, the skin, underwent several adaptions in the evolutionary process. In different animal species, unique and species-specific but also general structures and outgrowths have evolved in animals and humans, adapting their life to the environment.

Acknowledgment We would like to thank Prof. Erika Jensen-Jarolim, and crossip communications, Vienna, for kind support in the arrangement of illustrations. The work was supported by the Austrian Science Fund FWF grant SFB F4606-B19.

Further Reading

Bukowskiy LF (ed) (2010) Skin anatomy and physiology research developments. Human anatomy and physiology, 1st edn. Nova Science Publishers, New York, 30 April 2010, ISBN-10: 1607414694

Lindberg M, Forslind B, Forslind F (eds) Skin, hair, and nails: structure and function. Basic and clinical dermatology. Informa Healthcare, New York, Sept 2003, ISBN-10: 082474313X

Millington PF, Wilkinson R (2009) Skin. Biological structure and function books, 1 reissue. Cambridge University Press, Cambridge, 1 June 2009, ISBN-10: 0521106818

Body Messaging: The Endocrine Systems

8

Florian K. Zeugswetter and Erika Jensen-Jarolim

Contents

F.K. Zeugswetter (✉)
Small Animal Clinic, Clinical Department for Companion Animals and Horses, University of Veterinary Medicine Vienna, Vienna, Austria
e-mail: Florian.Zeugswetter@vetmeduni.ac.at

E. Jensen-Jarolim (✉)
Comparative Medicine, Messerli Research Institute, University of Veterinary Medicine Vienna, Medical University Vienna, University Vienna, Vienna, Austria

Division of Comparative Immunology and Oncology, Department of Pathophysiology and Allergy Research, Center for Pathophysiology, Infectiology and Immunology, Medical University Vienna, Vienna, Austria
e-mail: erika.jensen-jarolim@meduniwien.ac.at

E. Jensen-Jarolim (ed.), *Comparative Medicine*,
DOI 10.1007/978-3-7091-1559-6_8,

Abstract

Cells of multicellular organisms communicate via **chemical messengers** secreted into the extracellular fluid. This form of communication can be found in even the simplest animals like sponges and is phylogenetically older than neural transmission (Hartenstein 2006). Some of the diffusible signals travel short distances and influence cells in the vicinity (**paracrine signaling**), whereas others reach their target cells via the circulatory system (**endocrine and neuro-endocrine signaling**). Transported within body fluids, these highly specific messengers, called "**hormones**" (from the Greek word "hormaein," to excite), reach most cells throughout the body, where they bind to specific receptor proteins. Their interaction with target cells depends on the presence and sensitivity of these specific receptors. By binding to different receptor subtypes, most hormones are able to trigger more than one and in many cases even opposing responses. In interaction with the nervous system, hormones coordinate the activities of many different cells and regulate growth, metabolic processes, water balance, immune function, digestion, and reproduction. The ability to cooperatively respond to environmental (**exteroceptive**) or internal (**interoceptive**) stimuli enables the body to maintain a dynamic equilibrium, the so-called **homeostasis**.

8.1 Endocrine Tissues and Organs

8.1.1 Introduction

In general terms the endocrine system of highly derived taxa (vertebrates and invertebrates) can be divided into the **neuroendocrine system**, where specialized neurosecretory neurons synthesize, store, and secret the so-called neurohormones, and into the "**classical**" **endocrine system**. In regulatory circuits neurohormones commonly represent the interface between the central nervous system and the "classical" endocrine system.

8.1.2 Vertebrates

Many neurosecretory cells (NSCs) of **vertebrates** are located in clusters (nuclei) in specific areas of the hypothalamus (e.g., the paraventricular and supraoptic nucleus). Which anatomically forms the ventral part of the diencephalon. The axons of these cells either travel through the infundibular stalk and end near vessels of the neurohypophysis or release their neurohormones into the **hypothalamo-hypophysial portal system**. The latter provides a direct and quick path for neurohormones to stimulate (**releasing hormones**) or inhibit (**inhibiting hormones**) the secretion of "classical" hormones from the adenohypophysis (Table 8.1). The hypothalamus

Table 8.1 Systematic overview on hormone regulation

Hypothalamic hormone		Pituitary hormone influenced Adenohypophysis		Primary targets
Releasing hormones (stimulate)	Abbreviation	*Tropins (stimulate)*	Abbreviation	
Corticotropin-releasing hormone	CRH	Adrenocorticotropin or adrenocorticotropic hormone	ACTH	Adrenocortical cells (secrete glucocorticoids)
Gonadotropin-releasing hormones	GnRH	Gonadotropins (follicle-stimulating hormone; luteinizing hormone)	FSH, LH	Gonads (secrete sexual steroids)
Growth-hormone-releasing hormone	GHRH	Somatotropin or growth hormone	GH	GH-receptors, liver (secretes insulin-like growth factor 1 [IGF-1])
Prolactin-releasing hormone	PRH	Prolactin	Prolactin	Mammary gland, pigeon's milk (birds)
Thyrotropin-releasing hormone	TRH	Thyrotropin or thyroid-stimulating hormone	TSH	Thyroid glands (secrete thyronines)
Inhibiting hormones (inhibit)				
Growth-hormone-inhibiting hormone	GHIH	Somatotropin or growth hormone	GH	GH-receptors, liver (secretes insulin-like growth factor 1 [IGF-1])
Prolactin-inhibiting hormone	PIH	Prolactin	Prolactin	Mammary gland, pigeon's milk (birds)
		Neurohypophysis		
Antidiuretic hormone	ADH	Release (additionally stimulates ACTH secretion)		Kidneys (water balance), adrenocortical cells (secrete glucocorticoids)
Oxytocin	Oxytocin	Release		Uterus, mammary gland

together with the pituitary gland can be considered the master regulators of endocrine coordination.

The pituitary gland or hypophysis is an endocrine gland located below the hypothalamus and caudally of the optic chiasma (where optic nerves cross). In higher phyla it is connected with the hypothalamus by the slender infundibular stalk which consists of axons linking the hypothalamus with the neurohypophysis and of blood vessels (the portal system) linking the hypothalamus with the adenohypophysis. The adenohypophysis is separated by the residual cleft (remains of the Rathke's pouch) into a pars distalis and pars intermedia. Interestingly in contrast to hormones from the pars distalis, hormones produced by the mostly rudimentary pars intermedia are primarily controlled (inhibited) by dopamine. No pars

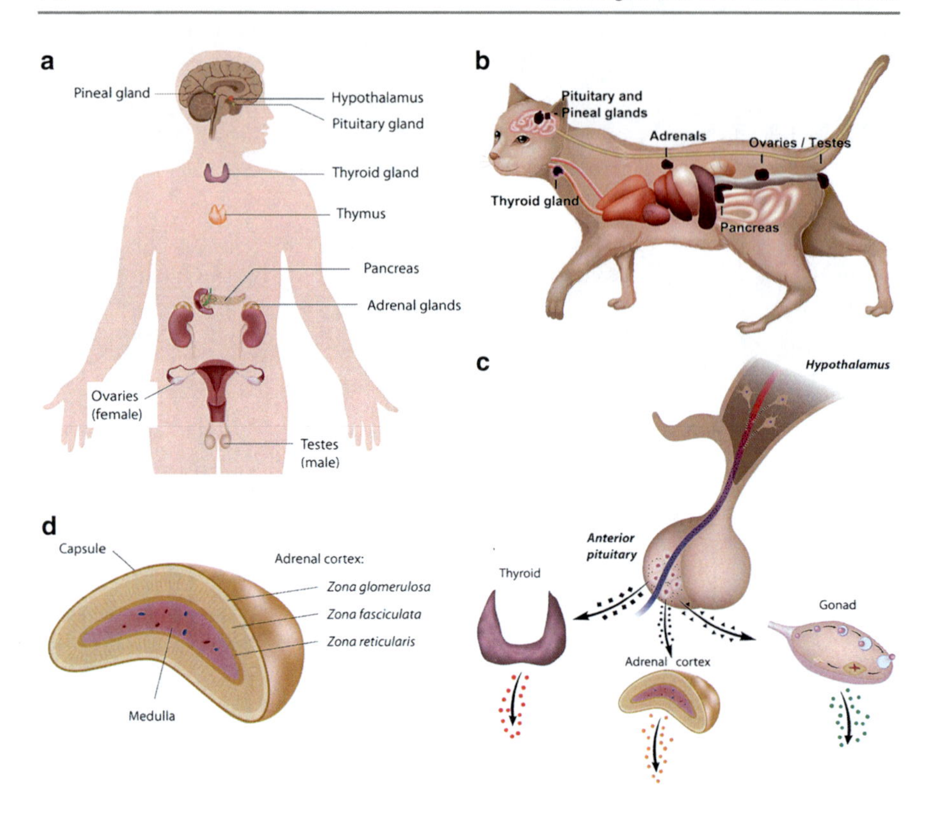

Fig. 8.1 The endocrine organs. (**a**) Human (© [Alila Medical Images]—Fotolia.com) and (**b**) cat (adapted from © [blueringmedia]—Fotolia.com). The highest organ in this hierarchy is represented by the hypothalamus, followed by the pituitary gland (hypophysis); at the third level, thyroid gland, thymus, pancreas, adrenal glands, and sexual glands (ovaries or testes) control diverse body functions, growth, and homeostasis. (**c**) As the most important hormone producer and regulator system, the hypothalamic-pituitary axis to exemplary endocrine glands is depicted (© [Alila Medical Images]—Fotolia.com). For instance, thyrotropin-releasing hormone (TRH) is released by the hypothalamus to induce secretion of the thyroid-stimulating hormone (TSH) from the pituitary gland, which induces thyroxine production and secretion from the thyroid gland. Further, upon hypothalamic stimulation by CRH (corticotropin-releasing hormone), adrenocorticotropic hormone (ACTH) is secreted from the pituitary to stimulate corticoid hormone production and release from the cortex of the adrenal glands. (**d**) The cortex of the adrenal gland is built up by cells of different function organized in three zones (© [Alila Medical Images]—Fotolia.com): the zona glomerulosa produces primarily the mineralocorticoid aldosterone (see Chap. 10), the zona fasciculate and the zona reticularis produce glucocorticoids (e.g., cortisol) and androgens (male sexual hormones), respectively (see Chap. 12). Principally, all hormones inhibit their own production by short- and long-distance feedback mechanisms (see example in Fig. 8.2d)

intermedia has been described in birds, elephants, and whales (Bentley 1998). Due to the different positions of the adeno- and neurohypophysis in various species, the terms "anterior," "intermediate," and "posterior" pituitary should be restricted for use in primates. In dogs and horses, for example, the adenohypophysis encloses the entire neurohypophysis (Nickel et al. 2004).

The different hormones secreted by the anterior pituitary can roughly be categorized according to their target cells. Whereas hormones, such as the milk secretion stimulating **prolactin** or the melanocyte-stimulating hormone (**MSH**), act directly on non-endocrine tissue, the "**tropins**" like adrenocorticotropic hormone (ACTH) or thyrotropin (thyroid-stimulating hormone, TSH) stimulate the secretion of other hormones.

Hormone-producing cells can be found with varying degree of association. They can be dispersed throughout other organs (the so-called **diffuse endocrine system** [DES]) and scattered as "**islets**" or are grouped in ductless organs called **endocrine glands** (Fig. 8.1a, b). Diffuse endocrine cells can be found, for instance, throughout the gastroenteropancreatic tract. Fourteen different specialized cells producing more than 100 different hormonally active peptides and amines have been described (Rehfeld 1989; Rindi and Bordi 2006). Unsurprisingly, the gut is considered the largest endocrine organ of the body. The classical hormones of the gastrointestinal tract regulating secretion are the peptides **secretin** secreted from S cells in the duodenum, **gastrin** released from G cells in the stomach, and **cholecystokinin** secreted from intestinal I cells. The traditional classification into the DES and classical endocrine glands is stressed by the fact that we now know that even the skin and fat tissue secrete factors with important endocrine functions (Kershaw and Flier 2004; Slominski et al. 2008).

Major discrete endocrine glands in the vertebrate phyla (Fig. 8.1c) are the **pituitary**, the **pineal gland** (absent in some animals, e.g., hagfish, crocodiles, and whales, Bentley 1998), the **urophysis** (in the tail of fish), the **thyroids**, the **parathyroids** (absent in fish, phyletic debut in amphibians), the **ultimobranchial glands** (produce calcitonin, in mammals incorporated into the thyroids), the **gonads**, and the **adrenals** (**interrenals** in fish). The adenohypophysis, the thyroids, the parathyroids, and the ultimobranchial glands originate from pharyngeal tissue. The morphology and positions in the body show significant interspecies variations. With the exception of the pituitary, which is stimulated and inhibited by the hypothalamus, most of these glands can be transplanted to other regions of the body without losing their physiological function. The possibility to reimplant (autotransplant) endocrine tissue is of great benefit after radical operations on the thyroid glands. In many cases the appended parathyroid glands also have to be removed, and the autotransplantation offers a great opportunity to avoid postoperative hypoparathyroidism (Lo 2002).

Even complex endocrine glands like the pituitary gland can be grown from stem cells cultured "in a dish." This opens up an attractive alternative perspective for treating hypophysial malfunction in the future (Suga et al. 2011).

Morphological studies suggest that discrete endocrine glands developed from dispersed cells of ancestral animals (Hill et al. 2008). One of the best-reviewed glands is the adrenal gland with publications going back to the nineteenth century. In fish the adrenal gland, also called the interrenal organ, consists of small islets highly dispersed throughout the kidneys, whereas in amphibians they are gathered either in groups (primitive amphibians like salamanders) or in cords (advanced amphibians like frogs) on the ventral surface of the kidneys. A further evolutionary step is observed in reptiles (except turtles), birds, and mammals where the adrenal glands are isolated encapsulated organs at the cephalic tip of the kidneys (Accordi et al. 2006) (Fig. 8.1d). In some birds they are even fused to a single adrenal gland (Bentley 1998).

8.1.3 Invertebrates (Mollusks and Arthropods)

Like in vertebrates most hormones of mollusks (snails and cephalopods) and arthropods are short polypeptides produced by scattered NSCs and epithelial cells.

In ***mollusks*** variable clusters of NSCs secreting neurohormones into the hemolymph have been described. The neurons of these cells are partly in contact with primitive peripheral nonneural endocrine organs such as the **dorsal bodies** and **lateral lobes** (snails) and **optic glands** (mollusks). These glands secrete amongst others the gonadotropic and ecdysteroid hormone (Hartenstein 2006).

In ***insects*** NSCs are located in the brain and ventral nerve cord, and their axons project towards the **corpora cardiaca**, the **corpora allata**, and the **prothoracic gland**. In Drosophila these organs are fused into a "**ring complex**." The corpora allata secrete the neurohormone **prothoracicotropic hormone** (PTTH, various types identified) and the **juvenile hormone** (JH). In larva and nymph PTTH stimulates the release of steroid hormone **ecdysone** from the prothoracic gland. Ecdysone, the "molting hormone," is peripherally activated to 20-hydroxyecdysone and stimulates the digestion of the old cuticle. This is an essential part of insect metamorphosis. In the presence of high JH concentrations, the insect molts into a larger juvenile form, whereas in absence of JH it triggers the development of an adult form. Both hormones are lipid soluble and bind to intracellular receptors. After molting, the JH also acts as a gonadotropin and stimulates the production of ecdysone which is now produced in the gonads and no longer in the prothoracic gland. In the adult form, ecdysone stimulates the synthesis of yolk proteins and the production of pheromones (Hill et al. 2008). Analogs of the juvenile hormones such as the carbamate fenoxycarb are currently used to control insect pests.

8.2 Endocrine Regulation

When hormones of one endocrine tissue stimulate the secretion of messengers from another tissue, it is called an axis. Depending on the hormones involved, we differentiate between **two-part** and **three-part axes**. A well-known example for a three-part axis is the hypothalamic-pituitary-adrenocortical axis (Fig. 8.1c). Under stressful conditions and at low levels of blood glucocorticoids, corticotropin-releasing hormone (CRH, corticotropin-releasing factor) is released into the hypothalamo-hypophysial portal system and stimulates the release of pituitary ACTH. Evolutionary, releasing hormone precursors might generally themselves play a role in the control of energy and water homeostasis (De Loof et al. 2012). ACTH is carried to the adrenal cortex where it stimulates the release of adrenocortical steroids, e.g., cortisol (alias hydrocortisone). Comparable to other hierarchically organized hormone systems, high concentrations of cortisol suppress its own secretion by inhibiting pituitary ACTH (**short-loop feedback**) and hypothalamic CRH (**long-loop feedback**) secretion. These **negative feedback mechanisms** modulate hormone concentrations and prevent that cortisol secretion is carried to excess. Knowing that cortisol- and ACTH-producing tumors commonly lose their ability to respond appropriately to hypercortisolemia is inevitable to understand dynamic suppression tests commonly applied in clinical endocrinology.

Most hormones are not released continuously, but in a **circadian** or **rhythmic, pulsatile manner**. The knowledge of these daily or annual cycles is of particular importance in the interpretation of endocrine tests.

8.3 Chemical Nature, Storage, and Cellular Response Pathways

Biochemically "classical" hormone molecules of invertebrates and vertebrates fall into three main classes—**peptides**, **steroids**, and **amines** (Fig. 8.2a). Exceptions are modified fatty acids like the prostaglandins (vertebrates) and the juvenile hormone (JH) of invertebrates. Vitamin D (calcitriol, 1,25-dihydroxyvitamin D) is now also considered a hormone and has similarities to the steroid hormones.

8.3.1 Peptide Hormones

All neurohormones from the hypothalamus, neurohypophysis, and urophysis and most hormones other than hormones from the adrenal glands, thyroid glands, pineal gland, skin, gonads, and placenta are **polypeptides**. As peptide hormones are structured from chains of three (e.g., thyrotropin-releasing hormone, TRH) to more than 200 amino acids (e.g., adiponectin), they vary enormously in size and structure (see examples in Fig. 8.2a, right side). Hormones with peptide chains of more than about 100 amino acids are called **proteo-hormones** (e.g., angiotensin, leptin, prolactin). Peptide and proteo-hormones are synthesized from larger **prepro**- and **prohormones** (mostly inactive) by proteolytic cleavage (posttranslational processing) and stored in membrane-bound vesicles attached to the protein neurophysin. Due to this enzymatic action and the storage in the same vesicle, some hormones are co-secreted. Examples are the cleavage products ACTH and β-endorphin from the preprohormone pro-opiomelanocortin (POMC) and insulin and C (connecting)-peptide from proinsulin. Related to their structure, polypeptides are **water** and not lipid **soluble** and as such cannot pass through the lipid bilayer of cell membranes. Accordingly they have to bind a **membrane-bound receptor**, which triggers a series of biochemical reactions leading to the intracellular generation of second messengers and **signal amplification** (Fig. 8.2b). A typical second messenger generated in the **G-protein-coupled receptor system** is cyclic

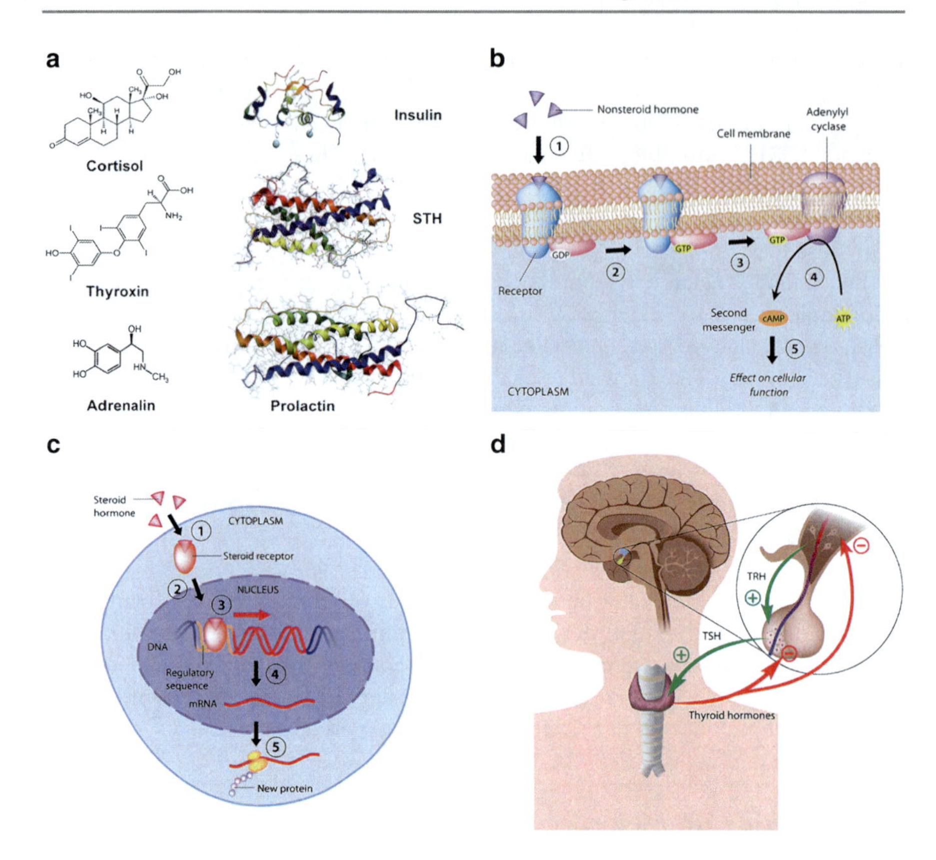

Fig. 8.2 Hormones and hormone actions. (**a**) Examples of important hormones of different substance classes (© [Vladimir Fedorchuk; 1 to 3; WikipediaCommons; ×3 Leonid Andronov]—Fotolia.com). (**b**) Nonsteroid hormone action is characterized by binding of the hormone to membrane-bound receptor. This triggers a series of biochemical reactions leading to the intracellular generation of the second messengers cyclic adenosine monophosphate (cAMP) generated from ATP. cAMP induces a cellular signal leading to effects on cellular function (© [Alila Medical Images]—Fotolia.com). (**c**) Steroid hormones enter the cytoplasm where they bind to a steroid hormone receptor which transports the hormone to the nucleus where the complex binds to DNA and modulates gene expression (© [Alila Medical Images]—Fotolia.com). (**d**) The hierarchy of thyroid hormone regulation: TRH from the hypothalamus induces release of TSH from the pituitary gland. TSH then binds to specific TSH receptors on thyroid cells from where triiodothyronine (T3) and thyroxin (T4) are released. The concentration of formed hormones is regulated by positive and negative feedback mechanisms (low concentrations of T3 and T4 induce, whereas high concentrations inhibit secretion of TRH and TSH from higher centers) (© [Alila Medical Images]—Fotolia.com). Any dysfunction of these mechanisms may cause hypo- or hyperthyroidism disease. The determination of T3, T4, and TSH concentrations is amongst others used to diagnose thyroid diseases

adenosine monophosphate (**cAMP**) generated from ATP. cAMP activates protein kinases which change the configuration and consequently the function of proteins by phosphorylation. Other common second messengers of the G-protein-coupled receptor system are inositol 1,4,5-trisphosphate (**IP**$_3$) and 1,2-diacylglycerol

(**DAG**) generated by the catalytic action of the membrane-associated enzyme **phospholipase C**. The major action of IP_3 is to open Ca^{2+} channels and increase cytoplasmatic Ca^{2+}, whereas DAG activates membrane-bound kinases. As polypeptide hormones act on preexisting proteins, their effects are rapid and start within minutes (Hill et al. 2008).

Peptide hormones in different animals often have a different amino acid sequence, and sometimes variants exist in the same species (e.g., four variants of growth hormone in cows). In many cases as for **antidiuretic hormone** (**ADH or vasopressin**), **oxytocin**, **ACTH**, **β-MSH**, **IGF-1**, **insulin**, **glucagon**, and **somatostatin**, the interspecies differences are small and apply for only a few amino acids. Suiformes including true pigs, for example, produce **lysine vasopressin** (nine amino acids) instead of **arginine vasopressin**. The difference is one amino acid in position 8. The change in pigs probably occurred 60 million years ago (Ferguson and Heller 1965). Vasopressin is released upon hyperosmolarity of the blood and causes water resorption by the kidneys. Thus, the endocrine master regulator center also has impact on volume regulation (see Chap. 10).

Oxytocin (nine amino acids) is derived from the hypothalamus and—at least in mammalians—stored in the posterior pituitary. It acts primarily on the smooth muscles of breasts producing milk ejaculation and on those of the uterus during orgasms and birth giving. Additionally it has gained attention in behavioral studies due to its important function for the social actions and bonding to partner and baby, as well as in the human-animal relation to pets and human-human relation, such as in autism (Chang et al. 2012). The equivalent hormone to oxytocin in amphibians, birds, and reptiles is **mesotocin** (differs by one amino acid) and **isotocin** in most bony fish (differs by two amino acids).

Other hormones show more profound differences, e.g., the bovine growth hormone (~200 amino acids) differs by 64 amino acids from human **growth hormone** (**GH**; or somatotropin; somatotropic hormone, STH; Döcke 1994). Knowing this, the importance of using validated species-specific assays is self-explanatory. It also clarifies why, for example, porcine and bovine growth hormones have no significant effects in primates (Ganong 2001). GH is released in a pulsatile fashion upon stimulation by **growth-hormone-releasing hormone** (**GHRH**) from the hypothalamus. An evolutionary ancestor of GHRH may have been glucagon-like as proposed from insect studies. Generally these insect RHs may fulfill other functions than the higher developed RHs (De Loof et al. 2012). Mammalian GHRH counteracts somatostatin by inducing the secretion of GH from the hypophysis. GH then acts through rapid catabolic (insulin antagonism) and slow anabolic actions (growth). The anabolic effects are mainly mediated via the **insulin-like growth factor 1** (**IGF-1**) which is produced in the liver and many other tissues. Prolonged GH excess in adult humans and animals (exogenous GH sources, somatotroph adenoma of the pituitary gland, progesterone induced in intact female dogs) may result in diabetes mellitus (common in cats) and acromegaly (bony and soft tissue overgrowth predominantly in the face, hands, and feet). Nonphysiological exposure before the closure of the epiphysial plates typically results in giantism. Currently, the only FDA-approved GH in animal breeding is bovine GH, which is primarily used to enhance milk production in dairy cows. This is possible due to its close relation to the peptide hormone prolactin which has a pronounced effect on lactation. In humans, GH is routinely used to treat growth disorders in children or to enhance vitality in elderly.

Parathormone (PTH) is a peptide hormone (84 amino acids) produced from the parathyroid glands. These glands located on the outside of, in, or near the thyroids (depending on the species under investigation) were already detected in dogs in the nineteenth century (Baber 1876). Low calcium levels in blood and in parathyroid cells stimulate the adenylate cyclase to convert ATP in cAMP. This leads to a release of PTH (secondary hyperparathyroidism). PTH then activates osteoclasts to free calcium from bones and calcium reabsorption in the kidneys. Further, PTH stimulates the kidneys to activate vitamin D by hydroxylation. This indirectly also leads to calcium absorption from the intestine. Unintended removal of the parathyroid glands during thyroid surgery or autoimmune gland destruction causes primary hypoparathyroidism and hypocalcemia. Clinical signs include tachycardia, cardiac arrhythmia, trismus ("lockjaw"), cramping and twitching of muscles or seizures, flaccid paralysis (reptiles), lethargy, and elevated body temperature. Notably, calcium is also involved as a cofactor in enzymatic reactions and important for neuromuscular signal transmission. For signs of hypocalcemia, see 8.5.3. Elevated PTH levels may also be due to autonomous hormonal active parathyroid adenoma, carcinoma, or hyperplasia (primary hyperparathyroidism). These patients are usually presented with polyuria, polydipsia, anorexia, and lethargy caused by hypercalcemia.

Calcitonin is a small peptide hormone (32 amino acids) derived from the parafollicular C cells within the parenchyma of the thyroid gland. Its effects on blood calcium levels were first suggested for dogs by Copp and Cameron (1961). Administration of large supraphysiological doses lowers elevated calcium concentrations by inhibiting osteoclastic, calcium-releasing macrophages in bones, inhibiting absorption from the intestine, and simultaneously enhancing Ca^{2+} secretion (and phosphate) from the kidneys. Further, it inhibits phosphate reabsorption. As calcitonin (endogenous or exogenous) does not lower serum calcium levels in normocalcemic patients, no biological effects are seen in patients with calcitonin-secreting thyroidal carcinoma, and calcitonin deficiency syndromes have not been identified, the physiological importance of this hormone has been questioned (Hirsch et al. 2001). Nevertheless, salmon calcitonin which is 20–100-fold more active than mammalian calcitonin (Bentley 1998) is commonly used in clinical endocrinology.

8.3.2 Steroid Hormones

Steroid hormones in vertebrates include the **glucocorticoids** (effects on glucose metabolism), **mineralocorticoids** (effects on electrolytes, see Chap. 10), and **sexual steroids** (see Chap. 12). Further, vitamin D with regulatory function on calcium levels is today considered a hormone (D-hormone) (see Chap. 10). Steroid hormones are synthesized from the precursor cholesterol, are stored in lipid droplets, and leave the cells by diffusion. They attain much higher plasma concentrations than peptide hormones. The biochemical pathways necessary to synthesize steroids and their role as signaling molecules are ancient and present in the earliest invertebrates. A well-known steroid in arthropods is **ecdysone** also called molting hormone. Classical corticosteroids (11- and 21-hydroxylated steroids) are unique to the vertebrates and can be traced back to the Cyclostomata (jawless fishes: hagfishes and lampreys, Denver 2009). As steroids are unable to dissolve in aqueous body fluids like blood or hemolymph, their transport greatly depends on the availability of **carrier molecules**. These specific transport proteins are synthesized in the liver, have different hormone affinities, and provide a reservoir for steroids. After having dissociated from their carrier, free hormones enter the target cells and bind to specific **cytosolic or nuclear receptors**. These hormone receptor complexes bind to DNA and modulate **gene expression** (Hill et al. 2008) (Fig. 8.2c). The nuclear receptors, absent in plants, are estimated to have arisen over one billion years ago (Denver 2009). As only **free steroids** are biologically active, their measurement (e.g., free cortisol) is favored in clinical endocrinology. The major drawback is that highly sensitive assays are needed and that appropriate species-specific filters to separate free steroids are rarely available. Traditionally, with emphasis on the classic delayed genomic theory, the action of steroid hormones was considered broadcast, slow acting, and long lasting. This concept has been flawed by the finding of **rapid, non-genomic effects** of arthropod and vertebrate steroid hormones mediated by novel, in most cases, unidentified cell-surface receptors (Sristava et al. 2005; Wendler et al. 2010).

Species differences even in mammals are manifold. Whereas corticosterone is the major glucocorticoid in rodents and some echidna, cortisol is the dominating hormone in cats, ruminants, guinea pigs, and humans. Due to the lack of the enzyme 17α-hydroxylase, cortisol is not produced in rats (Bentley 1998).

Cortisol (hydrocortisone) and/or corticosterone are released from the cortex of the adrenal glands upon ACTH stimulation. They counteract insulin, and their major function is to elevate glucose levels in blood in stress situations. This is done by gluconeogenesis from amino acids (protein from muscles and bones) and by glycogenolysis using up the hepatic glycogen storage sites. Cortisol and corticosterone further have lipolytic effects and elevate lipids in the circulation. Although cortisol binds to the mineralocorticoid receptor, the aldosterone target cells are protected by the enzyme 11-β-hydroxysteroid dehydrogenase that converts cortisol to an 11-keto analog. Only at very high concentrations cortisol acts like aldosterone and enhances peripheral volume by sodium reabsorption. It is also an immunosuppressant and therefore used in medicine and veterinary medicine to treat inflammation. Defects in the system can be caused from the primary site (adrenal glands), secondary site (pituitary), or tertiary site (hypothalamus) rendering hypercortisolism (e.g., Cushing's syndrome) or hypocortisolism (e.g., Addison's disease).

8.3.3 Amine Hormones

The catecholamines **norepinephrine** (syn. noradrenaline) and **epinephrine** (syn. adrenaline: Fig. 8.2a, left side, bottom) are **amine hormones** and synthesized from one single amino acid, namely, tyrosine. The activity of the enzyme that catalyzes the conversion of norepinephrine to epinephrine phenylethanolamine *N*-methyltransferase (PNMT) is highly glucocorticoid dependent and found primarily in the adrenal medulla and brain. Epinephrine contributes 98 % of stored catecholamines in rabbits, 83 % in humans, and 17 % in whales (West 1995). After hypophysectomy or in patients with hypocortisolism, epinephrine synthesis is decreased. Catecholamines are stored in granulated vesicles together with the protein **chromogranin A**. The function of this protein is unclear, but it is used as a marker in patients with endocrine or neuroendocrine tumors and as an index of sympathetic activity. Like peptides, catecholamines are water soluble and therefore bind to cell-surface receptors classified as $\alpha_{1,2}$- and $\beta_{1,2,3}$-receptors.

Thyroid hormones, including **thyroxine** (**T4**, Fig. 8.2a, left side, middle) **and triiodothyronine** (**T3**), are also amine hormones and like catecholamines are derived from the amino acid tyrosine. In contrast to them and most other amines, they are lipid soluble. As each molecule contains three or four iodine atoms, the adequate supply with iodine in the feed is a prerequisite for a normal synthesis. T4 is the major hormone secreted by the thyroids and has to be deiodinated to T3 to become active. The deiodination of the "prohormone" T4 takes place at target cells. Unique for hormones, iodothyronines are stored in large quantities extracellularly in follicles bound to the glycoprotein thyroglobulin. The follicles are hollow spheres of single-layered follicular cells (principal cells) surrounded by a basement membrane and are the basic functional units of the thyroid gland.

The thyroid architecture is highly conserved indicating that thyroid hormones are essential for metabolism and hence development of life. Thyroid hormones are ancient signaling molecules and can be traced back to the primitive chordate (Hazlerigg and Loudon 2008). For instance, thyroids were described in lizards (Ebanasar and Inbamani 1989), zebra fish (Elsalini et al. 2003; Bourque and Houvras 2011), and TRH discussed in insects (De Loof et al. 2012).

Synthesis and release of thyroid hormones is under strict hierarchic regulation below superior centers (three-part axis): TRH from the hypothalamus stimulates the pituitary gland to release thyroid-stimulating hormone (TSH), which acts directly on TSH receptors of the follicular cells in the thyroid. Consequently, T3 and T4 are formed and also released from storage. The released T3 and T4 exhibit a negative feedback effect on secondary (pituitary) and tertiary (hypothalamic) centers (Fig. 8.2d).

When released from the principal cells to the plasma, iodothyronines are bound to thyroid hormone-binding globulin (high affinity, low capacity; absent in non-mammals), transthyretin (or prealbumin; moderate affinity and capacity; not in cats, some marsupials, reptiles, amphibians, and fish), albumin (low affinity, high capacity), and lipoproteins (Feldman and Nelson 2004; Bentley 1998). As discussed with cortisol, only unbound hormones enter the cell and produce biological effects.

Generally, normal thyroid hormone levels in the periphery are called euthyroid, high levels hyperthyroid, and low levels hypothyroid. After entering the cell, thyroid hormones bind to nuclear receptors, act on T3-responsive elements (T3RE) gene expression, and increase RNA synthesis, protein, and enzyme expression.

Thyroid hormones enhance cellular metabolism by stimulating the Na^+-K^+-ATPase. They have positive inotropic and chronotropic cardiovascular effects by increasing the sensitivity to catecholamines (see stress reactions below) via β-adrenoreceptors; they enhance neuromuscular signaling speed, gastrointestinal mobility, and the turnover of tissues and bones; metabolic effects of T3 and T4 comprise glucose supply during stress reactions: gluconeogenesis and glycogenolysis are enhanced and they decrease cholesterol levels.

The vital effects of thyroid hormones T3 and T4 are best seen in settings of aplasia of the organ. This defect causes severe retardation of the child in terms of physical, sexual, and mental development, in all dwarfism specifically termed cretinism. Being dependent on iodine supply, any deficiency in iodine may result in compensatory enlargement of the thyroids (euthyroid goiter) to enhance the resorptive area for iodine uptake from blood and end in hypothyroidism upon decompensation. Endemic iodine deficiency in humans is prevalent in areas with low iodine supply in nutrition and drinking water, preventing an iodine uptake of 150 μg/day. Additionally, various nutritional "goitrogens" such as cyanogenic glucosides in cassava (manioc) or soybeans inhibit iodine uptake and may result in endemic cretinism or goiters (Delange et al. 1982). Hyperthyroidism, most often caused by autoimmune stimulation of TSH receptors (Graves' disease) in humans and thyroid adenoma in animals (extremely common in cats) results in pathological hypermetabolism, enhanced heart rate, sympathicotonia, and has a diabetogenic effect. In both situations, TSH concentrations are indirectly correlated to those of T3 and T4 due to the negative feedback regulation. This is exploited in diagnosis of thyroid function in human and veterinary medicine.

Melatonin (5-methoxy-*N*-acetyltryptamine) is secreted into the blood by the pineal gland (epiphysis), a small organ in the center of the brain at the roof of the third ventricle. This water-soluble hormone was named melatonin because of its action on melanophores in amphibians. Two high affinity G-protein-coupled membrane-bound receptor subtypes have been described (Hazlerigg and Loudon 2008). Melatonin is involved in biological rhythms (e.g., the circadian clock) and the synchronization with photoperiodic cycles. Its secretion is related to light and day length (Reece et al. 2008). In mammals the release of pineal melatonin is controlled by the suprachiasmatic nucleus in the hypothalamus that receives input from the retina of the eyes. Melatonin amongst others stimulates specific thyrotrophic cells in the pars tuberalis of the adenohypophysis. Increased TSH then activates TSH receptors in the hypothalamus and regulates seasonal reproductive responses (**reverse hypothesis**—the pituitary stimulates the hypothalamus, Hazlerigg and Loudon 2008). Interestingly in nonmammalian vertebrates such as amphibians, reptiles, and birds, the pineal gland additionally serves as a "deep brain" photoreceptor (Frigato et al. 2006).

8.4 Metabolism and Excretion

The time from the release of a hormone to its removal from the circulation is called "**clearance time**." This depends on many factors including binding to carrier proteins, cellular uptake, metabolic breakdown, and excretion with urine or bile. In general only very small amounts of authentic, unchanged hormones are excreted with bile or urine. Considerable interspecies differences exist (Bentley 1998).

Degradation of **peptides** is accomplished by specific amino-, carboxy-, and endopeptidases found throughout the body. An organ of utmost importance is the kidney where filtered peptides like insulin or glucagon undergo intraluminal cleavage by enzymes located in the brush border of tubular cells. The fragments either are then reabsorbed by endocytosis and undergo further degradation or are excreted. The excretion of intact hormones is negligible. An additional and in some cases even more important part of renal peptide metabolism is intracellular degradation after the extraction from postglomerular blood.

The principle organs involved in the biotransformation and clearance of **steroids** and **thyroid hormones** are species dependent and include the liver and the kidneys and the gastrointestinal tract, respectively. To give an example, after the injection of radioactive cortisol, 77 % is excreted via the kidneys into the urine in dogs and 82 % via the bile into the feces in cats (Schatz and Palme 2001). Deactivation of steroids includes reduction, hydroxylation, and oxidation (e.g., conversion of cortisol to the less active cortisone at target cells). Thyroid hormones are inactivated by deiodination in the liver, kidneys, and many other tissues. To increase their water solubility, the metabolites are usually conjugated as glucuronides or sulfates. The proportion of excreted conjugated and unconjugated polar metabolites is also strongly species dependent. Studies suggest that most steroid metabolites are excreted unconjugated in domestic livestock (sheep, ponies, and pigs) and conjugated or polar unconjugated in carnivores (Schatz and Palme 2001). Due to reabsorption, only very small amounts of steroids or thyroid hormones are excreted as the authentic free hormone.

8.5 Response of the Endocrine System to Various Stimuli

8.5.1 The Mammalian Stress System: An Example for Hormonal Synergism

The mammalian stress response the "fight and flight and freeze reaction" is an immediate response of the body to a threatening situation and includes a coordinated reaction of the autonomic nervous system and the hypothalamic-pituitary-adrenal axis. The first phase of this reaction includes the release of **catecholamines** from nerve endings and the adrenal medulla. In principle epinephrine binds to the same (adreno)receptors as norepinephrine, but its effects last about ten times longer (Wilson-Pauwels et al. 1997). The release of these amine hormones triggers an increase of heart (β_1-receptors) and respiratory rate, increases the blood supply to the muscles and lungs (β_2-receptors), and mobilizes energy from fatty tissue (fatty acids, α_1-, $\beta_{2,3}$-receptors) and the liver (glucose, β_2-receptors). The peripheral vasoconstriction (α_1-receptors) avoids blood losses in case of possible injury, and the aggregation of melanophores causing a color change of the skin of amphibians and fish helps to hide or intimidate a predator. Dilation of the pupils (α_1-receptors) and erection of body hair (α_1-receptors on pilomotor muscles) as impressively seen in stressed cats make the animal appear more dangerous than they actually are. In humans the same only leads to a rudimentary "goose skin" reaction. The activation of sweat glands via acetylcholine receptors helps dissipate muscle heat, but in the absence of muscle work causes "cold sweat." By inhibiting insulin and stimulating glucagon secretion (α_2-receptors), catecholamines cause hyperglycemia and a state of insulin resistance. This ensures sufficient glucose supply to the brain and working muscle, which can take up glucose by insulin-independent mechanisms. Several additional hormones amplify and support the stress signal. **Thyroid hormones T3 and T4** sensitize cells to adrenergic stimuli. **Glucocorticoids** reinforce the action of catecholamines especially on vagal tone and on energy. To avoid unnecessary energy expenditure, digestive, excretory, and reproductive systems are inhibited.

8.5.2 Glucose Homeostasis: An Example for Hormonal Antagonism

As glucose is used for the generation of ATP, the fuel for cellular respiration, glucose homeostasis is an essential process of life. The brain which uses glucose as the main source of energy relies on a continuing supply. This is warranted by food intake, gluconeogenesis, and glycogenolysis in the liver. Glycogen is a polymer of glucose molecules and the main form of glucose storage in the body. In case of glycogen depletion, gluconeogenesis from amino and fatty acids is of particular importance. The main hormones involved in glucose homeostasis are the two antagonistic peptide hormones **insulin** and **glucagon** secreted by the α- and β-cells of the pancreatic islets of Langerhans. The key site for the action of these hormones is the liver. Additionally important are the incretins (gastrointestinal tract), the adipokines (adipose tissue), the glucocorticoids (adrenals), the catecholamines (nerve endings and adrenals), and the growth hormone (adenohypophysis). When blood glucose levels drop, glucagon stimulates the breakdown of hepatic glycogen and gluconeogenesis. The glucose molecules are then released into the circulation and dispersed throughout the body. In case of pronounced hypoglycemia as after a prolonged fasting, in a stress situation, or after insulin overdosage, catecholamines, glucocorticoids, and growth hormone additionally stimulate glycogenolysis and gluconeogenesis. Accessorily they decrease insulin secretion and insulin receptor sensitivity. The rebound hyperglycemia also called Somogyi effect observed after insulin-induced hypoglycemia is a direct consequence of this mechanism. Stress or handling associated hyperglycemia is a diagnostic challenge in cats and pigs. In case of high glucose concentrations as seen postprandially, insulin is released from pancreatic β-cells (Fig. 8.3a) and promotes the uptake of glucose into nearly all body cells by promoting the expression of the GLUT4 receptor for glucose (Fig. 8.3b). Thereby, glucose can be utilized on demand in the periphery in healthy

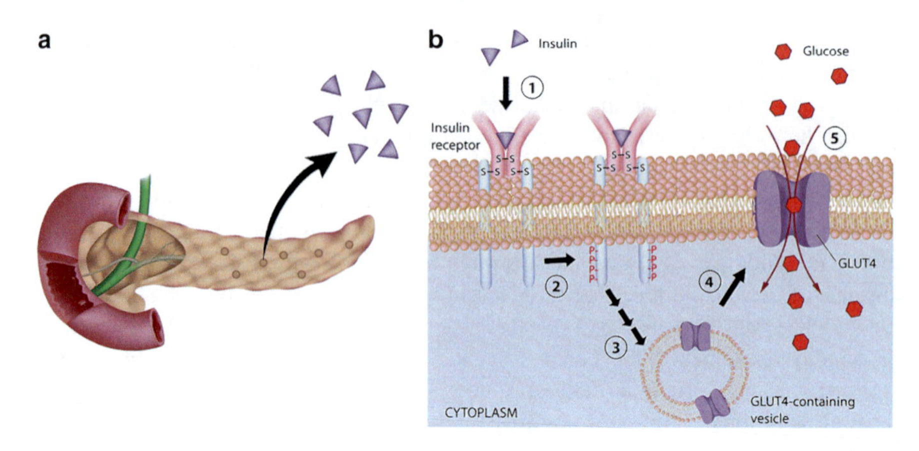

Fig. 8.3 The endocrine function of the pancreas. (**a**) Besides its excretory functions by secreting digestive enzymes into the duodenum, the pancreas also has important endocrine functions: the most important hormones are the antagonists insulin and glucagon, which are secreted by α-cells and β-cells, respectively, into the blood (adapted from © [Alila Medical Images]—Fotolia.com). (**b**) Insulin acts on peripheral cells where it binds to insulin receptor embedded into plasma membranes. This leads to the phosphorylation of many intracellular proteins which alters their activity (© [Alila Medical Images]—Fotolia.com). Consequently, the glucose transporter GLUT4 from cytoplasmic vesicles is inserted into the cell membrane, enabling glucose uptake. In diabetes mellitus either insulin secretion or the sensitivity of the insulin receptor may be impaired. This leads to starvation of the cells in spite of peripheral hyperglycemia

condition. Redundant glucose is stored as hepatic (most) or muscular (little) glycogen or fat. Interestingly the stimulus for insulin secretion is very different in various species. Glucose is the main trigger in humans and dogs, fatty acids in ruminants, and amino acids in cats, respectively. Birds are different in that glucagon is the dominating hormone in glucose homeostasis (Döcke 1994).

8.5.3 Calcium Homeostasis Controlled by a Complex Hormonal Network

Calcium, an essential natural element of live, is amongst others required for various enzymatic reactions, membrane stability, neuromuscular energy transition, muscle contraction, cell growth, blood coagulation, and bone formation. Its serum concentration is regulated by a complex interaction of calcium-elevating hormones (PTH and 1,25-dihydroxyvitamin D [calcitriol]) and calcium-lowering hormone calcitonin (physiological function not established; see 8.3.1). Exogenous factors are UV exposure in respect to calcitriol activation, as well as sufficient supply via food. Again, different pathologies are seen in developing versus adult organism. Whereas chronic hypocalcemia—caused by calcium or calcitriol deficiency during growth—results in softening of bones and consequently malformations (rickets, juvenile osteodystrophy), decalcification leads to osteomalacia (e.g., rubber jaws in dogs) with pseudofractures in adults. In domesticated reptiles such as turtles, lack of dietary calcium in combination with insufficient ultraviolet-B radiation causes skeletal deformity, compressible face and shell, paralysis, and prolapse of cloacal organs (Eatwell 2013). Hypocalcemic rickets may represent a breeding problem for chickens, resulting in splayed legs, deformity of long bones, and pathological fractures (Dinev 2012). This is due to the fact that a laying hen, producing approximately 300 eggs per year, deposits 24 times more calcium into her eggshells than into her bones. Aside from calcium and calcitriol deficiency, calcitriol receptor type 2 mutations can cause vitamin D-resistant hereditary rickets in dogs (LeVine et al. 2009). Systematic research on the effects of breeding on predisposition for specific diseases could help to better understand phenotypes and pathophysiologies originating from endocrinologic disturbances. For clinical signs of acute hypocalcemia see below.

Secondary hyperparathyroidism is a specific problem most often connected with kidney diseases in adults. Specifically cats and dogs (4:2) tend to develop renal insufficiency due to a low glomerular filtration rate. As renal phosphorus excretion declines, hyperphosphatemia inhibits vitamin D hydroxylation and consequently reduces serum calcium concentrations. This activates PTH secretion, resulting in secondary hyperparathyroidism. The reduced hydroxylation of vitamin D is later amplified by progressive loss of renal tubular cells.

8.6 Hormones, Aging, and Life Span

Vitamin D, insulin, glucocorticoid, and thyroid hormones are pleiotropic hormones with effects on several vital systems. Studies suggest that their levels correlate with the life span of small animals: two times higher levels of T4 were found in house mice with a lifetime of maximally 3.5 years than in Damara mole rats (15 years) and three times higher levels than in naked mole rats (28 years) (Buffenstein and Pinto 2009). The effects of female and male hormones are discussed in Chap. 12. Whereas estrogen reduction in postmenopausal women is correlated with aging including osteoporosis (see Chap. 3), in animals this is rarely seen. Only elephants experience the postmenopausal state due to their long life span. An additional falsification is that many pets are castrated, thus enter a nonphysiological state.

Acknowledgments The authors would like to thank crossip communications, Vinna, for kind support in figure preparations.

References

Accordi F, Chimenti C, Gallo VP, Liguori R (2006) Differentiation of chromaffin cells in the developing adrenal gland of Testudo hermanni. Anat Embryol 211:283–291

Baber EC (1876) Contributions to the minute anatomy of the thyroid gland of the dog. Phil Trans R Soc Lond 166:557–568

Bentley PJ (1998) Comparative morphology of endocrine tissues. In: Comparative vertebrate endocrinology, 3rd edn. Cambridge University Press, Cambridge, UK, pp 15–64
Bourque C, Houvras Y (2011) Hooked on zebrafish: insights into development and cancer of endocrine tissues. Endocr Relat Cancer 18:R149–R164
Buffenstein R, Pinto M (2009) Endocrine function in naturally long-living small mammals. Mol Cell Endocrinol 299(1):101–111
Chang SW, Barter JW, Ebitz RB, Watson KK, Platt ML (2012) Inhaled oxytocin amplifies both vicarious reinforcement and self reinforcement in rhesus macaques (Macaca mulatta). Proc Natl Acad Sci U S A 109(3):959–964
Copp DH, Cameron EC (1961) Demonstration of a hypocalcemic factor (calcitonin) in commercial parathyroid extract. Science 134:2038, PMID 13881212
De Loof A, Lindemans M, Liu F, De Groef B, Schoofs L (2012) Endocrine archeology: do insects retain ancestrally inherited counterparts of the vertebrate releasing hormones GnRH, GHRH, TRH, and CRF? Gen Comp Endocrinol 177(1):18–27
Delange F, Iteke FB, Ermans AM (1982) Nutritional factors involved in the goitrogenic action of cassava. International Development Research Centre Publications, Ottawa, pp 1–100
Denver RJ (2009) Structural and functional evolution of vertebrate neuroendocrine stress systems. Ann N Y Acad Sci 1163:1–16
Dinev I (2012) Clinical and morphological investigations on the incidence of forms of rickets and their association with other pathological states in broiler chickens. Res Vet Sci 92(2):273–277
Döcke F (1994) Veterinärmedizinische Endokrinologie, 3rd edn. Fischer, Jena, Germany
Eatwell K (2013) Nutritional secondary hyperparathyroidism in reptiles. In: Rand J (ed) Clinical endocrinology of companion animals. Wiley-Blackwell, Ames, pp 396–403
Ebanasar J, Inbamani N (1989) Histomorphology of thyroid gland in a ground lizard Sitana ponticeriana (Cuvier). ANJAC J 9:85–95
Elsalini OA, von Gartzen J, Cramer M, Rohr KB (2003) Zebrafish hhex, nk2.1a, and pax2.1 regulate thyroid growth and differentiation downstream of nodal-dependent transcription factors. Dev Biol 263(1):67–80
Feldman EC, Nelson RW (2004) Canine and feline endocrinology and reproduction, 3rd edn. Saunders, St. Louis
Ferguson GW, Heller H (1965) Distribution of neurohypophysial hormones in mammals. J Physiol 180:846–863
Frigato E, Vallone D, Bertolucci C, Foulkes NS (2006) Isolation and characterization of melanopsin and pinopsin expression within photoreceptive sites of reptiles. Naturwissenschaften 93:379–385
Ganong WF (2001) The pituitary gland. In: Review of medical physiology, 20th edn. McGraw-Hill, New York, pp 383–397
Hartenstein V (2006) The neuroendocrine system of invertebrates: a developmental and evolutionary perspective. J Endocrinol 190:555–570
Hazlerigg D, Loudon A (2008) New insights into ancient seasonal life timers. Curr Biol 18: R795–R804
Hill RW, Wyse GA, Anderson M (2008) Endocrine and neuroendocrine physiology. In: Animal physiology, 2nd edn. Sinauer Associates, Sunderland, MA, pp 391–424
Hirsch PF, Lester GE, Talmage RV (2001) Calcitonin, an enigmatic hormone: does it have a function? J Musculoskel Neuron Interact 1:299–305
Kershaw EE, Flier JS (2004) Adipose tissue as an endocrine organ. J Endocrinol Metab 89 (6):2548–2556
LeVine DB, Zhou Y, Ghiloni RJ, Fields EL, Birkenheuer AJ, Gookin JL, Roberston ID, Malloy PJ, Feldman D (2009) Hereditary 1,25-dihydroxyvitamin D-resistant rickets in a Pomeranian dog caused by a novel mutation in the Vitamin D receptor gene. J Vet Intern Med 23(6):1278–1283
Lo CY (2002) Parathyroid autotransplantation during thyroidectomy. ANZ J Surg 72:902–907
Nickel R, Schummer A, Seiferle E (2004) Lehrbuch der Anatomie der Haustiere 4, Nervensystem, Sinnesorgane, Endokrine Drüsen, 4th edn. Parey, Stuttgart, Germany

Reece JB, Urry LA, Cain ML, Wasserman SA, Minorsky PV, Jackson RB (2008) Hormones and the endocrine system. In: Wilbur B (ed) Campbell biology. Global edition, 9th edn. Pearson Education, San Francisco, CA, pp 1020–1041
Rehfeld JF (1989) The new biology of gastrointestinal hormones. Phys Rev 78:1087–1108
Rindi G, Bordi C (2006) Classification of neuroendocrine tumours. In: Caplin M, Kvols L (eds) Handbook of neuroendocrine tumours. Their current and future management. BioScientifica Ltd., Bristol, UK, pp 39–51
Schatz S, Palme R (2001) Measurement of faecal cortisol metabolites in cats and dogs: a non-invasive method for evaluating adrenocortical function. Vet Res Commun 25:271–287
Slominski A, Wortsman J, Paus R, Elias PM, Tobin DJ, Feingold KR (2008) Skin as an endocrine organ: implications for its function. Drug Disc Today Dis Mech 5(2):137–144
Sristava DP, Yu EJ, Kennedy K, Chatwin H, Reale V, Hamon M, Smith T, Evans PD (2005) Rapid, nongenomic responses to ecdysteroids and catecholamines mediated by a novel drosophila G-protein-coupled receptor. J Neurosci 25(26):6145–6155
Suga H, Kadoshima T, Minaguchi M, Ohgushi M, Soen M, Nakano T, Takata N, Wataya T, Muguruma K, Miyoshi H, Yonemura S, Oiso Y, Sasai Y (2011) Self-formation of functional adenohypophysis in three-dimensional culture. Nature 480(7375):57–62
Wendler A, Baldi E, Harvey BJ, Nadal A, Norman A, Wehling M (2010) Position paper: rapid responses to steroids: current status and future prospects. Eur J Endocrinol 162:825–830
West GB (1995) The comparative pharmacology of the suprarenal medulla. Q Rev Biol 30:116–137
Wilson-Pauwels L, Steward PA, Akesson EJ (1997) Autonomic nerves. B.C. Decker Inc., Hamilton

9 Alimentation and Elimination: The Principles of Gastrointestinal Digestion

Georg A. Roth
Hanna Schöpper and Kirsti Witter

Contents

G.A. Roth (✉)
University Clinic for Anesthesia, General Intensive Medicine and Pain Therapy, Medical University Vienna, Vienna, Austria
e-mail: georg.roth@meduniwien.ac.at

H. Schöpper (✉) • K. Witter
Institute of Anatomy, Histology and Embryology, University of Veterinary Medicine, Vienna, Austria
e-mail: Hanna.Schoepper@vetmeduni.ac.at

E. Jensen-Jarolim (ed.), *Comparative Medicine*,
DOI 10.1007/978-3-7091-1559-6_9, © Springer-Verlag Wien 2014

9.1 The Human Gastrointestinal Tract

Georg A. Roth

9.1.1 Abstract

The function of the human gastrointestinal (GI) tract is a continual supply of water, electrolytes, and nutrients to the human body; this provides the energy necessary for the uphold of our body's integrity and the performance of higher functions like talking, moving objects, or solving intellectual problems. To acquire this energy the body has to move the food through the GI tract, secrete digestive juices to digest the food, and absorb the nutrients during their passage through the GI tract. The anatomical and physiological basics of these processes are going to be reviewed in this chapter. The regulation of these processes is only partly under our voluntary control, which becomes particularly clear in times of emotional stress, where many individuals experience a "queasy feeling."

The GI tract does not only comprise organs like the stomach or the small intestine, also the oral cavity with the parotid and salivary glands belongs to the human GI tract. About 9 l of fluid in 1 day is secreted and reabsorbed in the human GI tract; its length is on average 9 m, which equates to the smaller side of a volleyball field. The surface of the GI tract estimates a football field; it harbors up to one thousand different bacteria species. It is estimated that up to 10^{18} bacteria are inside the human GI tract. The number of the bacteria and the surface of the human GI tract make the presence of immune-competent cells throughout the GI tract necessary; these immune-competent cells are distributed all over the GI tract in aggregations like the tonsils, adenoids, or the Peyer's patches.

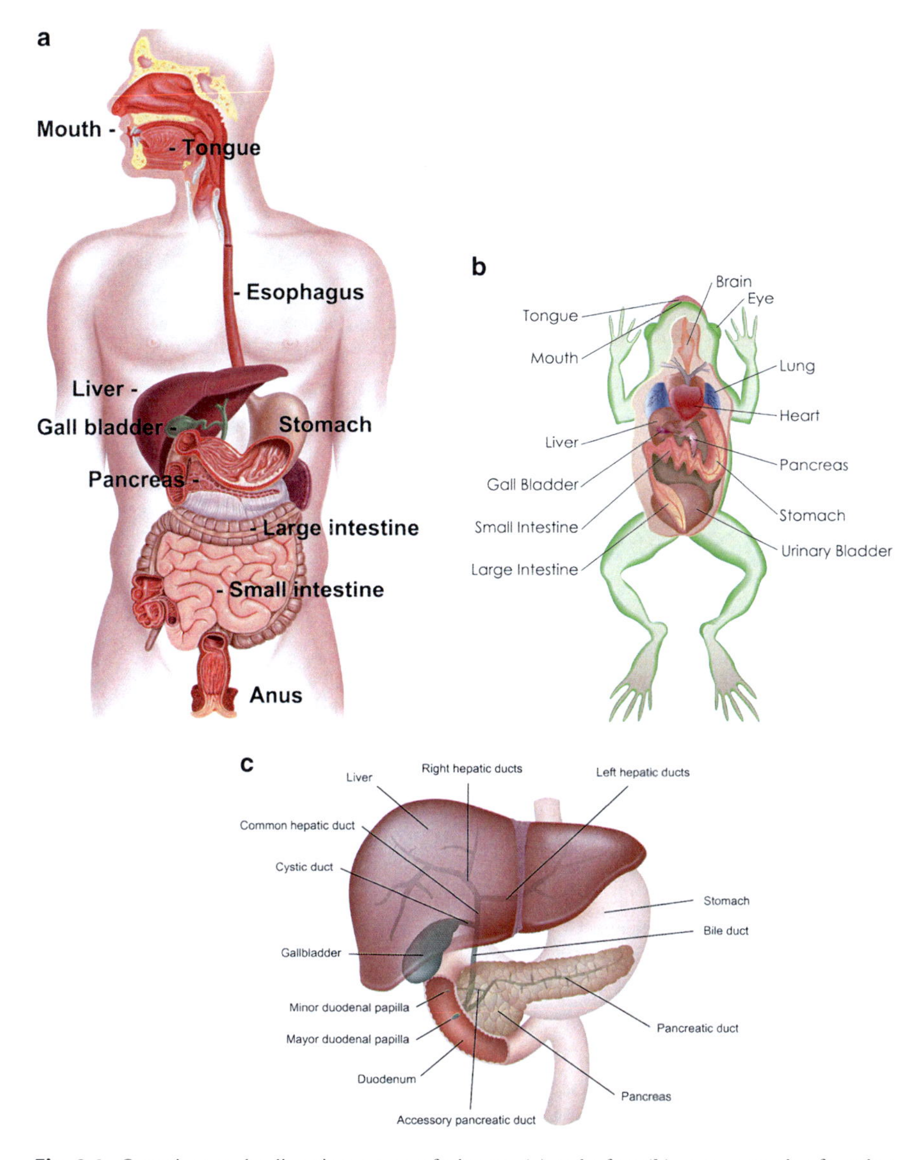

Fig. 9.1 Overview on the digestive system of a human (**a**) and a frog (**b**) as an example of another vertebran animal. Principally, the location of organs to each other is very similar in these species (© [pixelcaos] and © [snapgalleria]—Fotolia.com); (**c**) the anatomy of the liver, gallbladder, pancreas, and bile passage (© [peterjunaidy]—Fotolia.com)

9.1.2 Classification and Basic Anatomy and Histology

9.1.2.1 Classification of the Human GI Tract

The human GI tract consists of following organs and organ systems from oral to aboral (away from the mouth) (Fig. 9.1a):

- Oral cavity (teeth, salivary glands)
- Esophagus
- Stomach
- Duodenum
- Liver
- Gallbladder and extrahepatic bile ducts
- Pancreas
- Jejunum
- Ileum
- Colon
- Rectum/anus

Duodenum, jejunum, and ileum are sometimes summarized as small intestine, whereas the colon and rectum form the large intestine. Esophagus, stomach, and duodenum are designated as upper gastrointestinal tract; jejunum, ileum, colon, and the rectum form the lower gastrointestinal tract. The traditional boundary is the *ligamentum suspensorium duodeni* or ligament of Treitz; this small band connects the duodenum to the diaphragm. The ligament is named after the Bohemian pathologist Vaclav Treitz. In clinical practice upper from lower gastrointestinal bleedings are discerned by this anatomical landmark.

9.1.2.2 Peritoneum

The majority of the intra-abdominal organs of the GI tract are covered by the peritoneum, a thin membrane, which produces a serous secretion. Two layers of peritoneum can be discerned, the parietal peritoneum, the outer layer which is attached to the abdominal wall, and the visceral peritoneum, which forms the inner layer and is wrapped around the various organs. Its function is to reduce friction between the various organs and to connect the otherwise loose organs to the abdominal wall. The mesentery is a double layer of visceral peritoneum, which contains blood vessels as well as nerves and connects the various organs to the posterior wall of the abdominal cavity. If an organ lies inside the peritoneum, like, for example, the stomach, this is called intraperitoneal. The position in which an organ lies behind the peritoneum, like the kidneys, is called retroperitoneal.

9.1.2.3 Histological Structure of Human GI Tract

The histological or microanatomical composition is identical for all hollow organs of the human GI tract; from the inside to the outside the GI tract is build up by the:

- Mucosa
- Submucosa
- Muscularis externa
- Serosa/adventitia

The mucosa from the inside to the outside consists of the epithelium, which handles the secretion and absorption of the various substances. Directly underneath the epithelium lies the lamina propria, which is a thin layer of connective tissue. Next is the so-called muscularis mucosae, which is a thin layer of smooth muscle cells separating the mucosa from the submucosa. Its function is to facilitate the contact between contents of the lumen and the epithelium as well as the depletion of the various glands, which are situated in the wall of the mucosa.

The submucosa is a layer of connective tissue which joins the mucosa as well as the muscularis. It harbors blood as well as lymphatic vessels. Inside the submucosa also lies a network of intersecting nerves the submucosal or Meissner's plexus. These nerve cells innervate the smooth muscle cells in the muscularis mucosae as well as cells in the mucosa.

The muscularis externa is made up by two layers of smooth muscle cells, an inner circular layer and a longitudinal outer muscular layer. These two muscle layers are responsible for the peristalsis, the movement of contents in GI tract from its beginning to its end. In between these two layers of smooth muscle cells is the myenteric or Auerbach's plexus. Another network of nerve cells which controls the smooth muscle cells is muscularis externa. Together with the submucosal plexus they are a part of the so-called enteric nervous systems, which belongs to the vegetative nervous system.

The outermost part is formed by the serosa and adventitia, which is formed by several layers of connective tissue. The intraperitoneal parts are covered by the serosa, whereas extra-peritoneal parts are covered by the adventitia. The serosa consists of a layer of epithelial cells, which also secrete a serous fluid and underlying connective tissue. Its main function is to reduce friction between the various organs due to muscular movement. The function of the adventitia is to bind various structures together. The serosa is more or less the visceral peritoneum which covers the various organs of the GI tract.

9.1.3 Oral Cavity, Pharynx, and Esophagus

9.1.3.1 Oral Cavity

The oral cavity is the first part of the human GI tract; it harbors the necessary organs for the initial division of the food, the **teeth** (*dens*) and the **tongue** (*lingua*). In a human adult up to 32 permanent teeth (24 deciduous teeth in children) are found in the upper jaw (*maxilla*) and lower jaw (mandible). They can be discriminated into incisors, canine, premolar, and molar teeth and are built up by root and crone (Fig. 9.2). The tongue consists of four intrinsic muscles, which can change the shape of the organ, and four extrinsic muscles, which are able to change the position of the organ and are attached to a bone. The hard and the soft palate form the roof; the floor of the mouth forms the caudal boundary of the oral cavity. Frontal and lateral the oral cavity is confined by the teeth, the alveolar processes, as well as the cheeks. The tongue, which is attached to the floor of the mouth, takes up the major part of the oral cavity. The two minor salivary glands the *glandula sublingualis* and the *glandula submandibularis* lie beneath the tongue. The biggest salivary gland, the *glandula parotidea*, is located outside the oral cavity in the retromandibular fossa. The parotid duct enters the oral cavity in the vestibule of the mouth next to the maxillary second molar tooth. The secretion of the *glandula parotidea* is serous, whereas *glandula sublingualis* and the *glandula submandibularis* secrete serous-mucus. The secretion of these glands is the saliva; humans produce about 0.75–1.5 l

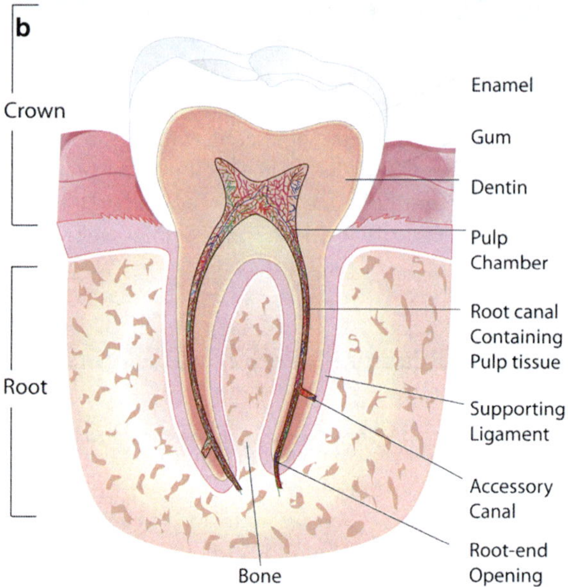

Fig. 9.2 Mammalian teeth are specialized to carnose, vegetable, or mixed food. (**a**) The chimpanzee has a very similar building-up of teeth like humans (© [Eric Isselée]—Fotolia.com). (**b**) The teeth are fixed by a root in the jaw bone whereas the crown is exposed to the oral cavity, saliva, and food. The form of teeth is adapted to grinding or cutting (© [snapgalleria]—Fotolia.com)

of saliva each day, with the majority being produced in the glandula submandibularis. Saliva consists mostly of water; its main function is to wet the food, to facilitate swallowing, and prevent mucosal surfaces from desiccation. It also contains proteins like secretory IgA, lactoferrin, or lysozyme, which work as natural disinfectants. With the production of alpha-amylase, which breaks down alpha bonds of polysaccharides, by the salivary glands, the oral cavity is the origin of the digestion of carbohydrates in humans. The transition of the oral cavity into the pharynx is marked by the pharyngeal isthmus, which is build by the palatopharyngeal arch.

9.1.3.2 Pharynx

The pharynx is located behind the nasal as well as the oral cavity. It is more or less a muscular tube with the length of approximately 12–15 cm, which extends from the skull base to the esophagus and the trachea. The lateral and posterior walls of pharynx are made up mainly by the pharyngeal constrictor muscles, which are essential in process of swallowing. In humans the airway as well as the digestive tract shares a common space in the oral cavity as well as in the pharynx. Therefore, the pharynx is frontally confined by the larynx and the epiglottis, which play an important role not only in respiration but also in phonation. The entry of pharyngeal secretions, food or drink, or stomach contents in the respiratory tract is called aspiration; to avoid this the epiglottis is moved over the larynx to seal it.

9.1.3.3 Esophagus

The esophagus is a hollow organ, which connects the pharynx with the stomach. Its main function is the transport of the food, which is crushed and grounded in the oral cavity to the stomach. The esophagus has three parts, the *pars cervicalis* from the neck until the begin of the thorax, *pars thoracalis* until diaphragm, and the *pars abdominalis* until its insertion into the stomach. The human esophagus has also three constrictions, the first is where the pharynx joins the esophagus, the second is at the aorta's crossing of the esophagus, and the last is where the esophagus pierces the diaphragm. In the human esophagus no relevant digestive processes are performed; it is merely a transportation organ.

9.1.3.4 Chewing and Swallowing

In humans four muscles of the head are relevant in the chewing process: the masseter muscle (*M. masseter*), the temporal muscle (*M. temporalis*), the medial pterygoid muscle (*M. pterygoideus medialis*), and the lateral pterygoid muscle (*M. pterygoideus lateralis*). The first three are responsible for closing of the jaw, whereas the lateral pterygoid muscle is responsible for the opening of the mouth. The masseter muscle is strong in humans like in all herbi- and omnivores, and is responsible for the bulk of the teeth's grinding of the food. The force, which is developed during chewing in humans, is about 100–250 N at the incisors and 300–650 N at the molars. The maximum force a human can generate with his/her jaws is approximately 1,900 N. All four muscles are innervated by the fifth cranial nerve, the *Nervus trigeminus* (see Chap. 6).

Swallowing can be differentiated into a voluntary oral phase and an involuntary pharyngeal phase. During the oral phase, food is transported into the pharynx, by backward and upward movement of the tongue against the palate. Soft palate closes the opening to the nasal cavity and the nasopharynx. During the pharyngeal phase the glottis is closed, respiration is paused, the larynx is moved upward and closes the airway, and epiglottis then seals the trachea and prevents aspiration. In the esophagus the food is transported through contractions of its longitudinal smooth muscle layer (peristalsis). Fluids need about 1 s for the passage of the esophagus into the stomach, pulp about 5 s, and solid particles about 9 s.

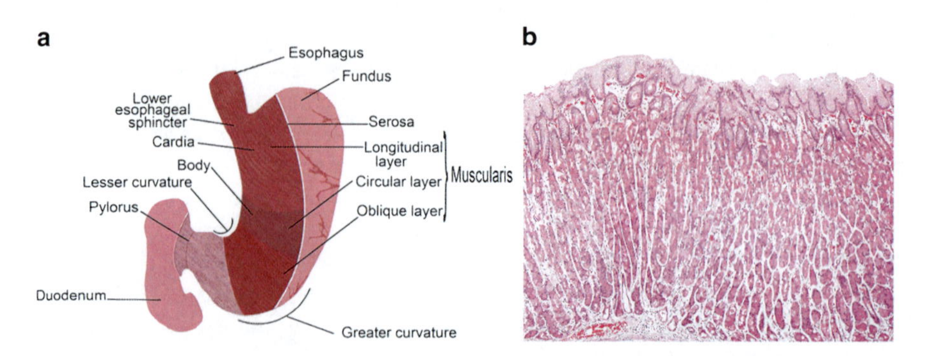

Fig. 9.3 Human stomach macro- and microanatomy. (**a**) The human stomach is divided into three major parts of different functions: When chymus enters the stomach from the esophagus through the upper sphincter, it reaches the cardia. Fluids may pass through quickly along the minor curvature, whereas solid food has to be salivated and enters the protein digestion in the fundus of the stomach at the side of the greater curvature. The procession of the food is supported by strong gastric smooth muscles with longitudinal and circular layers (© [Balint Radu]—Fotolia.com); (**b**) microscopic view of human stomach mucosa. The mucosa is equipped not only with mucous producing cells but also with HCl producing chief cells and pepsinogen producing parietal cells. Pepsinogen that is cleaved upon contact with low pH is converted to an active protease (© [Convit]—Fotolia.com)

9.1.4 Stomach

The stomach or *ventriculus* (*gaster*) is a hollow organ, which lies between the duodenum and the esophagus. Its form resembles a sack or a wineskin; four main parts can be described (Fig. 9.3a). The first is the **cardia**, which is the part where the esophagus leads into the stomach. Cranial and left from the cardia lies the **fundus**, which is the arched top of the organ. The main body of the stomach is called **corpus**; the part which leads into duodenum is called **pylorus**. The lateral boundaries of the organ are the **greater curvature** of the stomach on the left side of the stomach and the **lesser curvature** of the stomach on the right side.

The stomach's main function is to further downgrade the food. This is done by further homogenization and mixture with gastric secretions; this results in a semi-fluid mass, which is called **chyme** and which is further expelled into the duodenum. Food usually stays 1–6 h in the stomach; fluids usually leave the stomach within 10–20 min. The total volume of the stomach is about 0.8–1.5 l; its relaxation is controlled by a vagovagal reflex. In the human stomach only a very small amount of nutrients are absorbed. For solid particles a size of less than 2 mm is necessary to pass through the pyloric sphincter; 90 % of the particles, which pass through the pylorus, have a size of less than 0.25 mm. The tone of the pyloric sphincter is under hormonal control, which is again regulated by the tenth cranial nerve, the **vagal nerve** (*N. vagus*). The hormones gastrin, GIP (gastric inhibitory peptide), CCK (Cholecystokinin), and secretin increase pyloric tone, whereas motilin decrease the tone in the sphincter. A variety of cells exist in the stomachs mucosa, each with distinct functions (Fig. 9.3b). The foveolar cells or mucous neck cells are located all over the stomach; they produce mucus and bicarbonate ions to prevent the autodigestions of the stomach by hydrochloric acid and pepsin. The gastric chief cell produces pepsinogen, gastric lipase, and chymosin. Pepsinogen is the precursor of the enzyme pepsin, which degrades food proteins into peptides. Pepsinogen is activated into pepsin by the presence of gastric acid, which is produced by the parietal cells. The parietal cells also produce the intrinsic factor, which is required for the absorption of vitamin B_{12}. The **G-cells** and the D-cells in the antrum produce **gastrin**, which stimulates parietal cells and gastric motility, whereas the D-**cells release somatostatin**, which suppresses the release of gastrointestinal hormones.

9.1.5 Pancreas

In humans the pancreas is 15–20 cm long; it is glandular organ, with an exocrine and endocrine function (Fig. 9.1c). The pancreas is the largest digestive gland; in humans it produces about 1.5 l of serous secrete, which enters the duodenum by either one or two pancreatic ducts. The secrete contains **trypsin**, **chymotrypsin** to facilitate **proteins digestion**, **alpha-amylase** is responsible for the **breakdown of carbohydrates**, and the **pancreatic lipase** is responsible for the lipid digestion. The pancreatic enzymes, which are crucial for the degradation of the chyme in the small intestine, are inactive in the pancreas to prevent self-digestion, and not activated until the secrete reaches the duodenum. The production of these digestive enzymes is called **exocrine function** of the pancreas. The pancreas, however, also produces hormones in about a million cell clusters, which are also termed **islets of Langerhans**. The hormones produced include glucagon, which increases blood glucose (alpha cells), or insulin, which decreases the blood glucose (beta cells). The hormone production in the pancreas is also termed **pancreatic endocrine function** (see Chap. 8, Fig. 8.3).

9.1.6 Liver

9.1.6.1 Anatomy

The liver (*hepar*) is the largest internal organ in humans; it weighs about 1.5 kg (Fig. 9.1c). The liver's frontal side is split into the right and left lobe by the falciform ligament (*ligamentum falciforme hepatis*), which attaches the liver to the anterior body wall. From the posterior parietal side two more lobes, which are considerably smaller than the right and left lobe, can be distinguished, the more superior caudate lobe and the more inferior quadrate lobe. At posterior site of the liver lies the *porta hepatis*; at this site the portal vein (*Vena portae*) and common hepatic artery (*Arteria hepatica propria*) enter the liver and the bile duct leaves the organ. The liver receives about 75 % of the blood, through the **portal vein** and the rest through the common **hepatic artery**. The portal vein delivers blood from GI tract's other organs like the stomach, small intestine, etc. to the liver; it is therefore deoxygenated but rich in nutrients. The liver processes these nutrients, but also detoxifies this nutrient-rich blood. The hepatic artery delivers oxygen-rich blood. The blood from the liver is **drained via the hepatic veins** to the inferior vena cava; this blood is oxygen- and nutrient-depleted.

9.1.6.2 Function

The liver's main functions are synthesis, breakdown, and storage. The smallest functional unit in the liver is the liver lobule. In a human liver about 50,000–100,000 individual lobules can be found. Simplified the liver lobule is a hexagon with the central vein in the lobule's center. At hexagon's corners are portal venules and hepatic arterioles, which have their source in the portal vein and the hepatic artery. The central veins enter into the hepatic veins. Inside the hexagon lie the hepatocytes, which are arranged in a pattern, which resembles spokes in a wheel, with the central vein forming the wheel's center. The **hepatocytes** are lined by endothelial cells, to form the **hepatic sinusoids**. The blood flows from the outside of the hexagon through the liver sinusoids to the central vein. Adjacent to these sinusoids are **bile caniculi**, which converge to the bile ducts. The bile ducts run beside the portal venules and arterioles. The bile flow is exactly countercurrent to the blood flow, from the inside of liver lobule to its periphery.

The hepatocytes perform the various functions of the liver like the formation of glucose (**gluconeogenesis**), **cholesterol**, **ketone bodies** (ketogenesis as result of fatty acid breakdown), and **bile**. Moreover, **albumin**, **clotting factors**, and the **acute phase proteins** (see Chap. 13) are produced by the liver. The liver also converts **ammonia to urea** and **glucuronidates bilirubin**, the breakdown product of the hemoglobin from erythrocytes, facilitating its excretion into bile. The liver also plays a crucial role in **breakdown of drugs**. In the first phase the substance is modified, by, for example, oxygenation, reduction, or hydrolysis. In the second phase the target is conjugated to facilitate its excretion. This is done, for example, by conjugation with glutathione (GSH), sulfate or carboxyl (−COOH), hydroxyl (−OH) groups, etc. In the final phase the drugs then can be excreted by liver or the kidney.

The liver also plays an important role as **storage organ** in humans. **Glycogen**, **vitamins A and D**, **fat**, **iron**, etc. are all stored in the liver.

9.1.6.3 Gallbladder, Ducts, and Bile

The gallbladder or *vesica biliaris* (*vesica fellea*) lies on the posterior side of the liver, inside the gallbladder fossa (Fig. 9.1c). In humans the gallbladder is usually 8 cm long and 4–5 cm wide. It is a temporary storage organ for bile; the bile inside the gallbladder is concentrated by water withdrawal from the bile.

The bile ducts are the anatomical structure, which transport the bile from the liver into the duodenum. The usual classification is between intrahepatic bile ducts until the *porta hepatis* and extrahepatic bile ducts from outside of the liver to the duodenum. From the *porta hepatis* the bile duct is called common hepatic duct (*ductus hepaticus communis*). The cystic duct joins the gallbladder with the hepatic duct, which is called common bile duct (*ductus choledochus*), from its confluence. The common bile duct joins with the pancreatic duct to the hepatopancreatic ampulla, which is surrounded by the sphincter of ampulla or **sphincter of Oddi**, a muscular valve that controls the flow of digestive juices which enter the duodenum via the **major duodenal papilla**.

The bile consists of water (85 %), **bile salts** (10 %), mucus, pigments, inorganic salts, and cholesterol. The bile is bactericide which serves as surfactant and facilitates micelle formation and the resorption of fat-soluble substances as well as the excretion of bilirubin (hemoglobin endproduct).

9.1.7 Small Intestine

The small intestine is the part of the human GI, which lies between the stomach and the large intestine. It is about 6 m long in humans, and the largest part of the GI tract. Its three parts are the **duodenum**, the **jejunum**, and the **ileum**. The duodenum precedes the jejunum and is about 25 cm, the shortest part of the small intestine. The duodenum ends at the ligament of Treitz. It harbors the major duodenal papilla, where the secretes of the pancreatic and common bile ducts are released into the duodenum (Fig. 9.1c). The major duodenal papilla is about 7 cm below the pylorus of the stomach. The main function of the small intestine is the **absorption of the food's nutrients**, **minerals**, and **vitamins**. There is no clear distinction between jejunum and ileum; the most distinct features are **Peyer's patches** (aggregated lymphoid nodules) in the ileum (see Chap. 13). In the small intestine the digestion of carbohydrates, lipids, as well as protein is completed; its main task however is the absorption of the food's nutrients. The bulk of nutrients is absorbed in the jejunum; notable exceptions are: iron is mainly absorbed in the duodenum, whereas vitamin B_{12} and bile salts are absorbed in the terminal ileum. The nutrients can be absorbed by passive diffusion (e.g., water, lipids), facilitated diffusion (e.g., fructose), or energy-dependent active transports (e.g., sodium bicarbonate). Protrusions of the epithelium, the intestinal villi, increase the surface of the organ. The small intestine ends with the ileocecal valve, which prevents reflux from colonic material into the ileum.

9.1.8 Large Intestine

The parts of the large intestine are the cecum, where the ileum converges into the large intestine, the colon, which is the longest part and consists of the **ascending**, **transverse**, **descending**, and **sigmoid colon**, and the **rectum**, where the useless waste material is discharged from the body. The large intestine in humans is about 1.5 m long, and much shorter than the small intestine. Its main function is **water reabsorption** and transformation of the chyme into **feces**. With 5–70 h it has the longest transition time of all organs in the human GI tract. Unlike the small intestine, the large intestine possesses a huge **bacterial flora**. The number of bacteria in the human intestines exceeds the body's number of cells approximately tenfold. The bacteria have a lot of beneficial effects for the human body; for example, immune modulation, production of certain vitamins (especially in times of low dietary intake), production of short-chained fatty acids, stimulation of the peristalsis.

The defecation is the final act of the digestive process. The body disposes waste materials this way. Defecation is a polysynaptic reflex with involuntary and voluntary elements. Two muscles are involved, the internal and external anal sphincter. The internal anal sphincter consists of smooth muscle cells, which surrounds the anal canal as a muscle ring. The external anal sphincter is a skeletal muscle and, therefore under voluntary control, it surrounds the margin of the anus. To achieve defecation a relaxation of these two muscles is necessary; additionally the intra-abdominal pressure has to be increased; this is accomplished with the contraction of the diaphragm and abdominal muscles.

Acknowledgement Thanks to Prof. Erika Jensen-Jarolim, M.D., and crossip communications, Vienna, for kind support in the arrangement of illustrations.

9.2 Comparative Aspects of the Digestive System of Animals

Hanna Schöpper and Kirsti Witter

9.2.1 Abstract

Like humans, also all other animals are dependent on the intake of organic material and therefore called heterotrophs. The organic material is originally produced with the help of photosynthesis by autotrophs (e.g., plants and algae) that are solely dependent on inorganic material like carbon dioxide and water. In all heterotrophic species the digestive system is highly adapted according to the food. Three general groups can be distinguished: carnivores that feed on meat of prey, herbivores that feed on plants only, and omnivores consuming both meat and plants.

The main functions of the digestive system are **ingestion** of food, **digestion** (physical and chemical disassembly), transportation and temporal storage, **absorption** of nutrients, and subsequent **elimination** of undigested wastes. These functions are fulfilled by different parts of the digestive tract that are adapted to the dietary habits of the respective species: mouthparts, esophagus, stomach, and intestine. These organs are supported by salivary glands, pancreas, and liver.

As in the human, also for animals there are essential nutritional components that are to be provided with the food. Vitamin C, for example, is essential for guinea pigs, but not for most other mammals. An additional essential amino acid for cats is taurin. Cats can get a retinal degeneration if fed on dog food due to lack of taurin.

In the following we will mainly discuss characteristics of the vertebrate digestive system.

9.2.2 Mouthparts

The mouth and its associated structures like tongue, lips, teeth, and salivary glands fulfill different functions, such as ingestion of food, facilitating taste sensation, mechanical and chemical breakdown, transportation of ingesta, and also some non-digestive functions like defense or grooming.

9.2.2.1 Oral Cavity

All vertebrates feature a mouth that serves as the entrance to the body. The oral cavity is bordered by a palate above, separating oral and nasal cavity in air-breathing animals. In healthy mammals, the palate is always complete, whereas in birds, the lateral palatal shelves do not join. The caudal continuation of the oral cavity is the pharynx. With the exception of hagfish and lamprey, all vertebrates possess jaw joints that enable the animal to close the mouth.

9.2.2.2 Tongue

The multifunctional tongue is a vertebrate invention. In frogs and some salamanders it is attached in the rostral part of the oral cavity and shoots out if used to catch insects. In contrast, the tongue of mammals is attached to the bottom of the oral cavity in the caudal region leaving the rostral tip free. The tip of the tongue is highly flexible and built of numerous muscles. Lingual papillae serve mechanical functions in food transport and sensory functions by taste buds. Occurrence and number of papillae are species-dependent. Additional functions of the tongue are, e.g., grooming in cats.

9.2.2.3 Lips and Cheeks

Lips and cheeks adapted for sucking are a phylogenetic novelty in mammals. The lips are underlined with muscles and provide many sensitive nerve endings and vibrissae that give important information on potential food right before ingestion. Especially the upper lip can show additional adjustments for feed intake. In rodents, carnivores, and small ruminants, left and right upper lips are separated by a philtrum presumably enabling the animal to bite closer to a given surface. In pigs and cattle the upper lip is partly merged with the nose and used for digging and mechanical support during grazing. The upper and lower lips meet in the corner of the mouth and are continued by the cheeks. The space between lips, cheeks, and the row of teeth is called vestibulum. Lips and cheeks enable tight closure of the oral opening to suck milk or excavate the cheeks when chewing foodstuff. This room may be further enlarged to the neck region in cheek pouches of hamsters to temporarily store food.

9.2.2.4 Teeth

True teeth made of dentin, enamelum, and cementum are an achievement of vertebrates. Teeth are arranged in one or more rows as, e.g., in mammals and in the shark, respectively. Tooth row(s) are called dentition. Homodont dentitions comprising simple conical teeth can be found in lower vertebrates such as fish, some amphibians, and reptiles. Recent birds are toothless. True heterodont dentitions comprising differently shaped incisors (I), canines (C), premolars (P), and molars (M) are found in mammals.

For heterodont dentitions, tooth classes and the numbers of the according teeth are given in tooth formulae, describing one half of upper and lower jaw. The unreduced general formula of the permanent mammalian dentition is $\frac{I3.C1.P4.M3}{I3.C1.P4.M3}$. This tooth formula can be found in insectivores and the pig. Nearly all other mammalian dentitions are derived from this by reduction. Examples of maximal reduction are Pseudohydromys with the tooth formula $\frac{I1.C0.P0.M1}{I1.C0.P0.M1}$ and baleen whales with completely reduced teeth. Only few mammals have secondarily multiplied teeth, e.g., dolphins.

According to the requirements of mechanical breakdown of food, teeth can have different shape and period of growth. Teeth that develop, erupt, and grow until they reach their final size such as in humans, pigs, or dogs are called brachyodont. In brachyodont teeth, crown, neck, and root can be clearly distinguished. In contrast, some teeth of herbivores are ever-growing—they are hypsodont. Examples are the gnawing teeth of rodents and molars of horses, cows, and some rodents. In hypsodont teeth, instead of crown and neck, a corpus dentis is defined. If roots develop at all, they are short and form lately in life. The corpus of hypsodont teeth is usually covered completely with cementum and thus has a yellowish or brownish surface. The mantle of cementum and enamel can be folded into the tooth from occlusal and lateral surfaces. Since cementum, enamel, and dentin are of different hardness, abrasion during feeding causes an uneven occlusal surface and thus enhances grinding of hard food. If hypsodont teeth are not abrased constantly, they can reach enormous dimensions. If such cases appear in animals under human care, teeth have to be corrected by a veterinarian.

Not all teeth are arranged in form of a dentition on the edge of the jaw. An example is pharyngeal teeth of fish such as carps. In some animal species, the mechanical breakdown of fodder is not ensured by teeth. In birds, small stones ingested into the gizzard grind the feed into small particles. Another example is the radula of snails, a kind of small tongue with rows of chitin toothlets used for scraping food before ingestion.

9.2.2.5 Salivary Glands

Salivary glands produce saliva to lubricate ingesta in land living species. Animals that live in the water lack salivary glands. Large salivary glands are the parotid, mandibular, and sublingual gland; the small salivary glands are embedded in the mucosa and are named after the region they occur in. In addition to lubrication, saliva also acts as a solvent for gustatory substances and therewith enables sensory function like taste. Moreover, the fluid layer protects oral mucosa and teeth from drying out and effects of acids. Saliva may contain enzymatic substances as amylase in humans, pigs, elephants, and rabbits or lipase in calves. In some snakes saliva is poisonous and facilitates accelerated digestion if applied to prey before swallowing. Some biting insects and leeches have anticoagulatia in their saliva that prolonges blood flow of the victim. Quantity and pH of saliva differ between species, with cattle producing up to 150 l per day. This amount of saliva is produced not only for lubrication of ingesta but also to regulate the acid–base-balance of the forestomach by high contents of HCO_3^-. Interestingly, male mice have a mandibular gland that is twice the size of the females'—salivary glands play a role in pheromone production as well. Other functions of saliva not associated with digestion are thermoregulation by evaporation during panting (e.g., in dogs) or self-defense in camelids.

9.2.2.6 Pharynx

The pharynx is the crossroad of air and food way in lung breathing animals and major site for the swallowing process. In mammals, reptiles, and birds the pharynx connects to the middle ear via the Eustachian tube. With this connection between outer world and ear, adjustment of pressure on the tympanic membrane is possible. In equines the Eustachian tube is enlarged to the guttural pouch.

9.2.3 Esophagus

The esophagus connects the oral cavity to the stomach. The wall of the esophagus is built of mucosa, submucosa, muscularis, and surrounding connective tissue (adventitia). The inner epithelium of the mucosa is a more or less keratinized stratified squamous epithelium. Mucosal folds allow enlarging of the esophageal lumen during swallowing. In some species, glands are embedded in mucosa and submucosa to lubricate the surface. The muscle layer consists of varying portions of smooth and striated muscle. Striated muscle can be used in ruminants, some carnivores, and birds to bring back predigested food deliberately for additional grinding or to feed offspring. In birds, the esophagus is dilated to a crop. Its function is to store and lubricate ingesta. In pigeons the crop also produces crop milk—degraded epithelium rich in proteins and lipids. This process is triggered by prolactin as milk production in mammals. Other organs for temporal food storage are, e.g., cheek pouches in hamsters and the stomach in all mammals.

9.2.4 Stomach

Except some lower fish species like lamprey, all vertebrates possess a stomach to store and chemically break down ingesta. This expansion of the gastrointestinal tube is located between esophagus and intestine. The innermost mucosa of the gastric wall is followed by submucosa, muscularis, and peritoneum. The gastric mucosa contains glands that produce acid, enzymes for chemical digestion, and protective mucus. Except for ruminants, the stomach is not used extensively for absorption of nutrients; only water and ions may pass the stomach wall.

The stomach is more or less spindle-shaped with simplest forms in fish and amphibians. Crocodiles already show a bended stomach. The bird's stomach has two compartments: a proventriculus for chemical breakdown of ingesta, followed by a gizzard for mechanical breakdown. In mammals, one can distinguish species with a single-chambered stomach (monogastrics) and multi-chambered stomach (ruminants).

In monogastric mammals the stomach is still spindle-shaped and bended. In some mammals, the stomach is completely lined by glandular mucosa as in the human. In other species, there is an additional nonglandular region. In horses and mice, the nonglandular region covers almost half of the stomach. Functions of the simple stomach of animals are similar to that of the human, especially chemical breakdown of protein.

The multi-chambered stomach of ruminants is divided into three forestomach chambers without glands: rumen (paunch), reticulum (honeycomb), and omasum (bible), and one glandular chamber: abomasum. While the last is similar to the simple stomach of monogastrics, the three nonglandular chambers are an adaptation of the digestive tract to ruminating. The largest part is the rumen with up to 200 l in cattle. Reticulum in the cranial region and ball-shaped omasum are used for separation of differently sized food particles and liquid.

The forestomach makes virtually undegradable food components such as cellulose or nonprotein nitrogen (NPN) available for the ruminant via microbial fermentation by bacteria, fungi, and protozoa. The forestomach offers those microbes an environment with correct pH, water, right temperature, and sufficient substrates: complex carbohydrates such as cellulose and hemicellulose, but also proteins and other nutrients.

Cellulose but also all other carbohydrates ingested by the ruminant are split by microbes to pyruvate, which is further reduced to the short chain fatty acids (SCFA) acetate, propionate, and butyrate. Ruminants can use SCFA as humans use blood glucose: as an energy source—either directly or via storage as fat. All the glucose necessary for a ruminant is produced by the liver of SCFA (gluconeogenesis). Blood levels of glucose in ruminants are therefore considerably lower than in monogastric animals. On the other hand, a milking cow in high lactation produces about 5–7 kg SCFA per day.

Proteins, peptides, and amino acids are reduced in the forestomach to NH_4 which is resorbed by the rumen wall and transported to the liver for urea production. Urea is resecreted into the rumen, excreted in the kidneys, or transported to salivary glands for recirculation to the paunch by saliva. Microbes are able to use this internal urea to produce new protein. The microbes are eventually digested in the abomasum and used by the ruminant as a source of protein, but also of vitamins and other nutrients. In cattle, the abomasum secretes lysozyme active at low pH, which is believed to be involved in reduction of microbial cell walls. The process of recirculation of nitrogen is called ruminohepatic circuit. In nonruminant animals, great loss of energy is accepted by excreting nitrogen via the urine, while in ruminants being fed a low-protein diet, only 10 % of urea is lost this way.

During fermentation in the forestomach, large quantities of gas are produced, namely CO_2 (40–70 %) and methane (20–40 %), which has to be evacuated with the belch or ructus. A good milking cow discharges approximately 500 l methane each day. Since methane is considered a potent greenhouse gas, cattle farming has been criticized for the impact on climate change. Next to the gases that are eructed, volatile substances of rumen fermentation may be present after eating certain plants (wild onions or bad silage) and will be partly inhaled during ructus—then absorbed in the lungs, spread with the blood, and can be found as "off" flavors in the milk.

Rumination per se, i.e., the process of swallowing, predigestion in a forestomach, regurgitation (rejection) of the cud to the oral cavity in phases of rest to further mechanically reduce it, and addition of saliva, has evolved independently in at least three different groups of mammals: kangaroos, camelids, and ruminants (cattle, goat, sheep, antelopes, deer). In case of adequate diet cattle ruminate for about 8 h per day, with cyclic pouch contractions assuring rearrangement, selective retention, and discharge of gases by a ructus.

A short cut to avoid degradation of nutrients in the paunch, e.g., in milk-feeding calves, is the sulcus ventriculi, the shortest distance between esophagus and abomasum. The milk is transported directly to the abomasum that contains lab-ferment, bypassing the paunch and its microflora. Only with increased intake of raw fiber and affixed microbes the forestomach grows and takes over the main part in digestion.

An intact microflora is crucial for normal digestion in ruminants. Severe disruption can develop due to pH changes in the forestomach, e.g., as a result of increased intake of simple carbohydrates or of putrid feed. Antibiotics prescribed in case of health problems can affect the ruminal microflora considerably; oral administration is to be avoided. Other health problems can also develop due to disrupted ruminal digestion. Magnesium, for example, is solely reabsorbed from the forestomach—if transport mechanisms are impaired, hypomagnesemia may result, commonly known as grass tetany.

In addition to these potential health problems in ruminants, also other animals suffer from gastric problems: pigs, for example, tend to suffer from gastric ulcers if fed intensively and stressed by environmental and social factors like crowding. Horses and rabbits sometimes overload the stomach and as emesis is not possible, upcoming contractions may actually rip the stomach.

9.2.5 Intestine

All vertebrates possess an intestine as the main site of nutrient absorption and later elimination of undigested material. Its wall has the same general structure as the stomach. Except for fish, the intestine can be divided into small and large intestine. In the small intestine, digestion takes place: the chyme is mixed with fluids from different glands and further processed to smaller molecules, which are absorbed by the intestinal mucosa. The large intestine primarily absorbs water and electrolytes as well as eliminates undigested material. In order to prolong contact time of chyme with the mucosa and thereby accelerating absorption, the overall surface area of the intestine is enlarged in all vertebrates. Mechanisms of enlargement are increased length of the intestine with loop formation in the abdominal cavity, macroscopically visible plicae of the inner wall layers, mucosal villi, and microvilli of the epithelial cells themselves. Total length of the intestine depends on species and their dietary habits. Carnivores have a ratio of body length to intestinal length of about 1:5, herbivores like horses 1:12, and ruminants have the highest relative intestinal length between 1:20 (cattle) and 1:27 (small ruminants like goat and sheep).

9.2.5.1 Small Intestine

The small intestine of mammals is the major place of nutrient absorption and can be subdivided into duodenum, jejunum, and ileum. The duodenum receives pancreatic juice and bile from the closely associated pancreas and liver as well as from Brunner's glands in the intestinal wall itself. With the addition of those fluids, the pH changes from acidic to neutral and facilitates enzymatic activities. The longest section of small intestine is the jejunum, followed by the ileum that terminates at the ileocaecal valve as in the human.

The epithelium lining the intestine consists of highly specialized cells—enterocytes with interspersed mucous goblet cells and antigen presenting M-cells. The main function of enterocytes is the transfer of small molecules like amino acids, SCFA, or simple sugars from the intestinal lumen to the capillaries. Absorbed nutrients are collected in the venous blood system and passed via the *V. portae* to the liver; resorbed fat is transported by lymph vessels. With its function the intestine is a place of tight interactions of outer material (foodstuff) and inner system (mucosal tissue). Due to the great contact area host defense is vital and lymphoid tissue is visible as Peyer's patches along the small intestine.

9.2.5.2 Large Intestine

The large intestine of vertebrates can be subdivided into caecum, colon, and rectum with primary function of water reabsorption and elimination of undigested material as species-specific feces. The caecum shows great variation in number, shape, and length in different species. The pyloric caeca of some fish—up to a hundred blind-ending protruberances of the intestinal tube between stomach and small intestine—are not homologous to the caecum of mammals. Caeca are missing in amphibians and reptiles, whereas birds have commonly even more than one. In mammals the caecum is often a residence for intestinal bacteria. An appendix vermiformis as in the human can also be found in rabbits, whereas in most other species it is missing.

In close functional relationship to the caecum stands the colon. The colon is the longest part of the large intestine and can be subdivided into ascending colon, transverse colon, and descending colon as in the human. While the ascending colon shows major morphological differences between species, the transverse and descending colon remain relatively stable. In pigs the ascending colon is wound up to a cone shape, in cattle it is formed like a spiral, and in horses like a double U. Parts of the large intestine can show species-specific sacculations, called haustra, e.g., in horse, pig, rabbit.

In herbivores lacking a forestomach, caecum and colon are used as a chamber for microbial fermentation. The disadvantage of using the hindgut instead of the forestomach is that the main resorption is already carried out within the small intestine prior to the fermentation chamber in the large intestine. Therefore, not all microbial proteins or vitamins can be absorbed. Rabbits and guinea pigs, however, recover the nutrients produced by the intestinal flora by eating their own special feces from the caecum and allow a second passage of the nutrients and complete absorption in the small intestine. This process is called caecotroph nutrition.

The last part of the large intestine is the relatively short rectum which is continued by the anal canal ending in the anus. In the rectum, feces can be stored temporarily. Extension of the rectal wall leads to stimulation of receptors urging defecation. In this case feces pass the anal canal—a transition area where the mucosa is replaced by external skin. A smooth internal and a striated external sphincter muscle are providing tight closure that is just temporarily opened for defecation—a complex procedure of reflexes and conscious influence. Defecation behavior differs between species. For example, dogs defecate 2–3 times a day, but horses and cattle up to 20 times a day. Some carnivores like dogs, ferrets, and skunks have anal sinuses, filled with a secrete that is intense in odor, which is discharged with the feces and used for marking, intraspecific communication, and defense.

Originally, all vertebrates used a cloaca as a common exit for the digestive- and urogenital tract. Some fish and mammals except monotremes developed separate ways out secondarily. In birds and reptiles the cloaca is called vent. Turtles and some other species even have back passage, where urine can go back to the intestinal region and then being excreted together with feces. In birds, antiperistalsis occurs almost any time—in this way the caeca are filled and some of the urine is mixed with chyme. This process provides nitrogen for microbial metabolism in the caeca. Moreover, in the large intestine of birds, resorption of water from the urine is possible. In the roof of the cloaca in birds the bursa Fabricii is located—a lymphatic organ associated with the development of B-lymphocytes.

The anus or cloaca is not the only way of excretion. Indigestible substrates can also be eliminated orally, e.g., as hairballs in cats or the so-called pellets in owls or gulls.

9.2.6 Pancreas

In all vertebrates, the pancreas has a close relationship to the proximal intestine. In some fish, there are only nests of specialized cells scattered within or around the intestinal wall. In other fish species, the so-called hepatopancreas is developed. As in the human, the pancreas of other mammals has two major functions: to produce hormones like insulin and glucagon in the pancreatic islets (endocrine function) and to secrete pancreatic juice with enzymes (e.g., amylase, lipase, trypsin, and chymotrypsin) into the duodenum (exocrine function). The pancreatic juice is excreted via the one or two pancreatic ducts that are in close relationship with the bile duct. HCO_3^- in the pancreatic juice buffers the chyme.

9.2.7 Liver

Next to the skin, the liver is one of the largest organs. In all vertebrates it is located near the stomach. Its function is quite similar in all vertebrates: production of bile to facilitate fat resorption, storage and metabolic conversion of many substances that enter the liver via the *V. portae*, and detoxification. Its appearance differs slightly between mammalian species due to the degree of lobation. While carnivores show many lobes, in cattle they are largely reduced and the liver is displaced by the paunch. The gallbladder collects the bile produced in the liver; however, in some species the gallbladder is missing, e.g., in some fish, some birds, whales, rats, and horses. In those animals, bile is still normally produced and excreted to the duodenum, but without prior storage.

Further Reading

Arthur W (2011) Evolution—a developmental approach, 1st edn. Wiley-Blackwell, Oxford

Aspinall V, Capello M (2009) Introduction to veterinary anatomy and physiology, textbook, 2nd edn. Butterworth-Heinemann, Oxford

Hall J et al (2010) Textbook of medical physiology, 12th edn. Saunders, Philadelphia

Henry G (1918) Anatomy of the human body, 20th edn. Lea & Febiger, Philadelphia

Hickman CP Jr, Roberts LS, Keen SL, Eisenhour DJ, Larson A, L'Anson H (2011) Integrated principles of zoology, 15th edn. McGraw-Hill, New York

Hildebrand M (1974) Analysis of vertebrate structure, 1st edn. Wiley, New York

Johnson LR et al (2004) Physiology of the gastrointestinal tract, 4th edn. Elsevier/Academic, Waltham

König HE, Liebich HG (2012) Anatomie der Haussäugetiere, 5th edn. Schattauer, Stuttgart

Langer P (1988) The mammalian herbivore stomach: comparative anatomy, function and evolution, 1st edn. Gustav Fischer, Stuttgart

Lippert H et al (2006) Lehrbuch Anatomie, 7th edn. München, Urban & Fischer

Nickel R, Schummer A, Seiferle E (2004) Lehrbuch der Anatomie der Haustiere—Band II Eingeweide, 9th edn. Parey, Stuttgart

Pinnock CA et al (2002) Fundamentals of anaesthesia, 2nd edn. Greenwich Medical Media, London

Reece WO (2009) Functional anatomy and physiology of domestic animals, 4th edn. Wiley-Blackwell, Aimes

Reece JB, Taylor MR, Simon EJ, Dickey JL (2012) Campbell biology, 7th edn. Pearson, San Francisco

Schmidt R et al (2010) Physiologie des Menschen, 31st edn. Springer, Heidelberg

Sobotta J et al (2005a) Atlas der Anatomie des Menschen, Band I, 22nd edn. München, Urban & Fischer

Sobotta J et al (2005b) Atlas der Anatomie des Menschen, Band II, 22nd edn. München, Urban & Fischer

Stark D (1982) Vergleichende Anatomie der Wirbeltiere auf evolutionsbiologischer Grundlage, 1st edn. Springer, Berlin

Treuting PM, Dintzis SM (eds) (2012) Comparative anatomy and histology—a mouse and human atlas, 1st edn. Elsevier/Academic, Waltham

Von Engelhardt W, Breves G (2010) Physiologie der Haustiere, 3rd edn. Enke, Stuttgart

Volume and Clearance: Kidneys and Excretory Systems

10

Erika Jensen-Jarolim
Hanna Schöpper and Simone Gabner

Contents

E. Jensen-Jarolim (✉)
Comparative Medicine, Messerli Research Institute, Veterinary University Vienna, Medical University Vienna, University Vienna, Vienna, Austria
e-mail: erika.jensen-jarolim@meduniwien.ac.at

H. Schöpper (✉) • S. Gabner
Institute of Anatomy, Histology and Embryology, University of Veterinary Medicine, Vienna, Austria
e-mail: Hanna.Schoepper@vetmeduni.ac.at

E. Jensen-Jarolim (ed.), *Comparative Medicine*,
DOI 10.1007/978-3-7091-1559-6_10, © Springer-Verlag Wien 2014

10.1 Basic Physiology of the Human Kidney

Erika Jensen-Jarolim

10.1.1 Abstract

Kidneys and their evolutionary precursor organs are indisputable for the survival of mammalians to invertebrates like Drosophila. These organs are responsible for clearing of water-soluble exogenous and also endogenous metabolites with a potential for toxicity in settings of overload. Such substances are for instance ammonium ions (from protein metabolism), ureic acid (from nucleic acid metabolism), and creatinine (end product of muscle proteins). These substances are called urophenic. In situations of kidney failure these substances are retained and cause severe toxicity in the patients (uremia). Further, water and ions are secreted which have important functions in cells: Na^+ (sodium), K^+ (potassium), HCO_3^- (bicarbonate), and phosphate and calcium. Especially Na+ ions are recognized by specialized kidney tubular cells and indicate the volume state of the organism.

At the same time, the body must retain its homeostasis and retain its crucial fluids and molecules. The kidney is thus a major checkpoint for blood clearance, blood homeostasis, blood volume, blood pressure, and its pH.

10.1.2 Introduction

Whereas in mammalians the intestine preferentially secretes lipophilic substances, hydrophilic substances are preferentially secreted by the renal system, i.e., the kidneys. In the first part of this chapter predominantly the highly developed kidneys in humans will be discussed which are analogous to kidneys of other mammalian species. Basic principle No. 1 is filtration through **semipermeable membranes**. This means that water and small ions should be able to trespass, whereas molecules of a molecular mass above 68 kDa should be retained. This means that the major plasma protein albumin or other proteins only appears in the urine in situations where the permeability of the filter increases, like in inflammation. This is called proteinuria and is associated with loss of the oncotic pressure in the blood vessels. In consequence, fluid is lost to the interstitial space of the body and causes edema. The urine may be milky and/or cloudy. Further, immunoglobulins (160 kDa and more) and all cellular compounds (red and white blood cells) should be retained in physiological condition. In case the permeability increases further, red blood cells may be found in urine (hematuria).

However, also the increased amount of a substance in the blood may lead to urinary filtration. The renal threshold of glucose (RTG) is 160–180 mg/dL. Above that glucose is found in urine. Also the volume of the urine will be much higher due to the osmotic capacity of glucose. As a consequence, diabetic patients with insufficiently controlled blood glucose levels run the risk of exsiccosis.

The volume passing a filter is also dependent on the **pressure** on the membrane. The kidneys are located in short distance to the abdominal aorta from where they get oxygen-rich blood (Fig. 10.1a). In fact 20 % of the cardiac output, i.e., **1,500 L blood per day** run into the kidneys in an adult person. From this volume, 10 % are actually filtered to form the primary urine, i.e., 150 L/day. As the output urine, however, is only 1.2–1.8 or 2 L per day, significant concentration must take place in between. The ion concentration in the beginning is 300 mosm/L and is enhanced to 1,200 mosm/L in the end. This can be pursued by the so-called **countercurrent principle** in a specialized tubulus part—the **loop of Henle** (Fig. 10.1b, c). This is a special feature, which especially mammalian kidneys have developed in order to retain and control volume. The countercurrent principle relies on the active, energy-dependent transport of Na+ ions from primary urine into the renal interstitium. When water passes by in other segments of the tubes, it follows the osmotic gradient and thus can be reabsorbed. Concentration has taken place. In fact, 99 % of the primary urine is reabsorbed in healthy condition.

There may, however, be pathologic conditions where reabsorption is disturbed. This would result in **polyuria** (too much volume lost by urine—exsiccosis), or oliguria (too little urine is produced) or **anuria** (no urine). The latter will result in an intoxication of the organism with uriphenic substances.

10.1.3 Microanatomy of the Kidneys: The Nephron

So far, we have summarized the basic principles of kidney function, which are filtration and concentration. These are guaranteed by a highly specialized micro-anatomy. The smallest functional unit in the kidney is the **nephron**, of which 300,000 to 1 million exist in a human kidney; in mouse for instance 11,000 only (Fig. 10.1b). A nephron is the results of a collaborative action between one arteria and one urine-collecting tube, and their synapse is called a **glomerulum**. This is a small tulip-shaped bowel hosting the arteria, receiving the filtrate and forwarding it to a single tube. The incoming vessel (vas afferens) multiplies in several branches to increase the effective filter surface, before they assemble again to the vas efferens (outgoing vessel). An inner endothelial cell layer, a basal membrane, and so-called podocytes around compose each vessel. One can imagine that these multilayer ultrafilters can be easily hit by aggregates exceeding 70 kD, which stick to the filter and cause inflammation often affecting the whole organ—glomerulonephritis. Needless to say that secondarily the filter capacity may thereby be severely affected. In physiologic conditions, however, primary urine is filtered and then collected by a long single tube, finally assembling with others in the so-called collecting tubes. During its tube journey, urine is further concentrated. Whereas within the glomerulus the pressure is high (15 mmHg), the pressure in the tubal system decreases significantly allowing a number of selective or passive reabsorption and excretion processes.

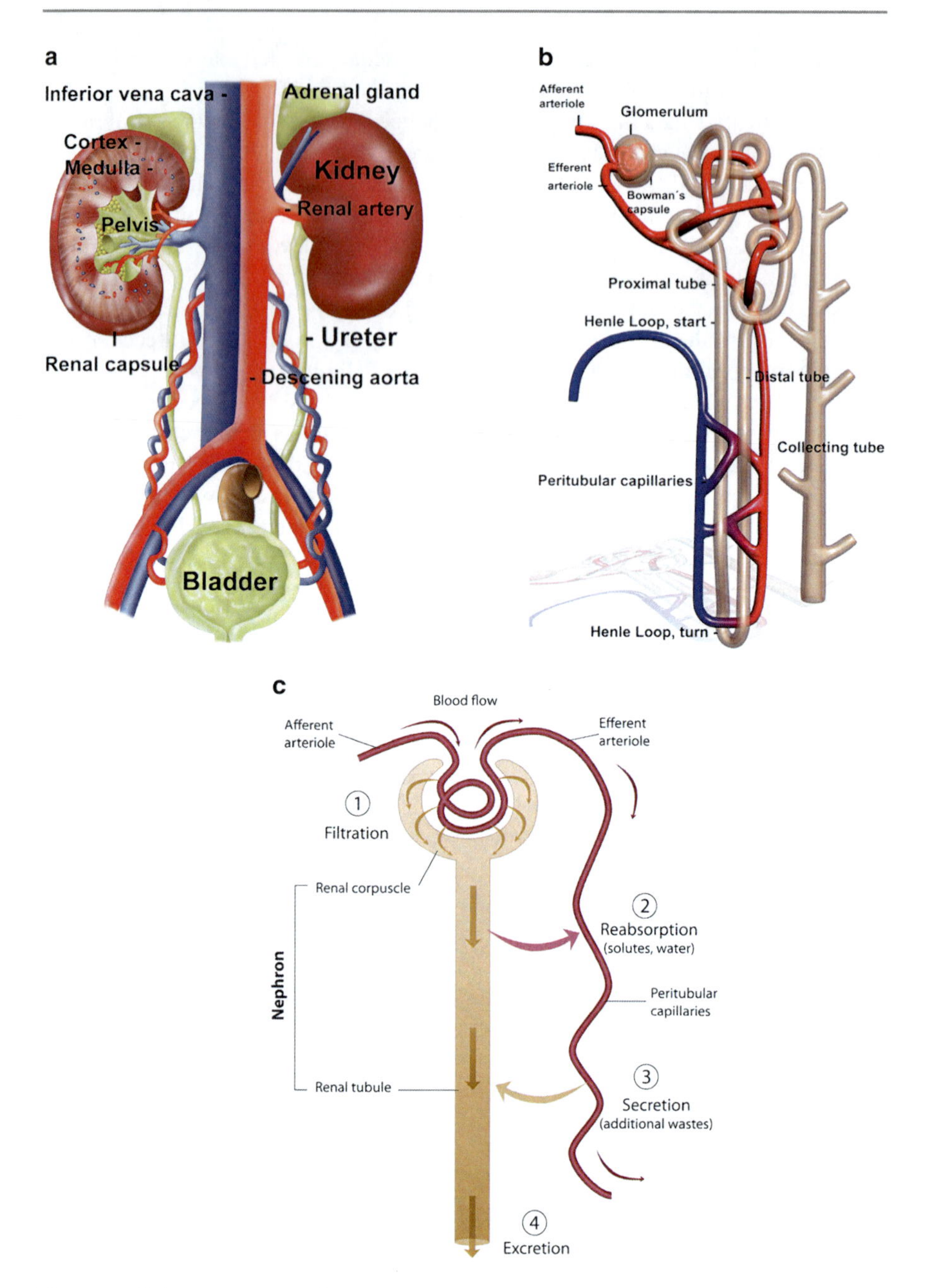

Fig. 10.1 Kidney anatomy and function. (**a**) The kidneys are tightly associated to the aorta where they may sense volume disturbances and immediately counteract them. Directly on top of the kidneys the adrenal glands are located which help in volume regulation by providing aldosterone. Further, regular blood pressure along the aorta is needed to produce primary urine, which is concentrated in the renal medulla by the countercurrent principle. The ureters transport the concentrated urine into the bladder as a transient reservoir (© [Martha Kosthorst]—Fotolia.

10.1.4 The Embryonic Kidney Development

From the early embryonic development on, soon after yolk sac and amnionic cavity have been established and cells are differentiated to ectoderm, endoderm, and mesoderm, the kidneys start to develop from the so-called **Wolffian duct** from mesoderm layer. The first developmental forms of it, termed **pronephros** and **mesonephros**, are transient in mammalians, birds, and reptiles. Only the third form, **metanephros**, persists and starts building up the excretory system from gestation week 5 on. The Wolffian duct on the caudal side fuses with the hindgut to form a primitive cloaca from where urine and products of the intestine are excreted. Urine production starts from week 10 on by the embryo.

The location of the metanephros is dorsal, along the neural tube and the aorta, but outside the perinatal cavity. The metanephros forms separated functional units, called **renculi** of which 10–20 exist in humans; in, e.g., guinea pig only one. This causes a lobulated surface typical for human embryonal kidneys, and existing also in other mammalian species (see below). From weeks 9–12 on the lobulated surface changes to a smooth one, and the developing kidneys ascent from the caudal to the lumbar region. During this journey which starts at week 7–9, they form loose multiple supply connections with the aorta which move cranially finally forming the renal arteria. The ureters are urine transporter tubes. They collect urine from the renal pelvis, where urine from the kidney nephrons is pooled. On the other end ureters make contact to the cloaca and later bladder. The ureters follow the kidneys during their ascent. As the urine system develops in parallel to the genital system they are together called urogenital system. During the ascent of the metanephros for instance, the male testes follow and later have to descend again along the ureters (see Sect. 10.2) in order to find their final position outside the body. The scrotum temperature of only 32 °C is optimal for the maturation of, at least human, semen. Some problems during the kidney and testes up- and down movements occur quite often causing malpositions and consecutive organ failure. Finally, after additional rotation the kidneys find their final position at term. Further, during the development a septum is formed to separate the hindgut/rectum from the urogenital sinus where primordial bladder with associated rudimental Wolffian duct and ureters are formed. In male mammalians the last distance of the urogenital system, where semen and urine are released, still indicates the common developmental history.

Fig. 10.1 (continued) com); (**b**) Scheme of the microanatomy of a nephron (© [krishnacreations]—Fotolia.com); (**c**) physiological function of a nephron: The afferent arteries transport blood into the glomerulum. The Bowman's capsule is used as a collector of primary urine which then enters the proximal tubes and Henle's loop, being surrounded by a tight capillary net. During its transport the urine is concentrated in the medulla and reaches via distal tubes the collecting tubes being connected to the renal pelvis and ureters (© [Alila Medical Images]—Fotolia.com)

10.1.5 Some Nephrons Participate in Volume Regulation by Autoregulation

As outlined above, a nephron is the minimal functional unit of the kidney. Glomerular capillaries and collecting tube system constitute it. There are two different types of nephrons with different functions. Type I is residing in the renal cortex where 99 % of the incoming plasma volume arrives. Therefore, this type of nephrons is an effective sensor of blood pressure and volume arriving. These cortical nephrons are also used for regulating the blood flow and thus the filtration rate in the kidneys. This is called myogenic **autoregulation**: Active change in arterial blood pressure may result in vasoconstriction (fall in renal plasma flow (RPF) ⟹ less filtration ⟹ retaining of fluid ⟹ elevation of blood pressure) or vasodilatation (higher filtration rate, lowering of volume and blood pressure). Vasoconstriction may occur due to nerval irritation, irritation of vasomotoric centers in the brain stem, hypoxia, stress (catecholamines such as adrenalin), or during work. This is reasonable as during body activity higher blood pressure and circulation rate are needed for full oxygen supply of organs. However, often after stress situations retained urine is released with satisfaction. Vasodilating conditions causing enhanced diuresis on the other hand are bacterial endotoxins causing fever (pyrogens) or protein-rich food.

The net release of urine is thus dependent on the capillary pressure, +50 mmHg, which is countered by the oncotic pressure of plasma proteins in the vessels (−25 mmHg), and the pressure of the **Bowman's capsule** (−10 mmHg) (Fig. 10.1c). This causes a net filtration pressure of +15 mmHg in healthy condition. Factors that influence that are (1) changes in permeability, for instance by precipitated immune complexes which block the filter, or (2) reduced filter area, for instance due to loss of parenchyma. In diagnostics it is important to have measures for the filtration capacity of the kidney. This can be achieved by measuring substances, which trespass the filter. Experimentally, para-amino hippuric acid (PAH) can be used to determine the arterio-venous difference, also called RPF, as it is excreted almost completely (90 %) and only to 10 % reabsorbed. Inulin is a small molecule which is freely filtrated through the glomeruli. It can thus be used to determine the glomerular filtration rate which should be 125 mL/min, resulting in 150–180 L primary urine per day. The formula would be

$$\mathrm{GFR} = \frac{V \cdot U_{\mathrm{In}}}{P_{\mathrm{In}}} = \text{Clearance Inulin}$$

However, for diagnosis it is more practicable to use body metabolites for determination which also are secreted easily, such as creatinine. For this, urine has to be collected from the patient during 24 h. The normal range of creatinine excretion is 500–2,000 mg/day for a healthy adult person, but depends also on sex and age.

10.1.6 Some Nephrons Participate in Volume Regulation by the Endocrine Network

The second type of nephrons are expanding from the cortex deep into the medulla. They are building up the osmotic gradient using Na+ ions. These nephrons are therefore important sensors of the Na+ contents in the urine. Excess Na+ ion levels at the level of distal tubes indicate that too little fluid is trespassing as a possible result of volume reduction by low fluid intake, volume loss by bleeding or other types of shock, or volume loss by excessive diarrhea or vomiting. This organ lying there is called the juxtaglomerular apparatus (juxta = close to). It consists of Na+-sensing cells of the distal tube forming the **Macula densa**, which are surrounded by vascular mesangium cells and in direct connection with juxtaglomerular endocrine cells. This is a major checkpoint for volume control: As a response to lowered GFR and hence lowered Na+ absorption, but also to sympathic stimuli (stress), these endocrine cells produce **renin**. It is released directly to the glomerular capillaries to direct their contraction state. Renin is further distributed to the whole organism to trigger a cascade of consecutive hormonal regulatory circles. Angiotensinogen from the liver is converted to angiotensin I by renin. The lung provides then angiotensin converting enzyme which produces **Angiotensin II**, which strongly enhances blood pressure, by multiple mechanisms: It augments sympathic activity, tubular Na+-reabsorption, and the release of **aldosterone**, a further regulatory hormone, from the adrenal gland, and enhances the secretion of ADH, the antidiuretic hormone from the pituitary gland. Together, this system is called the **renin–angiotensin–aldosterone system (RAAS)**. It is an important target of modern anti-hypertonic medications, such as the angiotensinogen-converting enzyme (ACE)-inhibitors. The highest level in this hierarchy is taken in by aldosterone.

Aldosterone is a product of the suprarenal glands' cortex and thus is ready to instruct the kidneys from short distance with immediate effect. It induces reabsorption of Na^+-ions (and hence water) from the tubes, especially proximal Henle's tubes. In exchange, K^+-ions are pumped out. In diseases with hyperaldosteronism, such as benign (Conns' disease) or malignant tumors of the adrenal glands, this leads to hypertonia with concomitant hypokalemia associated with muscular weakness and a risk for bradycardia and atrioventricular block in the heart. In addition, alkalosis occurs, as with the K^+ (potassium) ions also H^+ (protons) are lost (see below).

The antidiuretic hormone (ADH) induces water reabsorption by inducing H_2O channels in the collecting tubes which leads to a concentration of urine (antidiuresis). ADH is derived from the posterior pituitary gland (hypophysis) under the control of the corticotropin releasing factor (or hormone; CRH) from the hypothalamus (see Chap. 8). It is predominantly released upon signals from baroreceptors for low blood pressure, and signals from the hypothalamus for hyperosmolality (concentration, high ion strength) of the blood plasma, such as may occur in settings of extensive evaporation of body water, in heat or reduced drinking. Tumors in the hypophysis typically may lead to hyposecretion of ADH which leads to excessive diuresis of up to 20 L per day. This is a life threatening condition due to exsiccosis. From an evolutionary point of view, CRF is well conserved even in insects and may function itself as an ADH (De Loof et al. 2012).

10.1.7 Other Endocrine Functions of the Kidney

From the above it is clear that the kidney interacts with several other organs in regulating one of the most important body functions: volume regulation.

In addition to the secretion of the hormone **renin**, renal tubes are also a source of **prostaglandins**. They are formed from membrane phospholipids by the phospholipase A enzyme which first produces arachidonic acid which then can be transformed to either leukotrienes by the enzyme lipoxygenase or prostaglandins by the enzymes cyclooxygenase-1 or -2 (COX-1, COX-2). In the kidney, like in other epithelial cells (e.g., in the stomach, see Chap. 9), especially prostaglandin E has a regulatory function (Breyer et al. 1998). It inhibits the reabsorption of Na+ ions (and osmotically that of H_2O) in the renal tube and, thereby, enhances urine volume. This is one mechanism for enhanced diuresis in bacterial infections. Therefore, it counterregulates the RAAS and decreases blood volume and pressure. In therapy of pain and inflammation COX-2 inhibitors are used which counteract this and may lead to hypertension.

Erythropoietin (EPO) is a hormonal product directly of the kidneys. Its secretion by interstitial fibroblasts nearby peritubular capillaries is stimulated by hypoxia and consecutive release of hypoxia-induced factor (HIF) by the liver and the gastrointestinal tract. Therefore, several organ systems cooperate to protect the organism from hypoxic damage. When EPO reaches the bone marrow, site of blood cell production and maturation, it binds to EPO receptors of pro-erythrocytes (precursors of red blood cells). These cells still have nuclei and DNA and can thus enhance the production of hemoglobin. In consequence, hemoglobin-rich erythrocytes leave the bone marrow and are ready to transport more oxygen. This is a physiological mechanism to adapt to hypoxic conditions, such as stays in higher altitude. This phenomenon is used in sports to enhance the capacity of oxygen transport in athletes. This mechanism can also be misused by EPO doping.

The kidney takes also a critical role in **vitamin D3** metabolism by a hydroxylation step. Vitamin D3 (Calcitriol, 1α,25(OH)2-Cholecalciferol) is today mostly regarded as a steroidal hormone (D-hormone) because it is synthesized in the skin from 7-dehydro cholesterol upon UV irradiation (Präcalciol); besides it can be taken up by nutrition. Next, in the liver hydroxylation at position 25 takes place, followed by a second hydroxylation at position 1α in the kidney. The active 1α,25-Dihydroxy-vitamin D2 is from there distributed into the body where it specifically binds to vitamin D-receptor (VDR). It induces Ca^{2+} uptake from the intestine, supports the mineralization of bones and immune defense, and it suppresses malignant cellular growth. Reduced D-hormone levels are associated with a higher rate of colon carcinoma (Deeb et al. 2007). In case of chronic kidney failure such as during glomerulonephritis, the final activation of D-hormone is missing and results in low calcium levels. As a counterregulatory mechanism, the parathyroid gland secretes parathormone (PTH). This results in hyperplasia of the parathyroid organs, the so-called secondary hyperparathyroidism. Importantly, PTH activates osteoclasts to resolve calcium from bones. Chronic kidney failure thus typically results in renal osteodystrophic syndrome with massive decalcification of bones associated with pain and an enhanced risk for untypical fractures. The specific function of the osteoclasts in this context is discussed in Chap. 3.

10.1.8 The Kidney Controls the Acid–Base Balance

The urine is not only concentrated by the passage through tubes but also a pH shift takes place. Depending on the metabolism, pH values between 4.5 and 8.0 may be physiologic for human urine. This indicates that active processes may regulate the H^+ concentration. A key enzyme for this is **carboanhydrase**. This enzyme transforms H_2O and CO_2, which are present in all cells of the body, to H^+ ions and $HCO3^-$ (bicarbonate) ions. Thereby, two aims can be fulfilled: (1) excretion of protons to the luminal (urine) side (in exchange with Na+ of the proximal and distal tubular cells); (2) secretion of HCO_3^- ions back to the blood. Bicarbonate ion secretion is dependent on the function of the Na^+–K^+ ATPase (an energy-dependent ion pump). When one HCO_3^- leaves the cell to the blood, one K^+ potassium ion is taken up to the inside of the cell. Thereby, the kidneys contribute to the so-called buffer capacity of the blood: $HCO3^-$ ions released to the blood may collect free protons and be transformed to H_2CO_3. Carbonic acid is very instable and immediately dissociates again to H_2O and CO_2 again, two harmless substances. In case of renal insufficiency, when both the proximal and distal tubular functions are disturbed this leads to metabolic acidosis and hyperkalemia. Hyperkalemia leads to depolarization of the membrane potential of cells, most significant on neuromuscular cells. Hyperkalemia thus leads to atrioventricular conduction problems, resulting in arrhythmia up to asystole by AV-block.

10.2 Comparative Aspects of the Urinary System

Hanna Schöpper and Simone Gabner

10.2.1 Abstract

Main functions of the urinary system are **elimination of metabolic wastes** that otherwise accumulate and harm the integrity of the body, maintenance of **homeostasis and osmoregulation**, **acid-base regulation** and **hormone production**. Different species have developed different urinary sytems to adapt to the environment they are living in.

10.2.2 Basic Principles of the Elimination of Metabolic Wastes in Different Species

End products of metabolism that need to be eliminated are carbon dioxide, water, and species-specific residuals of the protein metabolism. After degradation of proteins to amino acids several potential destinies are feasible.

For production of energy amino acids can be metabolized in the citric acid cycle, with CO_2 (carbon dioxide) and H_2O (metabolic water) as end products. Carbon dioxide is exhaled via the lungs while water is either reused or eliminated as described in section homeostasis and osmoregulation.

Amino acids can also be used in the muscle metabolism with creatinine as final waste. Creatinine is apparent in all vertebrates and entirely filtered and eliminated by the urinary system.

Furthermore, amino acids can be incorporated as nuclear bases (pyrimidine and purine bases) in DNA and RNA which form species-specific end products during degradation. While pyrimidine bases are degraded to ammonia that is water-soluble and easily excreted in urine, purine bases are degraded in species-specific ways. While invertebrates, larval stages of amphibians, and most fish may also reduce purines to **ammonia**, adult amphibians and some other fish are only able to reduce purines to **allantoic acids**. Most mammals degrade purines to **allantoin**, while primates, guinea pigs, and humans together with reptiles and birds reduce purines only up to the stage of **uric acid**. Therefore, some residual energy is lost with these excretes.

The normal pathways are jeopardized in case of Dalmatians: dogs normally degrade their purine to the allantoin stage—however, this breed is affected by a genetic defect that inhibits complete degradation and purine metabolism ends with uric acid. This leads to pathologically high contents of uric acid, posing a risk to developing uroliths.

All other nitrogenous residuals are eliminated in a species-specific way. From all amino acids the nitrogen part has to be removed from the carbon skeleton before the latter can be used for energy production. The removed nitrogen is forming ammonia

(NH_3) which is toxic and therefore has to be eliminated as fast as possible from the body. In water living species the excretion is not a problem as ammonia is highly soluble in water and easily diffuses as **ammonia** or ammonium (NH_4) via the gills or other surfaces (**ammonotelic species**). In land living species excretion of ammonia in water stands against the attempt to save and reabsorb water and therefore only low amounts leave the body as ammonium ions with the urine, while for the majority detoxification is necessary. Detoxification is done by conversion to relatively safe substances like urea or uric acid. Amphibians and mammals that generally live in moderate habitats produce **urea** (**ureotelic species**). Birds, reptiles, and many insects that are also able to live in arid areas only excrete **uric acid** (**uricotelic species**). The advantage of producing uric acid is the possibility to excrete it in more or less solid form so only limited water is necessary. For fetal development, elimination of the nontoxic uric acid is advantageous too, as it can be stored in a space saving manner within the egg shell. However, the disadvantage is the great energy consumption for conversion and loss of nitrogenous compounds.

10.2.3 Possible Ways of Dealing with Unwanted Substances

There are different ways of removing unwanted substances from the metabolism: storage, **immediate excretion**, and **temporal storage with subsequent excretion.**

The simplest form of dealing with metabolic wastes is storing detoxified end products within the body. An example is the nucleobase guanine, which is stored as photonic crystal in the skin of fish and is responsible for the metallic look (Levy-Lioret et al. 2008). Sharks even store urea in the muscles and therewith become isotone with the surrounding sea water. In case the shark is fished, the urea transforms to ammonia and develops a bad smell that only slowly disappears by evaporation.

Immediate excretion is seen in aquatic species where substances can diffuse to the water surrounding the animal. Examples are gills of fish that are able to expel ammonia from the body or the highly permeable skin of amphibians like frogs that has an active transport system for sodium.

The simplest example of temporal storage is the contractile vacuole of unicellular organisms. Mainly water and soluble substances are stored in vacuoles that fuse with the cell membrane after a while and eliminate the content to the outer medium.

The most widely used way of excretion is via specialized organs for excretion. The general blueprint consists of convoluted canals with great surface area. In many cases temporal storage is performed in addition.

10.2.3.1 Excretion in Invertebrates

In invertebrates like annelids and molluscs there are dead end- (Protonephridium) and open-systems, which are in contact with the circulatory system (Metanephridium). Ciliated or flame cells produce a fluid flow in a canal system that opens with nephridopores to the outside from each segment of the body. In insects and spiders the Malphigian tubules are the main site of excretion. They are located in close relationship to the alimentary canal and are surrounded by hemolymph. Functionally, an active transport of hydrogenic and potassium ions is followed by a passive influx of water and nitrogenous wastes. Uric acid is precipitated in the intestine and water reabsorbed before excretion with the feces.

10.2.3.2 Excretion in Vertebrates

Vertebrates developed a highly specialized excretory organ: the kidney—with the nephron as functional unit for urine production. A nephron is composed of a renal corpuscle (glomerulus and capsule) and the adjacent tubular system. While the first is for **filtration** of blood to primary urine, the latter has the function of **secretion**, **reabsorption**, and **excretion**.

From comparative studies of development it is believed that the kidney of the earliest vertebrates extended the length of the coelomic cavity and showed tubules arranged in segments resembling the invertebrate nephridium. This ancestral kidney is called **archinephros** and is not apparent as a functional form in adult vertebrates. Nevertheless, archinephros can be seen up until now during development of embryonic hagfish. Each tubule opens to the coelomic cavity with a nephrostome and collects the filtrate in the archinephric duct. The adult hagfish already shows a **pronephros** that derives only from the cranial parts of the archinephros and shows a reduced number of tubules. The **mesonephros** is a functional kidney of embryos of reptiles, birds, and mammals, only seen during development and not occurring in adult animals. It further develops to an opisthonephros in fish and amphibians or a metanephros in amniotes.

The **opisthonephros** is based on the caudal aspects of the mesonephros and is built of glomeruli and many coiled opisthonephric tubules. **Fresh water fish** have larger glomeruli than fish from salt water. The opisthonephric duct conducts both urinary excretes and sperm derived from colocalized testis.

In contrast, the **metanephros** of reptiles, birds, and mammals that also derives from caudal parts of the mesonephros develops an additional and separated duct for transportation of urinary excretes only: the ureter. The ureter ends either in the cloaca or in the urinary bladder (mammals)—an organ for temporal storage.

The metanephric **kidney of reptiles** shows multiple lobulations and a multitude of branches of the ureter. As nephrons are only in the cortical region concentration of urine is only possible in isosmotic situation. Final water reabsorption is performed by the cloaca epithelium. For lizards number of nephrons is speculated to range between 3,000 and 30,000.

Kidneys of birds are normally lobulated in three parts and tightly secured underneath the synsacrum and ilium. Probably due to the high metabolic rate of birds and the resulting metabolic wastes, a high number of nephrons are prevalent: in chicken about 200,000. Avian nephrons can be classified into two types: reptilian-type nephrons located in the cortex and mammalian-type nephrons in the medullary regions. The latter are with about 10–40 % the minority and show larger, more complex corpuscle and a loop of Henle. The produced uric acid is passed into the cloaca and transported back up to the caeca. In the intestine more water can be reabsorbed before final elimination together with the feces.

The **mammalian kidney** is located retroperitoneal in the lumbar region, with the right kidney in a slightly more cranial position in many species (except humans).

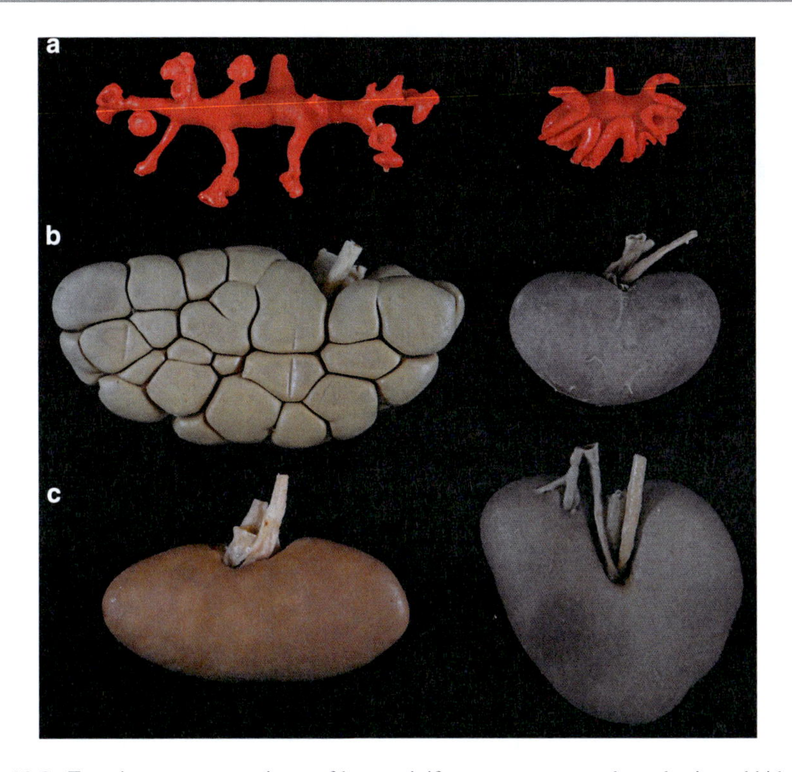

Fig. 10.2 Top view on cast specimen of large uriniferous structures and on plastinated kidneys. Color changes are due to original blood content of the specimen and fixation procedure. (**a**) Cast specimen of renal calices (cow) and renal pelvis (dog); (**b**) degree of fusion differs between species, *left*: lobulated (sulcated) bovine kidney, *right*: smooth canine kidney; (**c**) form also differs between species, *left*: oval-shaped porcine kidney, *right*: heart-shaped equine kidney. Preparations: Institute of Anatomy, Histology and Embryology, Vetmeduni Vienna

The kidney consists of a fibrous tissue capsule and a lobulated parenchyma, which can be divided into a cortical and a medullary region arranged in pyramids. The tip of each pyramid is called papilla renalis and collects the urine into the renal pelvis. The lobes of the kidney fuse to a varying level, which leads to a different kidney morphology in mammals (Fig. 10.2). When papillae are not fused the renal pelvis is partitioned in a multitude of calices (Fig. 10.2a). Nephrons are the functional units of the kidney and their basic structure is similar in all mammals. While corpuscles with glomeruli (network of capillaries) and Bowman's capsule as well as the contorted tubular parts are located in the cortex (Fig. 10.3c), the medulla mainly contains the loop of Henle for concentration purposes. The number of nephrons differs between species. Pig kidneys consist of 1,000,000 nephrons (comparable to humans) while cattle have up to 4,000,000 nephrons and mice only possess about 10,000–15,000. Not only number but also concentration capacities of nephrons differ. The psammomys, a gerbil, has only about 100 nephrons however, with very long tubules, which can most efficiently concentrate the urine as an adaptation to its

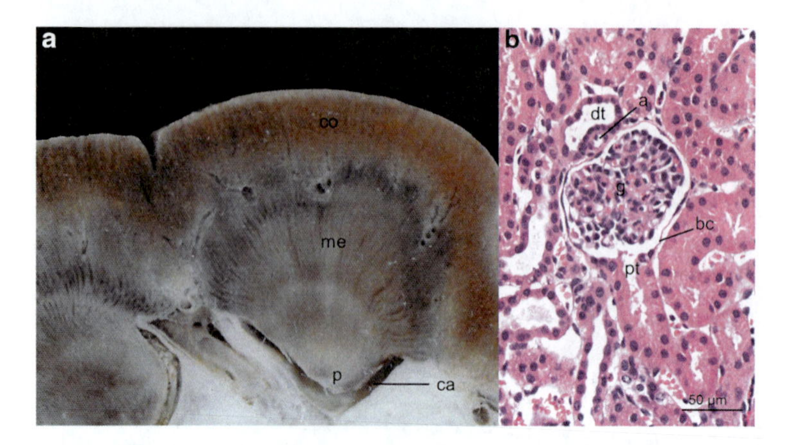

Fig. 10.3 Macro- and microscopic cross sections of the kidney. (**a**) Detail of midsagittal plane of a bovine kidney showing a lobe with the pyramid consisting of cortex (*co*) and medulla (*me*). Both portions are partially fused to the neighboring pyramid. Renal papil (*p*) ends in the renal calix (*ca*). (**b**) Histological section of the cortex of a rat kidney showing a glomerulum (*g*), Bowman's capsule (*bc*), the afferent arteriole (*a*), and proximal (*pt*) and distal (*dt*) tubules, H&E stain, 40× magnification. Preparations: Institute of Anatomy, Histology and Embryology, Vetmeduni Vienna

arid habitat. Concentration capacity is highest in the loop of Henle within the medulla due to the counterflow principle disposed here. In animals with especially long loops of Henle also the papilla is prolonged—reaching into the upper part of the ureter in some gerbils and white-tailed antelope squirrels. In all mammals the renal pelvis or calices join into the ureter and finally flow into the bladder.

The **bladder** is a hollow organ used for temporal storage of the constantly flowing urine until micturition. It can be divided into a corpus and a neck region. The composition of the bladder wall secures the great capacity of the bladder to shrink and expand, depending on the contending volume, as well as the deliberate micturition. The **urethra** connects the bladder with the outside and develops in a sex-specific fashion. Female animals have a very short urethra. In males a long urethra is present located within the penis. In male pigs and ruminants there is a sigmoid flexure of the urethra (see Chap. 12). The outer exit of the urethra is a lot smaller in males than in females, which increases the risk of urinary stones to be stuck in male individuals. Especially in small ruminants the ostium urether externum is situated on an elongation of the penis tip. This portion is called processus urethralis and increases the risk of problematic situations from urinary stones in male goats/sheep even more.

10.2.4 Homeostasis and Osmoregulation

Homeostasis is defined as the ability to maintain the inner system stable, independent of the environmental situation. Therefore, a regulation system and feedback mechanisms are needed.

Concerning homeostasis of water, the osmoregulation system controls uptake and loss of water and solved substances. The need of the animal to maintain this system to a constant level becomes clear by recapitulating the principle of osmosis: In a semipermeable membrane the water is diffusing to the compartment with higher concentration of solutes. Especially animals living in the water have to deal with this phenomenon because of diffusion of water and salt. There are two types of adaption: animals that conform to the surrounding situation are called osmoconformers, and those who regulate their inner osmolarity: osmoregulators. **Osmoconformers** are mainly invertebrates and some fish living in the deep sea, which try to adjust their inner osmolarity to the outside situation. To maintain isotonicity with the sea water sharks, for example, accumulate urea in their body fluids and muscles and only exchange certain ions with the surrounding water. As a constant internal situation is necessary for optimal functioning and their body is not able to react to changes very well, those animals are restricted to small habitats with relatively stable conditions. As the majority of habitats are inconstant in their conditions, also most vertebrates are **osmoregulators** that try to maintain a constant inner milieu independent of the environmental situation. Therefore, mechanisms have evolved to regulate inner water and solute concentrations. In aquatic species one has to distinguish between fresh and sea water animals: **Fresh water fish** are hypertonic in comparison to the water around and therefore water is constantly absorbed via the skin, gills, and together with food. To maintain high levels of salt concentrations fresh water fish have to take up ions by active transport systems and at the same time try to lose much water by highly diluted urine. **Salt water fish** are living in an outer milieu showing a high osmotic pressure. The internal fluids are lower in concentration than the outer sea water and additional salts are taken up with the food. Water is lost via diffusion through skin and gills. So there is great need to take up enough water and excrete ions in highly concentrated urine.

All land living animals are osmoregulators. They have limited access to water and need to reduce water loss. Mechanisms therefore are reduction of evaporation by body surface, reabsorption within the body, increase of water intake, or production of metabolic water. For the excretion of crystalline uric acid less water is needed compared to urea—therefore, reptiles and birds are more prone to live in arid regions. In some migrating bird species the metabolic water alone is sufficient for the function of kidneys. Another important aspect is to protect the fetus from evaporation of water—which is believed to be a very important prerequisite for the evolution of land living species. Eggs either are laid in moist surroundings, like in mosquitoes, or have an eggshell. In mammals amnions contain embryonic fluids and membranes to protect the embryo from drying out.

10.2.5 Acid–Base Regulation

Another function of the urinary tract is the homeostasis of the pH value. Two organ systems are responsible to maintain the acid–base homeostasis: the kidneys and the lungs. The lungs contribute by regulating the carbon dioxide through breathing. The kidneys have two very important roles. They pump the hydrogen in the urine and reabsorb the bicarbonate.

10.2.6 Hormone Production

A hormone of special interest in comparative aspects is **erythropoietin (EPO)** as it is present in all mammals and was recently demonstrated to be also involved in red blood cell development in fish and amphibians (Nogawa-Kosaka et al. 2011; Paffett-Lugassy et al. 2007). In mammals, three body sites produce EPO: In the central nervous system EPO is produced by the astrocytes and has also paracrine effects. The liver also contributes to the EPO production, but the majority of the plasma EPO is produced by the kidney. In case of hypoxemia, the kidneys produce EPO, which leads to increased erythrocyte production in the bone marrow and an increased erythrocyte content in the blood. In cats, there are many chronic diseases where the kidneys tempt to fail. As a consequence affected cats develop anemia. The cats show pale mucous membranes and have a decreased number of erythrocytes in the blood. A current treatment option for this condition is to transfer cat patients feline EPO genes (Beall et al. 2000).

Another important hormone affecting kidney function is **aldosterone**. It is derived from the adrenal glands (see Chap. 8), acts on the distal tubules and collecting ducts, and stimulates sodium reabsorption, water retention, and potassium secretion. In mice there is strain-specific, sex-specific, and generation-dependent aldosterone secretion (Spyroglou et al. 2012).

Parathyroid hormone from the parathyroid glands is upregulated if calcium levels in the blood are decreased. As already outlined above in Sect. 10.1, this hormone promotes the kidney to produce **vitamin D** to reabsorb calcium from the intestine. This is especially important for dairy cows and other lactating animals, because the daily body turnover of calcium increased threefold during lactation.

References

Beall CJ, Phipps AJ, Mathes LE, Stromberg P, Johnson PR (2000) Transfer of the feline erythropoietin gene to cats using a recombinant adeno-associated virus vector. Gene Ther 7: 534–539

Breyer MD, Zhang Y, Guan YF, Hao CM, Hebert RL, Breyer RM (1998) Regulation of renal function by prostaglandin E receptors. Kidney Int Suppl 67:S88–S94

De Loof A, Lindemans M, Liu F, De Groef B, Schoofs L (2012) Endocrine archeology: do insects retain ancestrally inherited counterparts of the vertebrate releasing hormones GnRH, GHRH, TRH, and CRF? Gen Comp Endocrinol 177(1):18–27

Deeb KK, Trump DL, Johnson CS (2007) Vitamin D signalling pathways in cancer: potential for anticancer therapeutics. Nat Rev Cancer 7:684–700

Levy-Lioret A, Pokroy B, Levavi-Sivan B, Leiserowitz L, Weiner S, Addadi L (2008) Biogenic guanine crystals from the skin of fish may be designed to enhance light reflectance. Cryst Growth Des 8:507–511

Nogawa-Kosaka N, Sugai T, Nagasawa K, Tanizaki Y, Meguro M, Aizawa AY, Maekawa S, Adachi M, Kuroki R, Kato T (2011) Identification of erythroid progenitors induced by erythropoietic activity in Xenopus laevis. J Exp Biol 214:921–927

Paffett-Lugassy N, Hsia N, Fraenkel PG, Paw B, Leshinsky I, Barut B, Bahary N, Caro J, Handin R, Zon LI (2007) Functional conservation of erythropoietin signaling in zebrafish. Blood 110:2718–2726

Spyroglou A, Sabrautzki S, Rathkolb B, Bozoglu T, Hrabě de Angelis M, Reincke M, Bidlingmaier M, Beuschlein F (2012) Gender-, strain-, and inheritance-dependent variation in aldosterone secretion in mice. J Endocrinol 215:375–381

Further Reading

Von Engelhardt W, Breves G (2010) Physiologie der Haustiere, 3rd edn. Enke, Stuttgart

Foelsch UR, Kochsiek K, Schmidt RF. Pathophysiologie. Springer

Hickman CP Jr, Roberts LS, Keen SL, Eisenhour DJ, Larson A, L'Anson H (2011) Integrated principles of zoology, 15th edn. McGraw-Hill, New York

Hildebrand M (1974) Analysis of vertebrate structure, 1st edn. Wiley, New York

Krück F (2000) Pathophysiologie/Pathobiologie, 2nd edn. Urban und Schwarzenberg, München

Nickel R, Schummer A, Seiferle E (2004) Lehrbuch der Anatomie der Haustiere: Anatomie der Vögel, Band V, 3rd edn. Paul Parey, Stuttgart

Reece WO (2009) Functional anatomy and physiology of domestic animals, 4th edn. Wiley-Blackwell, Aimes

Reece JB, Taylor MR, Simon EJ, Dickey JL (2012) Campbell biology—concepts & connections, 7th edn. Pearson, San Francisco

Stark D (1982) Vergleichende Anatomie der Wirbeltiere auf evolutionsbiologischer Grundlage, 1st edn. Springer, Berlin

Paulev P-E, Zubieta-Calleja G, eds. Textbook in medical physiology and pathophysiology. Essentials and clinical problems, 2nd edn. Chapter 30

Treuting PM, Dintzis SM (2012) Comparative anatomy and histology—a mouse and human atlas, 1st edn. Elsevier, Waltham

Breathing: Comparative Aspects of the Respiratory System

11

Hanna Schöpper, Cordula Bartel, and Krisztina Szalai

Contents

H. Schöpper (✉) • C. Bartel
Institute of Anatomy, Histology and Embryology, University of Veterinary Medicine, Vienna, Austria
e-mail: Hanna.Schoepper@vetmeduni.ac.at

K. Szalai (✉)
Comparative Medicine, Messerli Research Institute, University of Veterinary Medicine Vienna, Medical University of Vienna, University of Vienna, Vienna, Austria
e-mail: Krisztina.Szalai@vetmeduni.ac.at

E. Jensen-Jarolim (ed.), *Comparative Medicine*,
DOI 10.1007/978-3-7091-1559-6_11,

Abstract

The respiratory system of animals is composed of **cellular respiration** for provision of energy, **respiratory gas exchange** via a respiratory membrane to provide the body with substrates for cellular respiration, and a **transportation system** whenever necessary to bring the two together.

11.1 Cellular Respiration

The respiratory system is essential for the energy metabolism of individuals. Eukaryotes take up food stuffs as substrates for metabolism. The biochemical reactions during metabolism of substrates rely on oxidation to generate the needed energy. This process is called **cellular respiration** and takes place in eukaryotic cell organelles, called mitochondria—the power plants of the body (see Chap. 2). Here a biochemical reaction takes place with electrons being transferred from an electron donor like glucose to an electron acceptor like oxygen. An electron transfer always needs both oxidation (loss of electrons by a molecule) and reduction (addition of an electron to a substance); therefore the process is called redox reaction. During this descend of the electron, energy is released in the form of ATP (adenosine triphosphate) that can be further used. Despite the denotation of oxidation, the redox reaction is not exclusive for oxygen as also other compounds like nitrite, manganese, or iron may be used as electron acceptors during energy retrieval. Depending on whether oxygen is used or not, aerobic and anaerobic species can be distinguished. The majority of species is aerobic—using oxygen from either air or water in their mitochondria where the energy is produced. Anaerobic living species are mainly prokaryotes like some earth-bound bacteria. However, in 2010, it was reported that a small metazoan of the animal phylum Loricifera was discovered in the Mediterranean Sea (Danovaro et al. 2010). It could be demonstrated for the first time that a metazoan not only shows cellular adaptations to anaerobic habitat like hydrogenosomes instead of mitochondria, but undergoing all live cycles under completely anaerobic conditions. Not only provision of electron acceptors is necessary for energy conversion but also the elimination of end products. In aerobic species, oxygen has to be incorporated and carbon dioxide needs to be eliminated from the body. Both processes depend on gas exchange at a respiratory surface.

11.2 Respiratory Gas Exchange

Respiratory gas exchange is referred to as the process of oxygen and carbon dioxide crossing a respiratory surface by passive diffusion to reach another compartment. The respiratory surface needs to be moist in order to function properly. While carbon dioxide is released to the outer world, oxygen is being taken up into the adjacent cells. In animals that cannot provide every cell in the body with respiratory surface, a transportation system for oxygen is needed to supply peripheral cells that are too far away for simple diffusion. In case of vertebrates the water-soluble protein **hemoglobin** with central ion atoms is located in red blood cells and transports the oxygen and carbon dioxide. Arthropods and mollusks have a similar system; however, their transport protein **hemocyanin** uses copper as the central atoms. In the periphery the oxygen is then released to diffuse to target cells, while carbon dioxide is loaded up for return transportation to the primary respiratory surface and elimination to the environment. The respiratory gas exchange needs to be highly efficient to provide enough oxygen for the requirement of the respective animal. For this purpose different mechanisms have developed: enlargement of respiratory surface area, ventilation to maintain high-concentration gradient of gas between outer world and body cells, as well as different directions of flows of gas and body fluids. To cope with these different mechanisms, a variability of respiratory systems has evolved in different species.

11.3 Respiratory Systems

11.3.1 Skin

Only in small animals that live under moist conditions it is possible to use the outer skin as a respiratory surface in addition to the original protective purpose. In Protozoa, Coelenterates, Platyhelminthes, but also some annelids including Tubifex and Hirudinea, the whole body surface is used as a respiratory membrane. Right underneath the epithelium, a network of capillaries is located to assimilate oxygen from the outside and transfer it to the target cells. Some amphibians and fish use cutaneous respiration to support other respiratory systems. Especially, during winter time, when lakes are covered with ice, there is no possibility to reach the surface and breathe air for frogs. In succession they reduce their metabolism and then cutaneous respiration of oxygen from the water is enough to survive. This respiratory system is limited to small animals that live in moist environment. In most species gas exchange at the outer skin is not sufficient to supply the whole body with O_2, and therefore specialized organs with increased respiratory surface area developed for gas exchange.

11.3.2 Invertebrate Trachea

The trachea is a highly efficient open system of air-conducting tubes throughout the whole body to bring the oxygen-rich air to the target cells in terrestrial arthropods like millipedes, insects, and some spiders. As the respiratory surface needs to be moist for full functioning of cell membranes and efficient gas exchange, land-living animals try to protect their respiratory system from drying out. Several mechanisms have evolved for this purpose. The respiratory membrane is no longer on the body surface, but within the body, openings are minimized in size and additional provision of moisture is given. At spiracles along the thorax and abdomen, the tracheal system opens with a valve to the atmosphere and oxygen-containing air can enter and exit the branched tracheal system within the insect's body. Oxygen diffuses along the tracheal system that opens up in small tracheoles. Those ends are fluid filled and are in close relation with the cell membranes of each cell in the body. Organs that are highly oxygen demanding are in close relationship to additional air sacs that increase gas exchange. Oxygen transport via **hemolymph** is nearly unnecessary as each cell has a direct tubular system to transport and facilitate gas exchange directly. Nevertheless pigments in the circulatory system are apparent, but do not play a role in respiration. The tracheal system works passively, but ventilation can be intensified by body movements.

11.3.3 Gills

In fish and some amphibians, gas exchange takes place at **the gills**. Gills are extensions of the outer body surface, specialized on gas exchange in aquatic animals. There are two types of gills: **external** and **internal** gills.

External gills are present in some larval stages and amphibians as protuberances of outer skin with increased surface area. Also in some invertebrates like the star fish, aquatic worms, mussels, crustacea, and snails, simple forms of gills are exhibited. Sea stars display small **papulae**, and some worms feature **branchial tufts** which look like little clusters of gills at the parapodium to both sides.

Internal gills are characterized by enlarged surface area of respiratory membrane within the body. They are mainly exhibited in some molluscs, arthropods, and fish. The basis of gills is a row of several arches, each arch projecting two filaments with a multitude of lamellae. Also in gills, the respiratory surface is underlined by a network of vessels to enable oxygen and carbon dioxide diffusion to and from the capillaries. Highly superficial network of blood vessels accounts for the red color of gills of fish. The blood flow within the capillaries is counter current, meaning that the blood flows opposite to the movement of water past the lamellae of the gills. Like this the maximal concentration gradient of gas between water and blood is maintained and therefore gas exchange is optimized. In a countercurrent process more than 80 % of the dissolved oxygen from the surrounding water can be removed, while if blood and water would flow in the same direction, it would be only 50 % at the most. At the same time CO_2 is eliminated—so in contrast to air-breathing animals where there are inspiration and expiration—in water-breathing animals both necessities are performed at once. Due to the great vascularization, the gills are highly vulnerable to mechanical injuries. To avoid particles from blocking or violation of the gills, there is a filter as part of the gills in fish. In addition, protective structures have evolved in many species. Either gills can be retracted within the body or an external protection like the shell or the operculum (a movable flap in fish) shields the gills from harmful influences.

The advantage of surrounding water is that the respiratory membrane is kept moist and fully functional. In general gills could also take up oxygen from the air—however, if respiratory membrane is exposed to air, the moist will evaporate very quickly, so that gills dry out, stick together, and cannot exchange gas anymore and the fish has to suffocate. At the same time water as the surrounding element holds a huge disadvantage: in water only about 5–10 % oxygen is dissolved, while in air it is about 21 %. The oxygen concentration of water is further dependent on temperature and salt content. Therefore the situation is most difficult in warmer regions with high salt concentration. So in water-living animals, an even more efficient respiratory mechanism is needed to meet metabolic situation. Next to the countercurrent flow of blood, also constant change of medium at the respiratory surface—to keep up the gas concentration gradient—called ventilation, is a possibility to increase respiratory gas exchange. In simple forms like mussels exchange of water along the gill lamellae is achieved by drift only. In small, highly metabolic fish like herring ram ventilation is necessary, meaning the animal's movements within the water leading to constant water flow along the gills. If those fishes are kept in an aquarium—too small for rapid swimming—they might asphyxiate while in water! Water flow along the gills can even be more improved by active support of two muscle pumps—one in the oral cavity and one in the opercular cavity. Those muscles will lead to draw water into the mouth and then release it after passing the gills through the gill slits.

Even though adapted for living and breathing in the water, some fish can use the atmospheric oxygen in exceptional cases. Additional organs with potential respiratory function are air bladder and fish lungs. The air bladder of bony fish has developed from the upper gastrointestinal tract and is filled with air. It is primarily used to reduce the specific weight and enabling balance and floating in the water. However, in cases of low oxygen content of the water, it may also serve for some kind of additional respiratory function. Some very primitive lungfish developed adaptive organs comparable to air bladder to be able to survive on land in times of drought. These "lungs" enable gas exchange from air. Nevertheless lungfish needs to stay in moist surroundings like mud, but are believed to survive for years in case a pond is fallen dry. Also eels can survive outside the water short term, when wandering on land in rainy seasons, and they use cutaneous respiration during those circumstances. In both cases it is secured that the respiratory surface is lubricated—either by mucus, mud, or rain.

11.3.4 Lung

As gills and cutaneous respiration are unsuitable for breathing air and limited surface area of latter is not sufficient for larger animals, a highly adapted and specialized organ for breathing air is needed.

The lung meets the requirements of breathing air. Disadvantages of evaporation, vulnerability to mechanical impact, and limited surface area are addressed in dislocating the respiratory membrane into the visceral cavity—additionally protected by the rib cage and nearly infinite potential to increase surface area. Lungs are found in very many animal species—even some invertebrates like pulmonate snails, scorpions, and some spiders have lungs. In snails a pneumostome serves as the entrance through which air is "inhaled," by contractions of the mantle floor to the lungs. Generally, the lung is surrounded by a highly vascularized area so that oxygen can diffuse to the circulatory system. In tetrapod vertebrates the respiratory system is composed of a mouth/nose, trachea, and lung. Therefore oxygen-rich air is inhaled through either mouth or nose and led to the respiratory surface of the lung via the trachea. The tissue of the lung is surrounded by a capillary network that subsequently transports the exchanged gas to the rest of the body and back. The surface area is matched to the actual metabolic need of oxygen and therefore especially the inner buildup of respiratory surface differs between species.

There are additional functions to the respiratory function like moistening of the air, filtration of the air in the nasal cavity, temperature adjustment, sensory functions, and phonation. See for more details section "Lungs" below.

11.3.4.1 Amphibians and Reptiles

Frogs own a quite simple form of lungs with just few septa building subdivisions. Ventilation is a two-step movement: first lowering the base of the mouth produces negative pressure and therewith air is inhaled into the oral and buccal cavity. Afterwards, nostrils are closed and the base of the mouth is pushed against the palate to increase the pressure until positive and force the air to flow into the lungs. In contrary, in mammals and birds ventilation of lungs is performed by negative pressure only. In most species, the following passive retraction of tissue leads to the exhalation of inhaled air. **Reptiles** also own lungs for respiration, while some aquatic forms may also use cutaneous respiration additionally. The skin and also mucosal tissue of mouth and cloaca are able to exchange gas. In elongated forms like snakes, only the left lung is conserved, while the right lung is degenerated.

11.3.4.2 Birds

The avian lung is composed of air-conducting bronchi, gas-exchanging parabronchi with atria, and air sacs that are not involved in gas exchange. As the lung is nearly constant in volume, ventilation is performed by expanding and compressing the adjacent air sacs. When inhaling the air flows through the trachea, ventral parts of the lung for gas exchange, and finally into the caudal air sacs. During expiration, costal muscles compress the caudal air sacs and force the air through the lung and gas exchange a second time. Like this the lung is ventilated comparable to a bellow and with the cross-flow of blood and air even more efficient in gas exchange.

11.3.4.3 Mammals

The general structure of **mammalian lungs** and **respiratory system** is comparable in its structure and function. Therefore, the human respiratory system will be described in detail, and at the end of each section, a selection of comparative aspects from the animal world will be included.

11.3.4.4 Human Lung Function and Respiration

A newborn baby has a breathing frequency of 40–45 per minute, while every adult healthy person breathes 12–16 times every minute. Breathing is an automatically working process ensureing that the body takes up enough air to provide the cells with sufficient oxygen for energy metabolism.

11.4 Anatomy of the Respiratory Tract

The respiratory tract can be divided into two segments, based on the location of the organs:

- **Upper respiratory tract**: nose and nasal cavity, paranasal sinuses, pharynx, and larynx (Fig. 11.1).
- **Lower respiratory tract**: trachea, bronchi, and lungs (Fig. 11.2).

In another setup, respiratory organs can be grouped into conducting zone or respiratory zone based on their function. To the **conducting zone** belong the organs from the nose to the bronchi, conducting the inspired air, while the real **respiratory zone** takes place at the alveoli of the lungs:

- **Conducting zone**: nose, nasal cavity, larynx, pharynx, trachea, bronchi, and terminal bronchioles.
- **Respiratory zone**: alveolar ducts and alveoli.

11.4.1 Nose and Nasal Cavity

The nose is the entrance of the **nasal cavity** and located in the middle of the face and gives an own character of the face. It opens with the **nostrils**, where the outer air enters and continues its route to the nasal cavity. The shape of the nose is determined by the ethmoid bone, cartilages, and nasal septum, which also divides the nose into left and right noses. Other important structures in the nose are the **conchae** or also called **turbinates**, being a narrow bone shelf, protruding into the breathing passage of the nasal cavity. There are three conchae: superior, middle, and interior, increasing the breathing area of the nose.

The nose and nasal cavity are covered by the respiratory epithelium, which is composed of different cells with different functions:

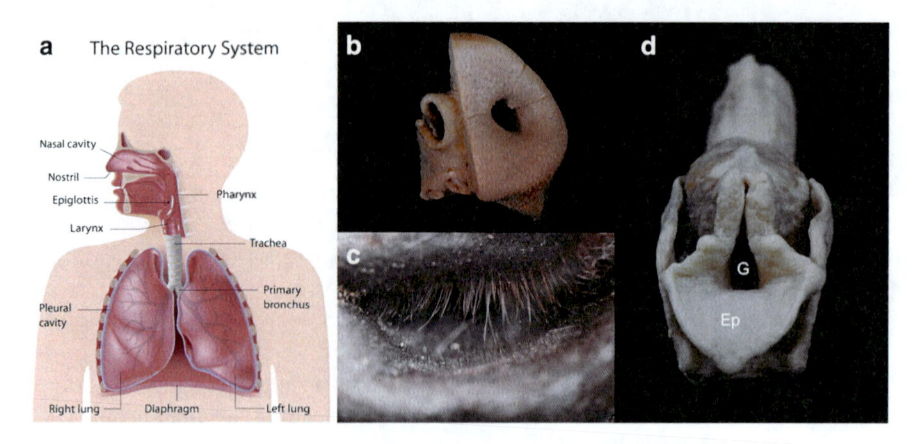

Fig. 11.1 Comparative aspects of the respiratory system. Pictures (**b**)–(**d**) are arbitrary representatives from other mammals than human. (**a**) Overview on human anatomy of the respiratory tract (© [Alila Medical Images]—Fotolia.com); (**b**) rooting disk of a pig with the nasal cartilage dissected on one side; (**c**) nostril of a horse protected by a multitude of hair; (**d**) larynx of a dog, view on the epiglottis (*Ep*) and subsequent glottis (*G*). Specimens are either dry or plastinate. Preparations of the Institute of Anatomy, Histology and Embryology, Vetmeduni Vienna, Austria

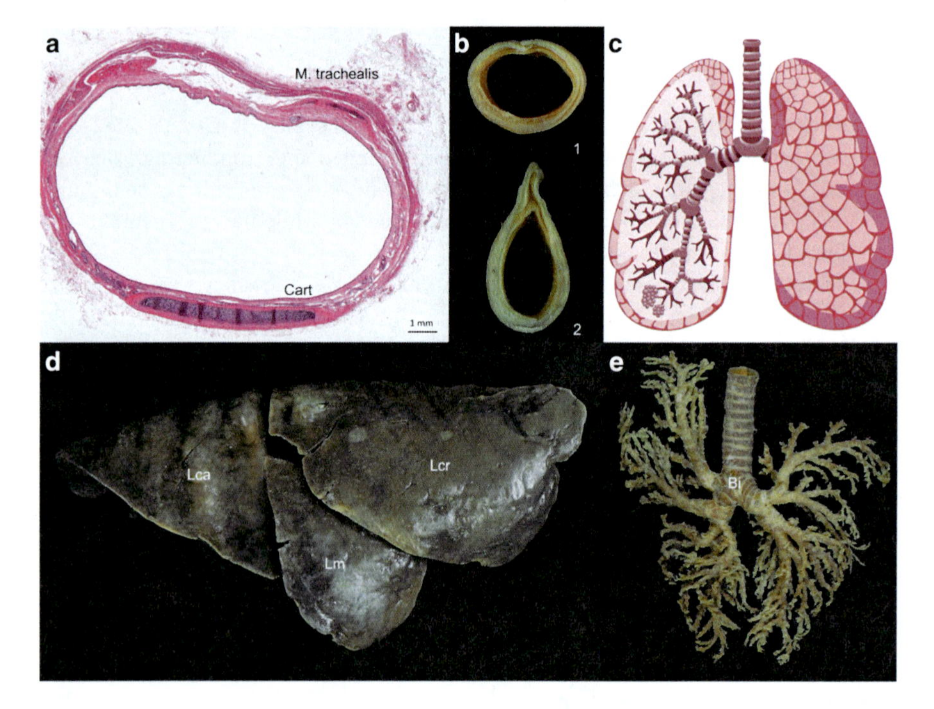

Fig. 11.2 Organs of the lower respiratory tract. (**a**) Histological cross section of a canine trachea showing cartilage tracheal ring (*Cart*) and the tracheal muscle (*M. trachealis*); (**b**) photographs of a porcine (*1*) and bovine (*2*) trachea to compare tracheal morphology; (**c**) overview on human lung anatomy (© [shurshusha]—Fotolia.com); (**d**) lateral view of the right lung of a dog: cranial (*Lcr*), middle (*Lm*), and caudal lobes (*Lca*) visible; (**e**) ventral view of the bronchial tree of a dog: bifurcation (*Bi*). Preparations: Institute of Anatomy, Histology and Embryology, Vetmeduni Vienna, Austria

- **Basal cells** are resting stem cells on the basal lamina of the epithelium and capable to divide into the different cell types in the nose.
- **Ciliated cells** are columnar cells, with densely packed apical cilia, to move the mucus along the surface of the epithelium. Ciliary motion occurs in the sol layer, covering the epithelial cells.
- **Goblet cells** are mucus-secreting cells, scattered individually within the epithelium. The produced mucus covers the whole epithelium.

The nose is the unique organ to inhale the air and for smelling. There are about 100,000 different smells existing in nature; however, a human individual can sense only 100–200 from it. Odorant molecules, which can be smelled, are dissolved in the mucus and bind to smelling receptors in the nose and nasal cavity. The sensing nerve of smelling, called **olfactory nerve**, builds up an intensive network within the conchae of the nasal cavity and provide the smelling information to the **bulbus olfactorius**, which enters into the brain, where the odor will be perceived.

Moreover, the nose has also the function to conditioning by humidifying and warming up (up to 37 °C) the inhaled air. Ciliated cells within the epithelium with their cilia filter the inhaled air from foreign and irritating particles, protecting the lung from these unwanted particles, which are then removed by the mechanism of sneezing.

The nose of mammals is highly adapted not only for inhalation of air and smelling, but also for additional purposes like usage as a tool in quadruped animals. The nose of the pig, for example, is formed as a rooting disk with a specialized prenasal bone as basis for digging and searching for food in the ground (Fig. 11.1b). In cattle the nose is merged with the upper lip to a planum nasolabial. The soft nose of horses shows a specialized alar cartilage that enables the nostrils to be widely flared and therewith increase airflow, while a multitude of hair prevents inhalation of particles (Fig. 11.1c). The olfactory region in the back of the nose is exhibited in a species-specific extend that is also reflected by the dimension of olfactory bulb of the brain (see Chap. 6).

11.4.2 Paranasal Sinuses

Paranasal sinuses are paired air-filled spaces, surrounding the nasal cavity and the eyes. In humans there are four subgroups, named by the bones within which sinus lies: frontal, ethmodial, maxillary, and sphenoidal. Sinuses are also covered with respiratory epithelium, therefore having similar functions to the nasal cavity: humidify and warm up the inspired air and participate in smelling. They also regulate the intranasal gas pressure and increase the resonance of the voice.

In other mammals the sinus system has the additional function to lighten the total weight of the head. While in two-legged species like humans the head is comparably easy to be held in a vertical position, in quadruped species more muscle work is needed to stabilize the head that is positioned in front of the body. As the skull is also the place of teeth and mastication muscles for accomplishing the grinding, a great bone volume is needed. So, on the one hand, the head should be light in order to minimize energy consumption of muscle activity, and on the other hand, the skull needs to be large enough to house large teeth and strong muscle packages for digestion of food. The paranasal sinuses seem to be the optimal answer to such diverging needs and are especially distinct in large herbivore species like the horse.

11.4.3 Pharynx

After passing through the nasal cavity and paranasal sinuses, the inhaled air continues its way to the **pharynx**, which is divided into three sections: naso-, oro-, and laryngopharynx with different functions. The air from the nose passes by the **nasopharynx**, while both food and air from the mouth pass into the **oropharynx**. The third part, the **laryngopharynx**, has the function to permit the airflow to the lungs. The pharynx is the crossroad for air and food and takes part in swallowing and also in the vocalization. The Eustachian tube connects the pharynx to the middle ear and is used for pressure equalization. In horses this duct is enlarged to a guttural pouch of a volume up to 300 mL each. As important blood vessels and cranial nerves are in close contact, an infection of the guttural pouch may have severe consequences.

11.4.4 Larynx

While also amphibians and reptiles show a simple version of cartilaginous larynx, only mammals show two characteristic structures: **the thyroid cartilage** and **the epiglottis** (Fig. 11.1d). **The larynx**, very commonly called **voice box**, is located in the neck and connects the lower part of the pharynx with the trachea. It participates in breathing and prevents disoriented food from ending up in the trachea or even worse in the lungs, where a pneumonia could emerge. The Larynx is the place of sound production, where pitch and volume are manipulated. In contrast to mammalian sound production, birds use a specialized organ called syrinx for vocalization. **The syrinx** is located at the bifurcation of the trachea and consists of membranes that are connected to elastic folds, regulated in their tension by muscles to modulate the pitch. The syrinx is enlarged by a bulla in some species like the drake.

11.4.5 Trachea and Bronchial Tree

The trachea, also called **windpipe**, is about 25 mm in diameter and 10–16 cm long in humans (Fig. 11.2c). It transports air into the lung, starting under the larynx, descending behind the sternum in front of the esophagus. It is composed of C-shaped cartilaginous rings (in humans 16–20) connected by the trachealis muscle, providing elasticity to the trachea (Fig. 11.2a,b).

At the sternal angle, the trachea divides into two smaller tubes (bifurcation), called **primary (main) bronchi**. The **right primary bronchus** is shorter, wider, and more vertical than the left in humans and enters into the right lung, while the **left primary bronchus** enters into the left lung. In the lung, the main bronchi are subdivided into smaller and smaller tubes, building up the **tracheal tree**. In humans the right primary bronchus subdivides into three secondary bronchi, which get into the lobes of the lung, and the left primary bronchus is split into two lobar bronchi. The secondary bronchi will further subdivide into tertiary, segmental bronchi entering into the **bronchopulmonary segments**. The bronchopulmonary segments are separated divisions of the lung, surrounded by a connective tissue.

The mucous membrane of the trachea consists of a basal membrane, supporting the stratified epithelium built up by basal, ciliated, and goblet cells, very similar as discussed for the nasal cavity above. The submucous layer is composed of connective tissue with blood vessels, nerves, and mucous glands.

Other mammalian species only show slight modifications of this basic structure of the trachea like differences in cartilage formation or muscle attachment. Birds, however, have closed cartilage rings and a prolonged trachea being sometimes looped within the ribcage.

11.4.6 Lung

The lung is the organ of respiration and gas exchange, has a weight of 2.3 kg, and includes 2,400 km of airways in humans (Figs. 11.1a and 11.2c). It has two sides, each side consisting of species-specific number of lobes (two lobes left and three lobes right in humans). Via **the trachea** and **the bronchial tree**, inhaled air gets into the lung and the bronchial tree continues into the real respiratory zone. The segmental bronchi get into primary bronchioles and further into terminal bronchioles, which then go on to divide into 2–11 alveolar ducts.

The alveolar ducts connect the terminal bronchioles with the alveolar sacs, which contain a group of alveoli (pulmonary alveolus has a diameter between 75–300 μm). A human lung contains approximately 700 million alveoli with the total surface of 70 m^2. The alveoli are composed of epithelial layer and an extracellular matrix and surrounded by tight network of capillaries, wrapping up to 70 % of the surface area. There are three major cell types in the alveolar wall, which are also called **pneumocytes**:

- **Type I cells**: squamous cells, forming the structure of the alveolar wall.
- **Type II cells**: surfactant cells, continuously producing pulmonary surfactant by exocytosis to lower the surface tension and increasing the capability of gas exchange.

Macrophages are important immune cells, participating in the immune defense by destroying foreign materials like bacteria.

11.4.7 Supporting Structures

11.4.7.1 Diaphragm

The **diaphragm** is built up by sheets of internal skeletal muscles and separates the organs of the thoracic and abdominal cavity. The diaphragm actively participates in breathing and moreover expels vomit, feces, and urine by increasing the intra-abdominal pressure.

11.4.7.2 Intercostal Muscles

The **intercostal muscles** are groups of muscles, located between the ribs, and help to form and move the chest wall, like by breathing. There are three different layers, external, internal, and innermost, with different functions. **The external group of muscles** takes part in the inhalation by expanding the dimension of the thoracic cavity. **The internal muscles** have the opposite function; by expiration they decrease the volume of the thoracic cavity. **The innermost intercostal** muscles are the deepest layer of the intercostal muscles, fixing the position of the ribs during respiration.

11.4.7.3 Pleura

The pleura is a serous membrane within the chest, forming the pleural cavity. The pleura contains two layers: the pleura parietalis which is connected to the chest wall and the inner pleura visceralis covering the lung. Within these two layers pleural fluid (serous fluid) produced by the normal pleura fills the pleural cavity, enabling the smooth movement of the lung within the chest cavity.

11.5 Physiology of the Respiratory System

In the previous section of this chapter, we described the anatomy of the respiratory tract. In the second section we discuss the function of the respiratory tract and go more into details of (1) **the mechanism of breathing** and lung volumes, capacities, (2) **pulmonary gas exchange**, and (3) **gas transport** in the capillaries.

11.5.1 Mechanism of Breathing

Via **inhalation** and **exhalation**, air is exchanged between the outer environment and the alveoli, depending on the pressure in the alveoli and lung volumes. In the respiratory system, air moves from high-pressure to the low-pressure areas. All pressures are relative to the atmospheric pressure, which is 760 mmHg at sea level.

Inhalation (inspiration) is the active part of breathing. The external intercostal muscles are contracted after the stimulus of nervous impulses and the diaphragm moves down (up to 1–10 cm) toward the abdomen, inducing an expansion of the thoracic cavity. The lung is via the pleura in connection with the chest wall, and the created negative pressure in the pleural cavity has the effect to hold the lung open and decrease the pressure in the alveoli.

Exhalation (expiration) is a passive process, where the elastic coil of the alveoli forces the air out from the lungs. In this mechanism, the internal intercostal muscles contract, via the ribs, and move back to the original position, but also the diaphragm moves back, reducing the space in the thoracic cavity and increasing the pressure in the lung.

During respiration, lung volumes and capacities can be measured and monitored with a spirometer (examples for the human are given here):

- **Tidal volume—TV** is the air volume being displaced during normal, unforced breathing between inspiration and expiration; 500 mL.

- **Inspiratory reserve volume—IRV** can be inhaled additionally after normal inspiration; 2,000–3,000 mL.
- **Expiration reserve volume—ERV is** the volume that can be exhaled additionally after normal expiration; 1,000–1,800 mL
- **Residual volume—RV is** the air that remains in the lungs after maximal expiration; 1,000 mL.
- **Vital capacity—VC (TV + IRV + ERV)** is the maximum volume of air that can be exhaled after maximal inspiration.
- **Inspiratory capacity—IRC (TC + IRV) is** the maximum volume of air which can be inspired after normal expiration
- **Functional residual capacity—FRC (ERV + RV) is** the volume of air remaining in the lung after the end of a normal respiration.
- **Total lung capacity—TLC (IRV + TV + ERV + RV) is** the total air volume filling the lung after maximal inspiration.

These parameters give important diagnostic hints for lung diseases.

In other mammals, lung volumes and capacities are highly dependent on body weight and metabolic rate of the animal. In rabbits tidal volumes of only 20 mL were measured, while horses move about 6,000 mL air during normal breathing.

11.5.2 Pulmonary Gas Exchange

Pulmonary gas exchange is the process where respiratory gasses (oxygen and carbon dioxide) are exchanged between the alveolar sacs (**alveoles**) and the blood, tightly wrapping the capillaries, building up the respiratory membrane, where the gas exchange takes place (Fig. 11.3). Via this very thin membrane, oxygen diffuses from the air-filled sacs into the blood where it binds with affinity to the hemoglobin of erythrocytes, while carbon dioxide from the blood diffuses into the alveolar sacs. Gas molecules either diffuse from the higher to the lower concentration or move from the higher pressure to the lower pressure area. There are several factors, which support efficient gas exchange of the lung: (1) the respiratory membrane of the alveoles is very thin, enabling fast diffusion of gas; (2) the air sacs are moist, so that gas can be dissolved before diffusing; and (3) last but not least, surfactant factor is produced locally in surfactant cells to support constant unfolding of the alveoles. In preterm babies, a deficiency of surfactant factor may inhibit regular unfolding of the lungs and breathing. Lung alveoli are enfolded to 70 % of their surface in healthy condition, providing a large area for respiratory gas exchange.

Oxygen exchange. Inhaled oxygen is dissolved in alveoli where it has a partial pressure of 104 mmHg. Via the respiratory membrane oxygen diffuses into the capillaries where the passing blood has a lower oxygen partial pressure (P_{O2} = 40 mmHg measured in humans). Here it immediately binds to the hemoglobin of erythrocytes, see below. Hence, oxygen-rich blood leaves the lungs to the periphery.

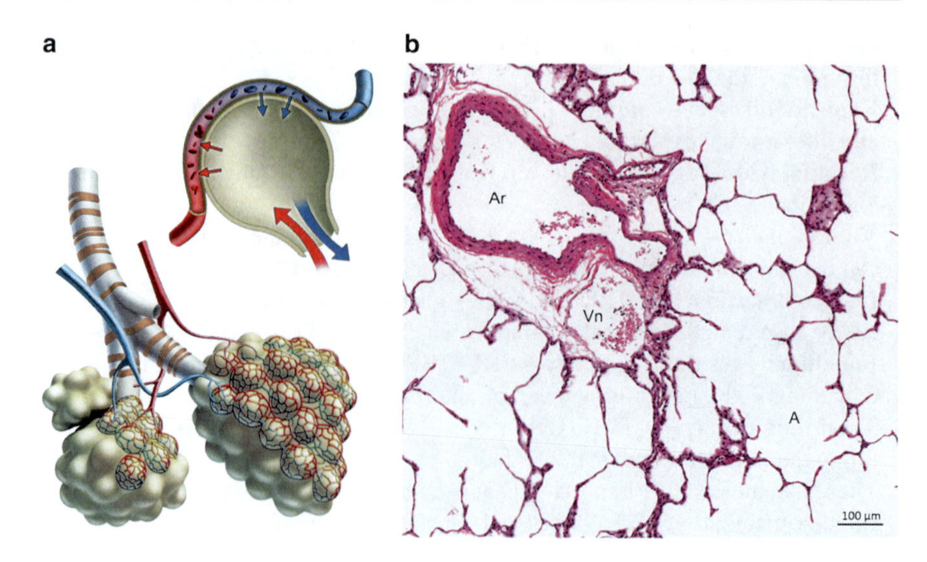

Fig. 11.3 The structure of lung alveoli and respiratory gas diffusion. (**a**) Schematic representation, representing oxygen (*red*) and carbon dioxide (*blue*) in the blood vessels or the air (© [Andrea Danti]—Fotolia.com); (**b**) histological section of lung parenchyma with air-filled spaces representing alveoli (A) and supplying blood vessels (*Ar—artery*; *Vn—vein*). Preparations: Institute of Anatomy, Histology and Embryology, Vetmeduni Vienna, Austria

Carbon dioxide has the opposite exchange direction. It is solved in blood and diffuses from the blood of arterial capillaries into the alveoli. Venous blood arriving from the periphery contains a higher concentration of carbon dioxide than the air inside the alveoli. Carbon dioxide diffuses from 45 mmHg in the blood to the lower 40 mmHg partial pressure in lung alveoli.

The respiratory gas exchanges are continuous and their intensity depends on the relative demand of the body, the supply of air, and the blood circulation.

11.5.3 Gas Transport

Respiratory gasses in the capillary blood are transported, oxygen to the tissues, while carbon dioxide from the tissues.

11.5.3.1 Oxygen Transport in the Capillaries

Almost all the *oxygen* (98 %) is transported in the capillary blood binding to hemoglobin in the red blood cells, while the other 3 % is dissolved in the plasma.

Hemoglobin is a globular protein with 4 polypeptide subunits. The stability of its structure is guaranteed by ionic and hydrogen bonds, hydrophilic interactions, and van der Waals forces. Each of the four subunits contains the **heme**, which provides the color to the red blood cells. Heme has a positively charged Fe ion (Fe^{2+}) complexed in the middle.

After the diffusion of oxygen into the blood, oxygen is dissolved in the blood and rapidly binds to the iron of hemoglobin, resulting oxyhemoglobin and transports oxygen to the tissue. Since one hemoglobin molecule contains four hemes, up to 4 Fe ions (Fe^{2+}) can attach to the 4 oxygen atoms. After the binding of the first oxygen atom, the structure of hemoglobin changes and enables the easier binding of the next oxygen. This binding is rapid and reversible. When hemoglobin binds up to three oxygens, hemoglobin is partially saturated, while binding all four oxygens is fully saturated. Hemoglobin saturation depends on the need of the body and influenced by several factors, like temperature, blood pH, partial pressure of oxygen, and carbon dioxide in the blood. Via blood transporting, hemoglobin loaded with oxygen reaches the different tissue, where oxygen is released and diffuses from the blood to the cells.

11.5.3.2 Carbon Dioxide Transport in the Capillaries

In human, blood from the tissues is rich in carbon dioxide which can be transported to the lung in three different ways. Most of the carbon dioxide (up to 70–80 %) diffuses into red blood cells where it reacts with H_2O resulting in instable **carbonic acid** (H_2CO_3). This reaction occurs slowly in the plasma, but the enzyme carbonic anhydrase catalyzes and speeds up the reaction. The produced **carbonic acid** then dissociates into **hydrogen ions** (H^+) and **bicarbonate ions** (HCO_{3-}). The resulted hydrogen ions are combined rapidly to **hemoglobin**, hindering the changes in the blood pH, while bicarbonate ions diffuse out from the red blood cells and enter the plasma. When blood passes the capillaries wrapping the lung alveoli, carbon dioxide diffuses from the blood with high carbon dioxide partial pressure to the lung alveoli.

Carbon dioxide (5–10 %) can be also transported via hemoglobin molecules, binding to its amino groups. Since the binding of the two respiratory gasses is independent from each other, both molecules can be transported at the same time.

Finally, 5–10 % of the whole blood carbon dioxide is dissolved in the blood plasma, which concentration is determined by the partial pressure of the gas. Higher P_{CO2} in the tissue results carbon dioxide in the blood.

In aquatic species using lung respiration (some reptiles, birds, mammals), there is no possibility to breathe during diving. Nevertheless, diving and hunting in the water represents a highly active state with increased oxygen demand and the necessity to eliminate carbon dioxide. Therefore oxygen has to be stored within the body and there are three possibilities to do so: (1) in the lungs (air), (2) in the blood (hemoglobin bound), or (3) in the musculature (bound to myoglobin). As in greater depth air is compressed, it is blood or muscle that is used for oxygen storage. In penguins it is mainly the breast muscles that are pumped with blood and here myoglobin is apparent in great amount and can even be recognized in necropsies by a grey to black color of these muscles. In addition the metabolism is reduced, while diving by lowering the action of the heart and circulation is what in return reduces lactate production. When surfacing the reverse is seen: heart activity and metabolism are increased to eliminate waste products as fast as possible.

11.6 Synopsis

The respiratory system is crucial for life as it provides the organism with the fuel for energy production and eliminates metabolic end products at the same time. Even though very similar in function, highly specialized organs evolved to meet the respective requirements of habitat and metabolic needs of species.

Acknowledgement The work was in part supported by the Austrian Science Fund FWF grant SFB F4606-B19.

Reference

Danovaro R, Dell'Anno A, Pusceddu A, Gambi C, Heiner I, Møbjerg Kristensen R (2010) The first metazoan living in permanently anoxic conditions. BMC Biol 8:30

Further Reading

Hickman CP Jr, Roberts LS, Keen SL, Eisenhour DJ, Larson A, L'Anson H (2011) Integrated principles of zoology. McGraw-Hill, New York

Hildebrand M (1974) Analysis of vertebrate structure. Wiley, New York

Kiran S (2011) Human anatomy. JP Medical Ltd, London

König HE, Liebich HG (2012) Anatomie der Haussäugetiere. Schattauer, Stuttgart

Nickel R, Schummer A, Seiferle E (2004) Lehrbuch der Anatomie der Haustiere—Band II Eingeweide. Parey, Stuttgart

Reece WO (2009) Functional anatomy and physiology of domestic animals. Wiley-Blackwell, Oxford

Reece JB, Taylor MR, Simon EJ, Dickey JL (2012) Campbell biology—concepts & connections. Pearson, San Francisco

Starck D (1982) Vergleichende Anatomie der Wirbeltiere auf evolutionsbiologischer Grundlage. Springer, Berlin

Treuting PM, Dintzis SM (eds) (2012) Comparative anatomy and histology—a mouse and human atlas. Academic Press (Elsevier), Waltham

Von Engelhardt W, Breves G (2010) Physiologie der Haustiere. Enke, Stuttgart

Propagation: Mammalian Reproduction 12

Christine Aurich and Isabella Ellinger

Contents

C. Aurich (✉)
Artificial Insemination and Embryo Transfer, Department for Small Animals and Horses, Vetmeduni Vienna, Vienna, Austria
e-mail: christine.aurich@vetmeduni.ac.at

I. Ellinger
Department of Pathophysiology and Allergy Research, Center for Pathophysiology, Infectiology and Immunology, Medical University Vienna, Vienna, Austria
e-mail: isabella.ellinger@meduniwien.ac.at

E. Jensen-Jarolim (ed.), *Comparative Medicine*,

DOI 10.1007/978-3-7091-1559-6_12,

Abstract
Reproduction is the process by which organisms create offspring. This process can be either sexual or asexual. The majority of the multicellular organisms such as plants and animals including mammalians generate new individuals by sexual reproduction. Mammalians have an obligatory sexual reproduction, while asexual reproduction remains an additional option for a variety of multicellular organisms (e.g., lizard). The major advantage of sexual reproduction is the formation of genetically novel individuals by combination of the genes from two individuals. This increases genetic diversity and produces new material on which natural selection can operate. Overall, sexual reproduction improves the chances of the population to face environmental challenges.

12.1 Introduction

Mammalian sexual reproduction involves two major steps (Fig. 12.1a):

1. The formation of haploid female (oocyte or egg) and male (spermatozoon or sperm) gametes by **meiosis**. The principle steps of meiotic cell division are illustrated in Fig. 12.1b.
2. The fusion of the gametes to produce a diploid, genetically novel cell (zygote) in the process termed **fertilization**. In placental mammalians, fertilization occurs within the female reproductive system.

In mammalians, the gametes are produced by distinct male and female individuals, have distinct male and female forms (spermatozoon and oocyte) to avoid self-fertilization, and are made in specific male and female organs, the gonads (testis and ovary). In sexually reproducing organisms, all cells have at least two copies of every gene (alleles), each derived from one parental individual. Genes are arranged on chromosomes, in humans 23 chromosomes ($n = 23$). Somatic cells have two copies of each chromosome per cell and are therefore diploid (humans: $n = 46$). The gametes, in contrast, have only one haploid set of chromosomes (humans: $n = 23$).

After fertilization, a new individual is formed from the zygote by cell division and differentiation. In mammalians, the growing fetuses are maintained and nourished within the female reproductive tract.

The zygote and all derived cells, however, will subsequently reproduce mitotically, always generating two genetically identical offspring cells. In a strictly controlled cell cycle (Fig. 12.2), these cells grow and duplicate their chromosomes during interphase, producing two identical chromatids (2C) per chromosome (n) joined at the centromere ($=2n$, 4C), and then divide mitotically (2 daughter cells, each $2n$, 2C).

The principle steps of **mitosis** are illustrated in Fig. 12.1b. The diploid somatic cells will either continue to divide or eventually enter a resting (G_0), nonreproducing state until new cells are required. This mitotic division is the fundamental process used not only during asexual reproduction of eukaryotes but also during embryonic development, growth of pluricellular beings, as well as tissue renewal.

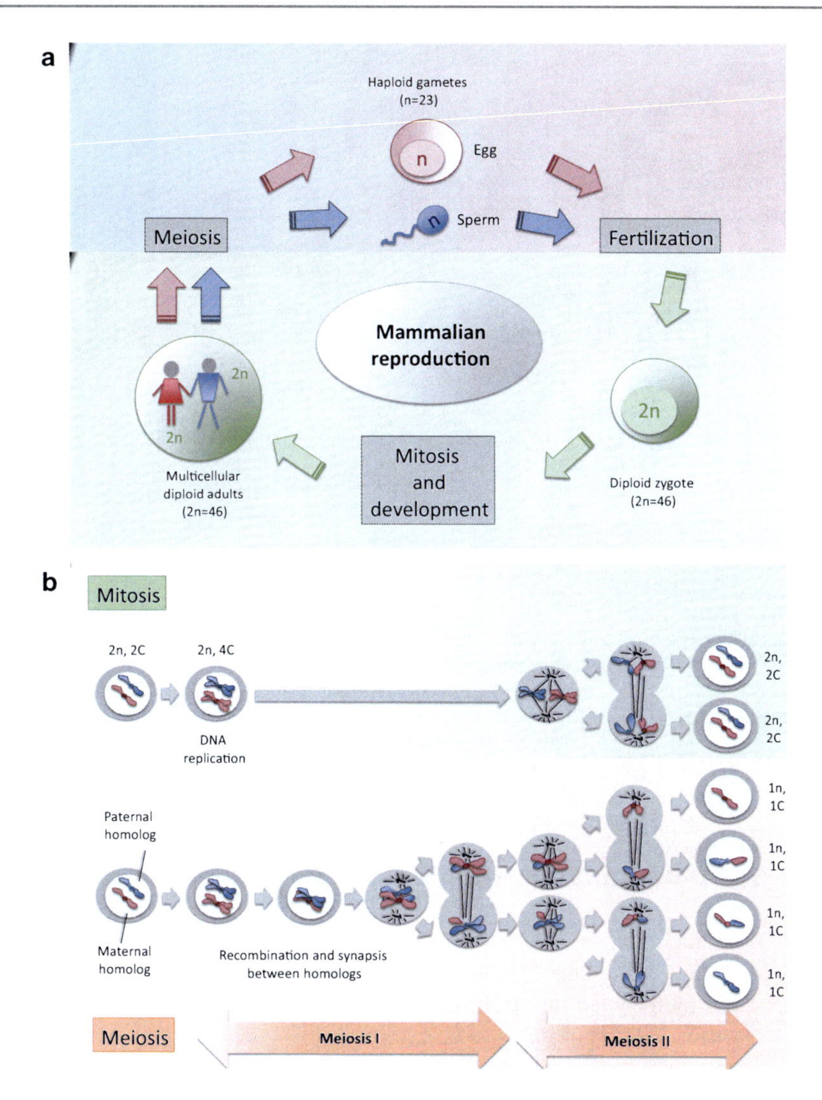

Fig. 12.1 (**a**) Two major steps of mammalian reproduction: formation of haploid male and female gametes and fertilization. Gametes are produced by meiotic cell division, while all other mammalian cells divide mitotically. (**b**) Comparison of mitotic and meiotic cell division. During meiosis, chromosome duplication is first followed by recombination and synapsis between the homolog chromosomes. Then, two events of chromosome separation occur. During meiosis I, the homologs separate, while in meiosis II the sister chromatids are separated. In mitosis, no pairing of homologs occurs and the sister chromatids are separated during a single division. *n* number of chromosome set, *C* number of chromatids per pair of chromosomes

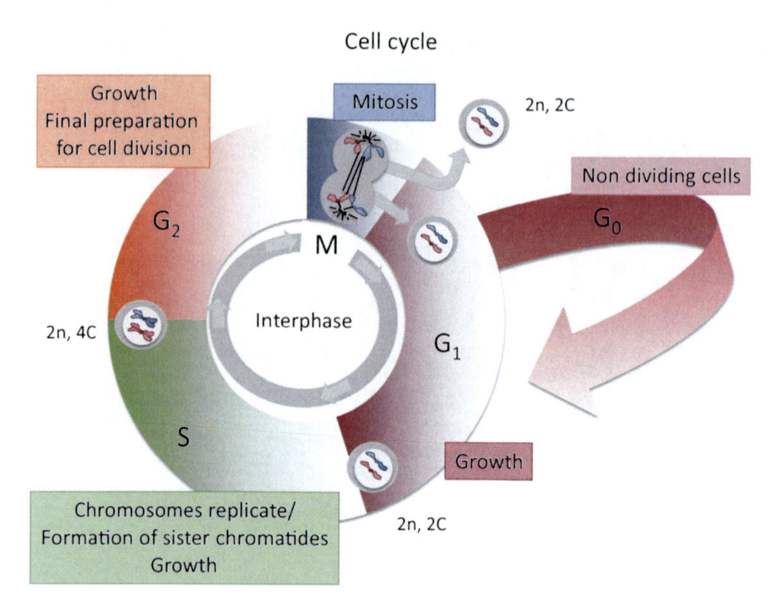

Fig. 12.2 Schematic illustration of a cell cycle, which consist of G1-, S (synthesis)-, G2-, and M (mitosis)-phase. After cell division, cells have two alleles per chromosome (2*n*, 2C). During interphase (G1-, S-, and G2-phase), cells grow and replicate their chromosomes, producing two identical chromatids (2C) per chromosome (*n*) joined at the centromere. During mitosis, the sister chromatids are separated and distributed to the daughter cells. If no mitotic signals from the environment are received, cells enter a nondividing Go state. *n* number of chromosome set, *C* number of chromatids per pair of chromosomes

12.2 Sex Determination and Gonad Development

Sexually reproducing mammals have two sexes and their development has a genetic basis. While, e.g., in humans, 22 chromosomes exist as identical alleles ($n = 22$, 2C) (one from mother and one from father), the sex chromosomes (female = X, male = Y) differ significantly in size and function of their genes. Sexual differentiation is usually induced by one main gene. In all mammalian species investigated so far, this is the gene encoding the **testis-determining factor** (**TDF**), which is located on the Y chromosome. Female-derived gametes contain only X chromosomes, while male-derived gametes can have either an X or a Y chromosome. Upon fusion of gametes, the combination of XX will determine a female individual, while the combination XY determines a male offspring.

In the early embryo, primordial (primitive) germ cells develop which are the gamete precursor cells. They originate from the inner lining of the yolk sac (extraembryonic tissue) and finally invade the undifferentiated gonads of the embryo, which are located in the genital ridge. They will eventually form male or female gonads depending on the sex chromosomes of the primitive germ cells.

In the gonads, the primitive germ cells stimulate the formation of primitive sex cords. In the male embryo, they produce TDF which stimulates development of the male reproductive system. In contrast, the absence of TDF results in development of a female reproductive system. Besides the gonads, in males this includes the epididymis and ductus deferens which originate from another embryonic organ, the mesonephric or **Wolffian ducts**. In contrast, in females, the oviducts, uterus, cervix, and part of the vagina originate from the paramesonephric or Müllerian ducts. In males, the anti-Müllerian hormone inhibits development of the paramesonephric ducts.

Also the primordial germ cells experience different fates in the male and female gonads. The male germ cells (spermatogonia) can go through mitotic cell cycles during the entire adult reproductive life, thereby constantly renewing the stem cell pool and enabling almost unlimited production of haploid gametes by meiosis. In contrast, all ovarian (female) germ cells cease their mitotic division either before birth (e.g., human, cow, sheep, mouse, goat) or shortly thereafter (e.g., rat, pig, cat, rabbit, hamster). These female germ cells enter their first meiotic division (see Fig. 12.1b) to become primary oocytes. Meiosis then stops in a prophase state of Meiosis I and the oocyte is arrested in a state called dictyotene for up to 50 years. Stromal cells surround the oocyte and condensate, resulting in formation of primordial follicles. Further development of the follicles will continue after entering puberty. However, most female germ cells die around the time of birth or until puberty, reducing the number of female gametes (e.g., from the original 7,000,000 by 75 % in humans). Overall, the female mammalians develop only a limited number of gametes, which cannot be replaced anymore from stem cells.

During puberty, the gonads mature and start to perform their two major functions, which are (1) production of haploid gametes and (2) production of sex steroid hormones. These sex hormones stimulate the tissues of the body to generate distinctive male and female somatic phenotypes. In addition, they affect behavior and physiology to ensure that mating will occur between different sexes at time of maximum fecundity. Finally, in females, they also prepare the body for pregnancy and lactation.

12.3 Reproductive Organs

Female and male reproductive organs (human: Fig. 12.3) are involved in producing, nourishing, and transporting either the oocyte or spermatozoon.

12.3.1 Female Reproductive Organs

The **female reproductive system** (human: Fig. 12.3a) is composed of the internal organs **ovaries**, **oviducts** (or Fallopian tubes), **uterus**, and **vagina**. The **vulva** is an external female reproductive structure, which enables the sperm to enter the body and which protects the internal genital organs from infectious organisms.

Additional tissues related to mammalian female reproduction are the mammary glands and—in placentals—the placenta. Under hormonal control (estrogen,

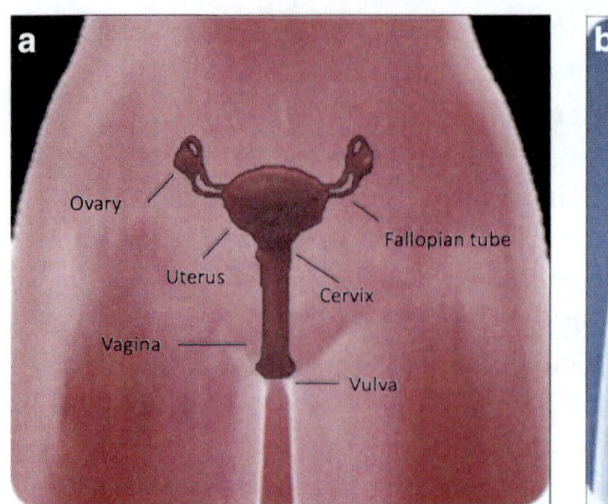

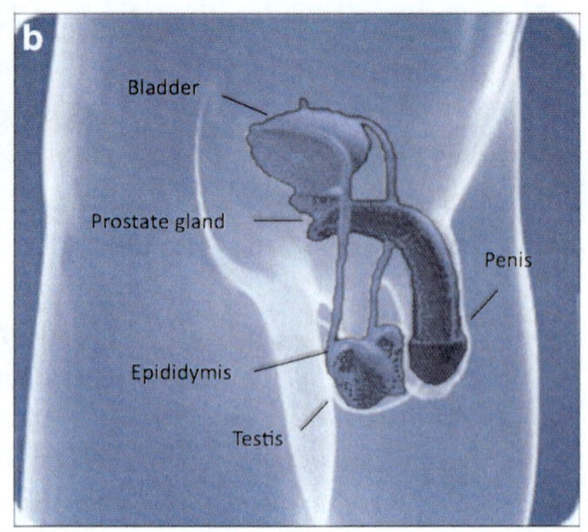

Fig. 12.3 Schematic illustration of the human (**a**) female and (**b**) male reproductive system

progesterone, prolactin, oxytocin), **mammary glands** produce and release milk to feed the offspring. Depending on the species, the location and number of mammary glands can vary. While in humans, primates, and elephants, the mammary glands are located in the thoracic region, they are located in the inguinal region in ruminants and horses or form a bilateral ridge covering parts of the thoracic as well as the whole inguinal region (rat, mouse, rabbit, dog, cat, pig). **Placentas** are organs which develop in female mammals during pregnancy, but exhibit large functional and structural variability among species. A plethora of fetal functions (gas exchange, nutrient transfer, immunologic functions, and many more) are partially or completely accomplished by the placenta during pregnancy as long as embryonic or fetal organs are still immature.

12.3.1.1 Ovaries

The **ovaries** are the female gonads, i.e., the primary reproductive organs (Fig. 12.4). Two ovaries develop in the absence of the TDF gene. They are small, oval-shaped glands that are located on either side of the uterus. Structurally, in most domestic animals, ovaries are divided into the vascular zone or **medulla**, which contains blood vessels, lymphatic vessels, and nerves in the connective tissue, and the parenchymal zone or **cortex**, which contains ovarian follicles of various sizes and at different stages of development as well as corpora lutea (Fig. 12.4a). In the horse, the anatomy of the ovary differs as the parenchymal zone is surrounded by the vascular zone (Fig. 12.4b). The **follicles** are spheric and differ in size depending on the species and their developmental stage. They consist of **one oocyte** that is **surrounded by the cumulus cells**. It is attached to the follicular wall that is formed of granulosa cells and synthesizes estrogens. After ovulation, granulosa cells develop into the corpus luteum that produces progesterone (Fig. 12.4c).

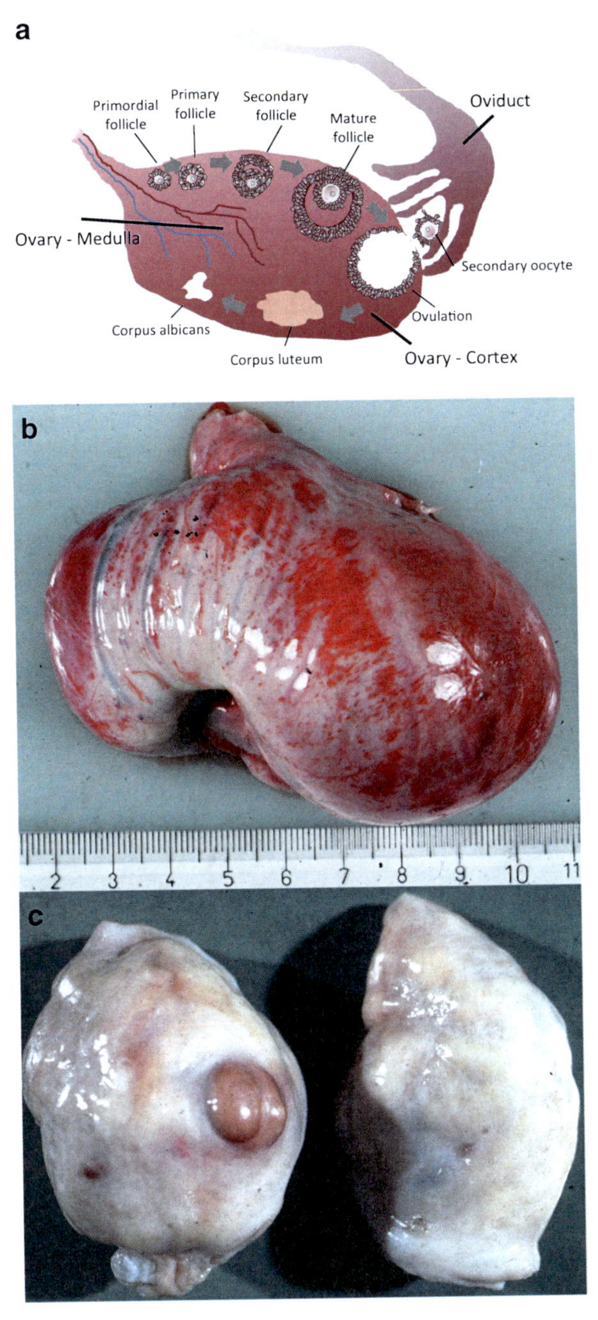

Fig. 12.4 Structure and comparison of the ovaries. (**a**) Illustration of a human ovary. Multiple follicles at different stages of development are located in the cortex. Upon ovulation, the released oocyte is captured by the fimbriae of the oviduct and migrates in the Fallopian tube towards the uterus, (**b**) ovaries from a mare, and (**c**) a cow (with corpus luteum on the left ovary)

Table 12.1 Characteristics of the estrous cycle in domestic animal species and the menstrual cycle in humans

Species	Cycle length (days)	Duration of estrus	Ovulation	Remarks
Human	24–32	–	Days 11–15	Menstrual cycle
Cattle	21	12–18 h	24–30 h after start of estrus	
Horse	22	4–9 Days	1–2 Days before end of estrus	
Sheep	17	24–36 h	24–30 h after start of estrus	
Pig	21	48–72 h	36–42 h after start of estrus	
Dog	31 weeks	6–8 Days	End of estrus (several days)	Endometrial bleeding during proestrus

12.3.1.2 Cyclic Ovarian Activity

From puberty onwards, ovaries produce the haploid oocytes and exhibit endocrine activity. The endocrine activity is coordinated with the gamete production. Though this holds also true for testicular function, there are significant differences between the functions of the two organs. First, only few mature oocytes are released during the reproductive years of a woman (about 400), and second, this release of oocytes (ovulation) occurs in cycles. The period before ovulation is characterized by **estrogen dominance**, while after ovulation **progesterone** is the dominant sex hormone. Due to this cyclic release of the steroids, the female body, and in most species also the behavior of the females, exhibits cyclicity. In most **animals**, this is called **estrous cycle**, while **higher primates** experience **menstrual cycles** (Table 12.1).

In most animal species with an estrous cycle, the females are only willing to mate around the time of ovulation and at this time, exhibit marked changes in their pattern of behavior. Day 1 of the estrous cycle is the first day where the female shows permissive behavior to a male. In canines, this is paralleled by bloody vaginal discharge. In contrast, higher primates show little evidence of estrous and the manifestation of ovarian cyclicity is the shedding of bloody endometrial tissue via the vagina (menstruation). Day 1 of the menstrual cycle is the first day of menstruation.

The reason of the changes occurring during female cyclicity is the dual function of the female reproductive tract. Under estrogen dominance, the body is prepared to receive the male gametes and enable fertilization, while under progesterone dominance, the body is prepared for implantation and nourishment of the conceptus.

12.3.1.3 Fallopian Tubes

The Fallopian tubes are narrow tubes which are attached to the upper part of the uterus and are used by the oocyte to travel from the ovaries to the uterus. Within the tubes, the oocyte will eventually be fertilized, becomes a zygote, and starts cell division (see below).

12.3.1.4 Anatomy and Function of the Uterus

During development of the female, part of the paramesonephric ducts develops into the uterus (womb). Depending on the mammal species, the paramesonephric ducts fuse to a different extent. In some species (e.g., rabbits), they will only fuse in their distal parts and form a common vagina, while in the proximal part, two separate uterine bodies exist, each having its own cervix that opens into the vagina (duplex uterus). In other species (e.g., ruminants, equids, and pigs), a so-called bicornuate uterus develops, which has two uterine horns and a small uterine body that has one cervix which opens into the vagina. The length of the uterine horns shows pronounced differences between species (Fig. 12.5). In humans, due to a high degree of fusion of the paramesonephric ducts, a uterus simplex is formed. The morphology of the cervix is species specific and highly variable. It is important for closure of the uterine lumen during pregnancy. Irrespective of the species, the function of the uterus is to enable implantation of the fertilized egg (zygote), house the growing conceptus, and expulse the fetus during delivery.

The uterus has a triple-layered wall, composed of the perimetrium (outer serosal layer), the myometrium (muscular layer), and the endometrium. The **endometrium** is a mucous membrane consisting of a basal layer (stratum basale, in humans 1 mm) and a functional layer (stratum functionale, in humans 5 mm). The structure, thickness, and state of the endometrium undergo marked changes during the estrous cycle and are most pronounced during the menstrual cycle (Fig. 12.6a). In case an oocyte has been fertilized in the Fallopian tubes, it travels to the uterus developing into a blastocyst, which implants into the endometrium. In primates, implantation transforms the endometrium into its pregnant state, the decidua. If implantation does not occur, the superficial part of the endometrium is shed during the (hemorrhagic) menstrual phase of the uterine cycle. This endometrial shedding does only occur in primates, but not in domestic animal species. In these, implantation occurs later than in humans, and—depending on the species—the contact between the conceptus and the endometrium is much less intense (see also placentation) and does therefore not require as profound structural changes of the endometrium as in humans.

12.3.1.5 Cyclic Reproductive Activity in Females

Human females enter cyclic reproductive periods with the first menstrual bleeding (menarche) at an average age of 13 years. The initiation of puberty is the activation of mechanisms in the hypothalamus, which result in delivery of the peptide gonadotropin-releasing hormone (GnRH) to the anterior pituitary. The pituitary responds by production of the glycoproteins luteinizing hormone (LH) and follicle-stimulating hormone (FSH) which in turn lead to production of estrogen and progesterone in the ovaries, thereby establishing the **hypothalamic-pituitary-gonadal axis** (Fig. 12.6b), which is characterized by positive and negative feedback

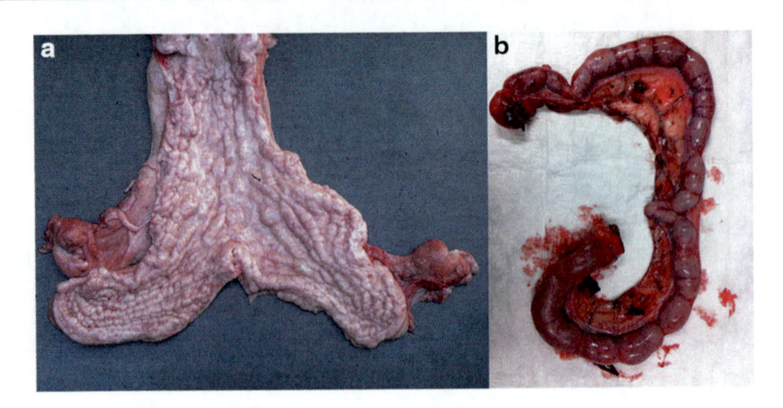

Fig. 12.5 Pathologic preparation of the uterus with the ovaries. (**a**) From a mare (the uterus opened, therefore endometrium visible) and (**b**) a uterus from a bitch. Note the different length of the uterine horns

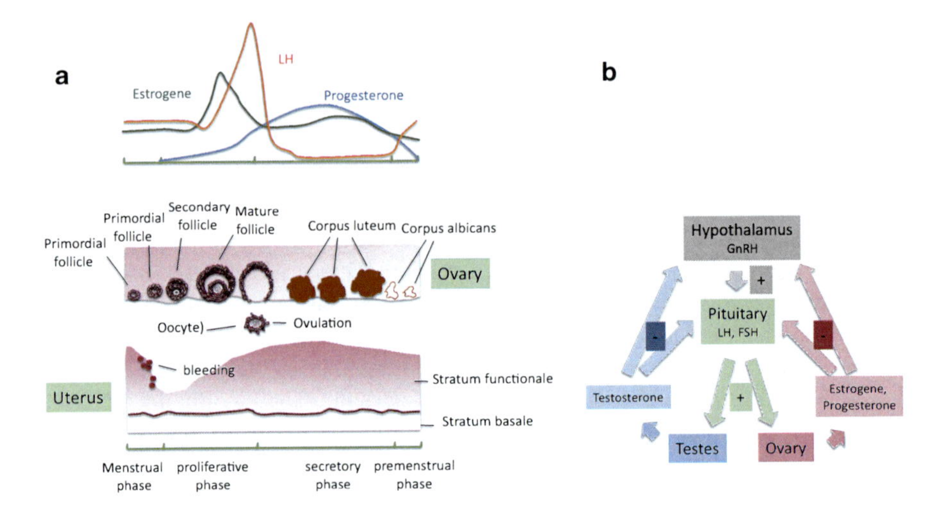

Fig. 12.6 The ovarian cycle is regulated by endocrine loops. (**a**) Illustration of serum hormone levels (*top*), ovarian folliculogenesis (*middle*), and alterations of the endometrium of the uterus (*bottom*) during the human menstrual cycle. *LH* luteinizing hormone; (**b**) schematic depiction of the hypothalamic-pituitary-gonadal axis. *GnRH* gonadotropin-releasing hormone, *LH* luteinizing hormone, *FSH* follicle-stimulating hormone

mechanisms but can also be modulated in various species by external factors such as environmental influences or proximity of reproductive partners.

The cessation of the menstrual cycles is termed menopause and occurs at an average age of 52 years in humans. It is caused by the declining number of ovarian follicles as well as their reduced responsiveness to FSH and LH. In most domestic animal species, females may enter a similar stage where ovarian activity ceases due to age, but the majority of individuals stay fertile until death. During each menstrual

Table 12.2 Involvement of steroid hormones in the regulation of female reproduction

	Estrogens (mainly estradiol-17ß)	Progesterone
Ovary	Follicle: oocyte maturation	Indirect effect: LH ⇓ → progesterone⇓ (negative feedback)
Endometrium	Edema	Secretory function (uterine milk⇑)
	Proliferation	
	White blood cells ⇑	
Cervix	Dilatation/edema	Fixation
	Mucus: viscosity ⇓	Mucus: viscosity ⇑
Myometrium	Enhanced contractions	Quiescence
Vagina	Edema	Mucus: viscosity ⇑
	Mucus: viscosity ⇓	
Behavior	Interest in males	No interest in males
	Searching for males	Dismissive behavior to males
	Acceptance of male contact and mating	

(human) or estrous (domestic animals) cycle, the ovaries, oviducts, and the uterus undergo structural and functional changes (Table 12.2).

The cycle is divided in three main phases: the **follicular phase**, **ovulation**, and **luteal phase** (Fig. 12.6a). During the follicular or proliferative phase, several follicles containing an oocyte start to grow in the ovaries ("follicular wave"). One dominant follicle is selected, while all others die. The dominant follicle secretes estrogen that induces proliferation of the endometrium in the uterus. The increasing estrogen level that normally represses LH and FSH release reaches a threshold value that induces LH release, which in turn stimulates the ovulation, i.e., the release of the oocyte from the ovary. The oocyte enters the Fallopian tube, where it should become fertilized with a sperm cell within the next 12–24 h. The cells of the Fallopian tubes are also influenced by the hormones estrogen and progesterone. They secrete mucus, which helps to nourish oocyte, spermatozoa, and zygote and promotes capacitation of spermatozoa. Once fertilized, the zygote undergoes embryogenesis and starts to divide within the Fallopian tube. It enters the uterus at the late morula or blastocyst stage (e.g., human, 70–100 cells). In the luteal or secretory phase, the remaining cells of the follicle, which has released the oocyte, transform into lutein cells and the collapsed follicle develops into the corpus luteum. This transformation is initiated by the high LH level around ovulation. The corpus luteum now produces significant amounts of the hormone progesterone, which prepares the endometrium for implantation. LH is required to maintain the corpus luteum. However, increasing progesterone concentrations inhibits LH synthesis and secretion by the pituitary gland. In humans, this finally results in atrophy of the corpus luteum (corpus albicans). The falling levels of progesterone then initiate the shedding of parts of the endometrium (menstruation) and the menstrual cycle starts again. If, however, the oocyte becomes fertilized, trophoblast cells derived from the zygote secrete the hormone human choriogonadotropin (hCG), which replaces LH and helps to maintain the corpus luteum as well as progesterone production.

While in humans luteal regression is passive and independent from the endometrium, in most domestic animals species, the luteal phase ends due to the secretion of another hormone, prostaglandin (PG) $F_{2\alpha}$, from the endometrium. $PGF_{2\alpha}$ is secreted into the uterine veins and—via a countercurrent system—travels to the ovaries where it initiates luteolysis (death of the corpus luteum) in the nonpregnant animal. Due to luteolysis, progesterone production ceases and thus the next follicular phase can start. In contrast, the presence of a conceptus in the uterine lumen (pregnancy) results in a mechanism called maternal recognition of pregnancy. This inhibits luteolysis and is thus responsible for maintenance of the corpus luteum.

12.3.2 Male Reproductive Organs

The **male reproductive system** (human: Fig. 12.3b) is composed of the male gonads (**testes**) producing spermatozoa and sex hormones, the **reproductive ducts** mediating sperm transport, and several accessory **reproductive glands**, which produce secretions that become part of the ejaculate. They are important for motility of spermatozoa and protect them against environmental challenges in the female reproductive tract.

12.3.2.1 Testes

The male gonads are located outside the body in the scrotum, a sac of skin. Spermatogenesis requires keeping testicular temperature 4–5 °C lower than the abdominal core temperature in most mammal species except e.g., elephants and dolphin species. Though the gonads develop in the upper lumbar region of the embryo, in adult individuals, the testes have descended from the abdominal cavity to the scrotum. Depending on the species, testicular descend occurs at different times during fetal, neonatal, or prepubertal life. Already during the fetal life, the testes secrete the hormones androgens and Müllerian-inhibiting hormone, which are the messengers of male sexual differentiation and induce the formation of the male genitalia. Structurally, each testis is divided into lobuli, each containing one or more seminiferous tubules, which are the functional units of the testis and end in the rete testis.

These seminiferous tubuli are the place of spermatogenesis, the development of the male gametes. They contain two types of cells: the **germinal cells** that give rise to the spermatozoa and **Sertoli cells**, which nourish and protect the developing sperm cells during the stages of spermatogenesis. These "mother cells" provide both secretory and structural support for spermatogenesis. With their tight junctions, Sertoli cells form the "blood-testes barrier." This allows for a specific milieu necessary for spermatogenesis. Between the tubules, the **Leydig cells** or interstitial cells are located, which produce and release the testicular hormones, mainly androgens (e.g., testosterone), but also estrogens and some other hormones. These hormones are involved in the regulation of male reproduction (Table 12.3).

Table 12.3 Involvement of steroid hormones in the regulation of male reproduction

	Androgens
Testis	Sertoli cell function
	Spermatogenesis
Epididymidis	Maturation of spermatozoa
Penis	Erectile function
Accessory sex glands	Production of seminal fluid
Behavior	Aggressive behavior (against males)
	Herding behavior/territorial behavior
	Mating behavior (precopulatory/copulatory/postcopulatory)

12.3.2.2 Male Reproductive Ducts

The reproductive ducts are composed of the epididymis, ductus deferens, ejaculatory duct, and urethra. The **epididymis** is a coiled tube (the uncoiled length in humans is 6 m) on the posterior surface of each testis. Within the epididymis, the sperms complete maturation and achieve fertilizing ability as well as motility. Epididymal maturation takes 10–12 days. In the cauda epididymidis, spermatozoa are stored until the next ejaculation occurs. Due to the acidic milieu, the sperms are immobile. Upon ejaculation, smooth muscles in the wall of the epididymis move the sperm to the ductus deferens. The **ductus deferens** extends from the epididymis in the scrotum into the abdominal cavity through the inguinal canal, ending near the prostate gland. In humans, it has a length of about 40 cm. Its function is the transport of sperms from their storage place to the urethra. The smooth muscle layer of the ductus deferens contracts in peristaltic waves during ejaculation. The **ejaculatory ducts** then receive the sperms as well as the secretions of the seminal vesicle and pass the semen into the urethra. The **urethra** is the last part of the urinary tract and is a passage both for urine and for the ejaculation of semen. It traverses the corpus spongiosum and its opening—the meatus—lies on the tip of the glans penis.

12.3.2.3 Male Accessory Sex Glands

The male accessory sex glands comprise the seminal vesicles, the prostate gland, and the bulbourethral glands. Species-specific differences exist with regard to their presence, size, and form. In all species known so far, the function of the accessory sex glands depends on the presence of androgens. In castrated males, they will degenerate. Together with the epididymis, the accessory sex glands contribute to the liquid, noncellular part of semen which is called seminal plasma. In humans, a pair of **seminal vesicles** is located posterior to the urinary bladder and is derived from the ductus deferens. They secrete fructose to provide an energy source for sperm as well as semenogelin, which is a major secretory protein that supports a gel-like consistence of the ejaculate (70 % of ejaculate is provided by seminal plasma) which is specific to humans. The duct of each seminal vesicle joins the ductus deferens on that side to form the ejaculatory duct. The **prostate gland** is a muscular gland that surrounds the urethra. The smooth muscle of the prostate gland contracts during ejaculation and thereby contributes to the expulsion of semen from the urethra. In humans, the secretions of the prostate gland contribute about 30 % to the ejaculate. Two major proteins derived from the prostate gland are PSA (prostate-specific antigen) and prostate-specific acid phosphatase, which are used to screen men for prostate cancer. The **bulbourethral glands** are located below the prostate gland and likewise empty into the urethra. The alkalinity of the seminal fluid helps to neutralize the acidic vaginal pH and permits sperm mobility in an otherwise unfavorable environment.

12.3.2.4 Penis

The size and form of the penis are again variable with regard to species. The penis consists of three major parts: the base of the penis, the shaft (main portion), and the glans penis. The morphology of the glans penis shows pronounced differences between species. While some species (e.g., ruminants and pigs) have a **fibroelastic** penis with a sigmoid flexure that allows elongation during sexual arousal, the penile shaft in stallions, dogs, and men has large corporal sinusoids that fill with blood during sexual stimulation (**corpus cavernosum and corpus spongiosum**). The unerected glans penis is covered with the foreskin (**prepuce**) that again shows a great variety among species.

An erection, which is required to place the ejaculate in the female reproductive tract, depends on changes in blood flow within the penis. Due to sexual arousal, specific nerves cause penis blood vessels to expand. More blood flows in and less flows out of the penis, hardening the tissue in the corpus cavernosum and spongiosum. In bulls, boars, rams, stallions and camelids, erection requires also relaxation of the retractor penis muscles. In dogs, the penis contains an os penis.

12.3.3 Reproductive Behavior

Reproductive behavior is an obligatory component of reproduction. It is driven by the reproductive steroid hormones, testosterone in male and estrogen and progesterone in female. In postpuerperal female domestic animals, sexual receptivity occurs only during estrus, i.e., in the absence of progesterone. It is characterized by **attractivity**, **proceptivity**, and **receptivity**. Besides, physical activity increases due to increased locomotion with peak activity occurring at the time of ovulation. This is also true for humans. Attractivity serves to attract males. This may include vocalization, specific postures, and the secretion of chemical cues. Proceptivity is characterized by behavior exhibited towards males that stimulates copulation. This may also be seen between females if they are in estrus. Finally, receptivity is a specific female behavior aimed to ensure insemination. It includes immobility, deviation of the tail, and backing towards the male. During the luteal phase of the cycle as well as in pregnant animals, progesterone influence diminishes estrus behavior in females of most domestic animal species. However, the female horse may accept male interaction and copulation also during pregnancy.

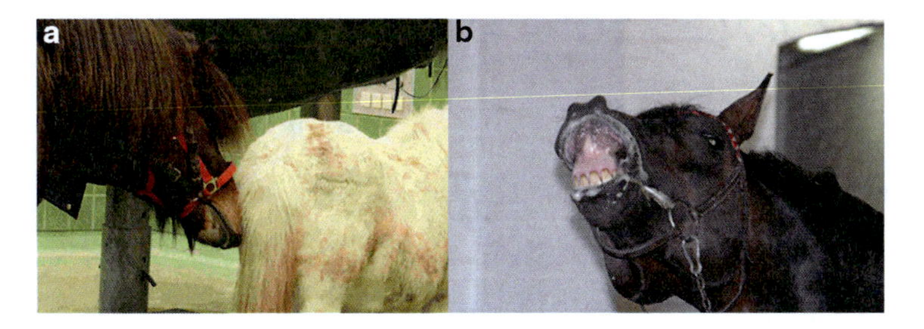

Fig. 12.7 Typical precopulatory behavior in stallions: (**a**) sniffing of the vulva and (**b**) flehmen

In contrast to the female, the male is capable of initiating reproductive behavior at any time after puberty. Three distinct stages of reproductive behavior are seen: **precopulatory**, **copulatory**, and **postcopulatory behavior**. Precopulatory behavior includes search for females, courtship (e.g., sniffing of the female vulva, flehmen, chin resting, increased vocalization, Fig. 12.7), sexual arousal, erection, and penile protrusion. Copulatory behavior includes mounting of the female, intromission, and ejaculation. Its duration varies significantly among species (from a few seconds in ruminants to approximately 20 min, e.g., in dogs or camelids). By the use of specific devices, e.g., artificial vaginas, semen can be collected during the copulatory phase with the male mounting either an estrous female or a dummy. During the postcopulatory stage, the male dismounts. This stage is further characterized by refractivity, i.e., a certain period of time during which a second copulation will not take place.

In males as well as in females, sexual experience may influence appropriate reproductive behavior. Negative experience will decrease enthusiasm. Therefore, in domestic animal species, management of breeding males should always aim to provide the male with positive stimuli.

12.4 Formation of Gametes

12.4.1 Oogenesis: The Formation of the Female Gametes

Oogenesis is the formation and maturation of the oocyte (Fig. 12.8). As detailed above, these process starts during the embryonic phase and is divided in **a proliferative period** (mitosis, this ends before birth) and **a maturation period** (meiosis). Already during the fetal development, oocytes become part of follicles (primordial follicles), in which they are accompanied by granulosa cells (epithelium of follicle). Follicles of later stages (secondary and tertiary follicles) are, additionally, surrounded by stromal cells (theca cells).

During the proliferative period, primordial follicles are generated. First, the stem cells, called oogonia, proliferate. Proliferation stops and the cells enter their first meiotic division (see Fig. 12.1b) to become primary oocytes. Meiosis is then halted

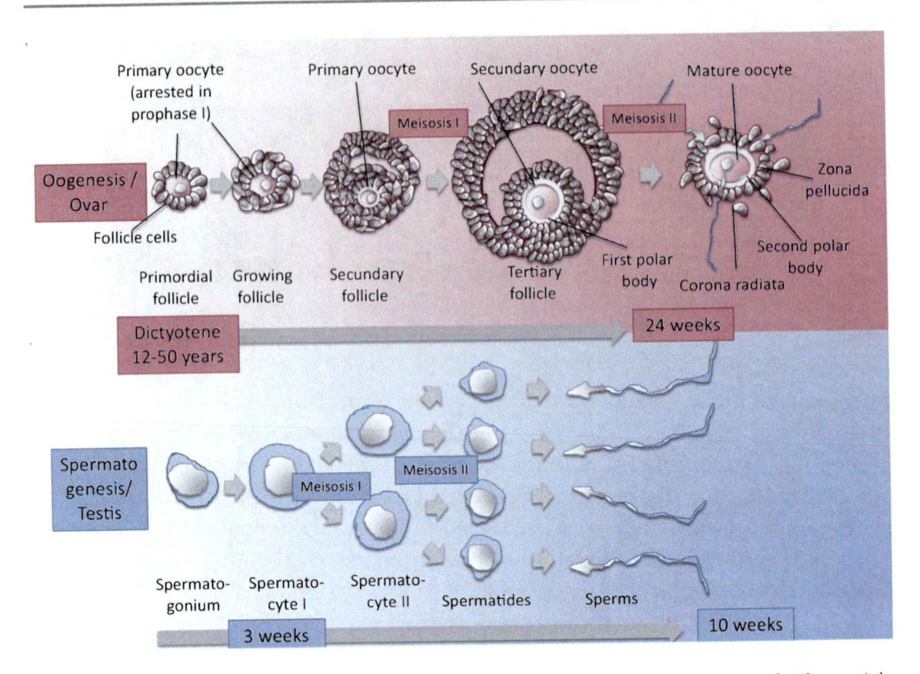

Fig. 12.8 Major steps during oogenesis/folliculogenesis (*top*) and **spermatogenesis** (*bottom*) in the female and male gonads

in a prophase state of meiosis I and the oocyte is arrested in the dictyotene until it is recruited for further growing. This period can last for up to 50 years. From the original 7 million primordial follicles in a 5-month fetus, one million are available around birth, 300 000/ovary at entry into puberty, and only 300–400 oocytes ovulate during the reproductive period.

The maturation period starts with puberty. During each follicular phase, 20–30 follicles are selected and grow in an FSH-dependent way to become tertiary follicles. During this period, the oocyte as well as the granulosa cells grow and theca cells (stromal cells) attach to the follicle. Even before puberty, follicles can grow and become tertiary follicles. However, these follicles all die by follicular atresia. During each cycle, only one tertiary follicle matures. Rarely, two or more follicles may develop and be fertilized, thus resulting in the conception of twins, triplets, etc. The size of this oocyte is <200 µm, and the size of the mature follicle in humans is <25 mm (e.g., cow 10 mm, mare 40 mm). Shortly before ovulation, meiosis I is terminated. The follicle antrum contains liquor follicularis. The oocyte is covered by glycoproteins (zona pellucida), is surrounded by several layers of granulosa cells (corona radiata) and theca cells, and remains attached to one side (cumulus oophorus) of the follicle cells. Due to LH secretion, the secondary oocyte is released during ovulation. It completes meiosis II only upon fertilization.

Table 12.4 Duration of spermatogenesis in different mammalian species (development from spermatogonia to spermatozoa)

Species	Duration of spermatogenesis (days)	Daily sperm production	Length of spermatozoon (μm)
Human	64	0.2×10^9	55
Cattle	61	6×10^9	75
Horse	57	$5\text{–}10 \times 10^9$	60
Sheep	47	10×10^9	65
Pig	39	16×10^9	55

12.4.2 Spermatogenesis: The Formation of the Male Gametes

Spermatogenesis (Fig. 12.8) takes place within the seminiferous tubules. The process can be subdivided into three phases. The first is called the **proliferative phase** (mitotic divisions of spermatogonia). During this phase, also stem cells are renewed that allow the continuous production of spermatogonia during the whole life of the male. The second phase is termed **meiotic phase** where primary and secondary spermatocytes are produced by the first and second meiotic division. DNA replication guarantees genetic diversity. At the end of the meiotic phase, haploid spermatids are formed which are still spherical and have to acquire their typical highly specialized morphology during the **differentiation phase**. This morphology—independent from species—includes a head with nuclear material, a flagellum including a midpiece with a number of mitochondria for energy supply, as well as a principal piece. As already described, spermatogenesis occurs in direct contact to the Sertoli cells. At the end of differentiation, spermatozoa are released into the lumen of the seminiferous tubules. This is a continuing process. Duration of spermatogenesis and daily sperm production are species specific (Table 12.4).

12.5 Fertilization and Development of the Embryo

Fertilization requires a complex series of events outside and inside the female reproductive tract.

From the male seminiferous tubules to the female oviduct, the **spermatozoa** overcome a distance, which is about 100,000 times their own length. Only one sperm in a million will complete this journey, and for fertilizing the oocyte, the spermatozoa must undergo **maturation in the male tract** as well as **capacitation** and the **acrosome reaction in the female tract**.

Spermatozoa require a fluid vehicle for an optimal transport to and within the female reproductive tract. This is the seminal plasma provided by the male accessory sex glands. The volume of seminal plasma can range from 3 mL in humans to up to 0.5 L in boars, but independent of species, it contains nutritional factors such as fructose and reducing agents to meet oxidative stress and provides buffering capacity to meet the acid pH of vaginal fluids. **Maturation** of spermatozoa occurs in the epididymis and involves coating of membrane with glycoproteins, several structural changes (e.g., nuclear condensation), metabolic alterations (e.g., use of fructose for energy production), and the acquisition of the potential to swim.

The semen (spermatozoa + seminal fluid) needs to be deposited in the female reproduction tract at coitus. Depending on the species, semen is either ejaculated within the vagina (e.g., humans, sheep, cows) or directly deposited in either the cervix or the uterus (e.g., mouse, rat, pig, dog, horse). In humans, only 1 % of the deposited spermatozoa enter the cervix. In this species, the capability of the sperm to penetrate the cervical mucus depends on the absence of progesterone and is usually highest around the time of ovulation. The lifetime of male gametes within the female genital tract is species specific but in most species averages 28–48 h. Spermatozoa reach the oviducts, the place of fertilization, most likely by their own propulsion. Secretions of the uterus as well as the oviduct, which contain proteolytic enzymes and high ionic concentrations, induce capacitation of the sperm. **Capacitation** has mainly two effects. The spermatozoa enter a hyperactivated motility state, and, furthermore, the plasma membrane properties change, making the sperm sensitive to signals encountered in the close proximity of the oocyte. These signals, when received, induce the acrosome reaction. However, capacitation leads to a metastable state and results in death of the sperm unless the oocyte is met rapidly.

Meanwhile, the oocyte and its surrounding cells have ovulated from the ovary (Fig. 12.9a, step 1). The fimbriae of the oviduct pick up the oocyte, and the beating cilia of oviductal cells transport the oocyte towards a certain region of the oviduct (the ampulla), where fertilization is supposed to occur. In most species, the ovulated oocyte may survive 6–24 h without fertilization.

Once capacitated spermatozoa and oocyte meet (Fig. 12.9a, step 2), the sperms bind to molecules in the zona pellucida. The molecular interaction is species specific and prevents cross-species fertilization. The binding induces the **acrosome reaction**. The acrosome is a modified lysosome-like structure of the spermatozoa containing digestive enzymes. Due to the acrosome reaction, these enzymes are released and digest a way through the zona pellucida. The sperm follows, employing hyperactivated tail movements.

Once the sperm has passed through the zona, the plasma membranes of the sperm and oocyte bind to each other and fuse. The nucleus and—depending on the species—parts of the tail are internalized by the oocyte, which thereby is transformed into a **zygote**.

In the next steps, the zygote has to ensure **diploidy**. For that purpose, calcium is released from intracellular stores of the oocyte following fusion of the two gametes. A pulsatile 5–10-fold raise of free intracellular calcium levels lasting for several hours is observed. The elevated calcium levels causes a modification of the zona pellucida by enzymes released from oocyte granules. This "zona reaction" prevents further binding and entry of other spermatozoa. In addition, high calcium levels trigger the oocyte's completion of the second meiosis, by destabilizing protein complexes responsible for the meiotic arrest. Having finished the second meiotic division, the second polar body is expulsed. The formation of a euploid zygote

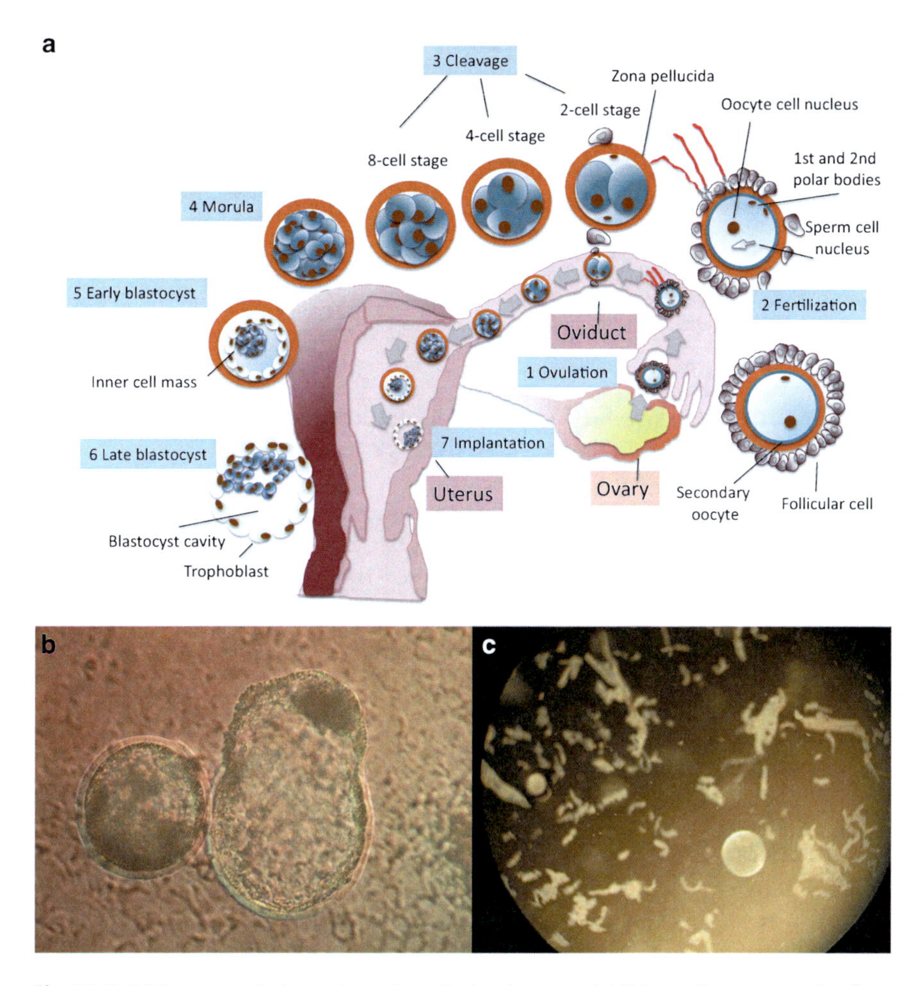

Fig. 12.9 Major events during early embryonic development. (**a**) Schematic representation from ovulation to implantation; (**b–c**) blastocysts derived from in vitro culture: (**b**) bovine blastocyst is hatching from its zona pellucida and (**c**) equine blastocyst collected on day 7 after ovulation from an embryo donor mare

based on fusion of one maternal and one paternal set of haploid chromosomes is essential for subsequent normal development of the conceptus. Aneuploidy causes embryonic abnormality but can also result in development of tumors (hydatidiform mole, choriocarcinoma), which is life-threatening for the mother.

Next, both a female and a male pronucleus form and the chromosomes in these pronuclei duplicate their DNA. The pronuclei approximate, and, after the breakdown of the pronuclear membranes, the gametic chromosomes come together in a process called syngamy. The first mitotic division follows, and thereby, the **one-cell zygote** has become a **two-cell conceptus**.

Table 12.5 Pregnancy outcome and fetal compartments in humans and domestic animal species

Species	Number of fetuses	Duration of pregnancy (days)	Allantoic fluid (Volume per fetus in mL)	Amniotic fluid (Volume per fetus in mL)
Human	1 (2–3)	270–290	No data	400
Cattle	1 (2)	280	9,500	3,500
Horse	1	340 (320–360)	4,000–10,000	3,000–7,000
Sheep	1–3	150	500–1,500	500–1,200
Pig	6–20	114	No data	40–150
Dog	3–8 (or more)	63	10–50	8–30

12.6 From the Two-Cell Conceptus to Implantation and the Establishment of the Placenta

The conceptus remains in the oviduct for several days, and this time length is characteristic for each species (Table 12.5). A four-cell state in humans is reached after about 2 days. The conceptus is transported by the cilia of the oviduct and enters the uterus after about 3.5 days. In some species, the transport, e.g., oviductal transport, depends on hormone secretion of the conceptus.

During this time, the cells of the conceptus divide in a process termed "cleavage divisions," where the size of the individual cells reduces progressively (Fig. 12.9a, step 3). As a result, the total size of the conceptus remains the same. These early cell divisions are still dependent on the molecules available in the cytoplasm of the maternal oocyte. Any deficiency that has occurred in oocyte maturation will consequently result in a failure of early development and can result in failure of pregnancy. Between the 4- and 8-cell stage in humans, however, a marked increase in transcription of the genes of the conceptus occurs which is accompanied by an increase in biosynthetic capacity. Further development is under the genetic control of the conceptus.

The formation of the **morula** is the next important step (Fig. 12.9a, step 4). In most species studied, it happens around the 8- to 16-cell stage. At this stage, the cells transform their phenotypes from symmetric to epithelioid and the offspring cells then develop into different cell types. At the 32- to 64-cell stage, the **blastocyst** is formed in most species (Fig. 12.9a, steps 5 and 6), where two distinct cell types are clearly visible. The **trophoblast cells** surround a blastocoelic cavity filled with fluid. Trophoblast cells are the first extraembryonic tissue and will help to establish the contact zone between the mother and fetus. Additionally, there is an inner group of pluripotent cells. Derivatives of these **pluriblasts** will be hypoblasts, amniotic ectoderm, and extraembryonic mesoderm, which contribute to extraembryonic tissues required to establish materno-fetal exchange as well as embryonic endoderm, mesoderm, and ectoderm that will form the embryo. The separation of the extraembryonic tissues (amnion, chorion, yolk sac, allantois) from those that will give rise to the embryo occurs mainly during the **embryogenic phase**. The subsequent time during which the basic body plan is laid down and the various embryonic cell types differentiate is called **embryonic phase**. The last phase of pregnancy is the **fetal phase**. The length of these phases is species specific, being, e. g., very similar in humans (14/36/220 days) and cows (16/25/240) and different in mice (6/5/9) or opossum (7/4.5/1.5).

The conceptus enters the uterus during transition from morula to blastocyst (Fig. 12.9a, steps 4 and 5) and aims to implant at a site that is characteristic for each species. The implantation site is especially important in polytocous species (several conceptuses, bicornuated uteri) to reduce the competition between conceptuses. Immediately before implantation, the zona pellucida is lost, or—in some species, e.g., the bovine—the blastocyst hatches actively from the zona (Fig. 12.9b, c). Implantation is required to guaranty sufficient nutrition and oxygen supply for all cells in the growing conceptus. While **histiotrophic nutrition** (e.g., by oviduct and uterine secretions or destroyed cells) is sufficient to nourish embryonic cells until towards the end of the first trimester in humans, the embryo has to develop its own blood vascular system that distributes metabolites to embryonic tissues and exchanges them via extraembryonic surfaces with the maternal blood circulation. Most importantly, an intimate contact zone between maternal and fetal blood circulation has to be established, which is called the **hemotrophic placenta**.

The process of implantation of the conceptus is species dependent, but always starts with attachment to the surface of the uterus. **Implantation** (Fig. 12.9a, step 7), which follows, either can be **invasive** (humans, most primates, dogs, cats, mice, rabbits), meaning that the conceptus will break through the surface epithelium of the maternal uterus and invades the underlying stroma or can be **noninvasive** (pig, sheep, cow, horse), integrating the uterine epithelium in the placenta. Noninvasive conceptuses usually attach later than invasive conceptuses and grow to a much greater preimplantation size than invasive conceptuses (free-living phase in sheep, 15 days; pig, 18 days; cow and horse, 30–40 days). Here, it is mainly the extraembryonic tissue that grows to establish a large surface area for exchange of metabolites upon attachment. Often, attachment occurs at multiple sites resulting in diffuse (mare, pig) or cotyledonary (cow, sheep) types of placentas. In contrast, invasive conceptuses have a short free-living phase (mouse and rat, 4.5 days; rabbit and human, 7–8 days). Their trophoblast cells contact the maternal epithelium and this leads to a stromal transformation known as the **decidualization reaction** (endometrium becomes decidua). Cell morphology and extracellular matrix components in the endometrium change, accompanied by an increased number of capillaries. During the subsequent invasive phase, uterine tissue is partly destroyed by the trophoblasts and used for histiotrophic support. The depths of invasion as well as the degree of proximity between maternal and fetal circulation can vary largely among species. In humans and guinea pigs, the conceptus invades the stroma so deeply that the uterine surface epithelium is restored over it (interstitial implantation). Other species (dog, cat, rat) may invade the stroma only partially and project into the uterine lumen (eccentric implantation). This can result in a contact to other sites of the uterine lumen and additional placental development (e.g., bidiscoid placenta in rhesus monkeys, zonary placenta in dogs and cats). While in some species the maternal tissue remains relatively intact (dog, cat), meaning that

Table 12.6 Progestin sources during gestation in humans and domestic animal species

Species	Gestation length (days)	Corpus luteum	Placenta	Other sources
Human	270–290	+ (Activity ⇓)	+ (>Day 56)	
Cattle	280	+	+ (>Days 160 to <240)	
Horse	340	+ (<150)	+ (>60)	+ Corpora lutea accessoria (>days 37 to <120)
Goat	150	+		
Sheep	150	+ (Activity ⇓)	+ (>day 70)	
Pig	114	+		
Dog	63	+		

trophoblast cells contact maternal capillary endothelium, in other species, trophoblast cells also invade the maternal endothelium and ultimately bathe in maternal blood (human, rabbit, rat, mouse). Further differences exist among these species with respect to the numbers of trophoblast layers (humans have one, rabbits have two, rat and mouse have three trophoblast layers separating their blood circulation and maternal blood). Interestingly, the efficiency of placental transfer cannot directly be related to the numbers of tissue layers that separate the fetal and maternal blood circulation. Most likely, it is related to the type of transport mechanisms employed within the cell layers that may vary among species.

In summary, in humans only trophoblast cells (syncytiotrophoblasts) and fetal endothelial cells finally separate maternal and fetal blood circulation and mediate the exchange of nutrients, gases, and many other substances required to ensure proper growth and development of the fetus. Many of these mechanisms are not fully characterized yet. However, among all other tissues, the placenta exhibits the largest degree of species specificity, which renders it difficult to compare the metabolic exchange processes among species or use other species as model systems to analyze human transplacental transport processes.

The duration of pregnancy is species specific (Table 12.5). Pregnancy and maintenance of the fetus require a specific endocrine milieu. In all mammalian species studied so far, the presence of progestins, mainly progesterone, is necessary. Depending on the species, progestins are produced either by the corpus luteum that has developed after ovulation of the oocyte or from the placenta (Table 12.6). Besides progestins, the placenta produces also a variety of other hormones which are necessary for stimulation of ovarian function, maintenance of pregnancy, modulation of fetal growth, stimulation of mammary function and—at the end of pregnancy—regulation of **parturition**. In humans, the placenta very early after ovulation produces human chorionic gonadotropin which is responsible for maintenance of the corpus luteum and thus progesterone production. Similarly, in the horse, the chorion produces equine chorionic gonadotropin (eCG) between days 37 and 120 of pregnancy. This hormone will stimulate the formation of accessory corpora lutea on the ovaries and thus contribute additional sources for progestin production. While in some species (e.g. goats, pigs, dogs) progesterone production

Table 12.7 Stages and duration of parturition in humans and some domestic animal species

Species	Stage I (= myometrial contractions + cervical dilatation)	Stage II (= expulsion of fetus)	Stage III (= expulsion of fetal membranes)
Human	7–13 h	Up to 60 min	Less than 15 min
Cattle	2–6 h	30–60 min	6–12 h
Horse	1–4 h	5–20 min	1 h
Sheep	2–6 h	30–120 min	5–8 h
Pig	2–12 h	150–180 min	1–4 h
Dog	6–12 h	6 h	Simultaneously with neonates

is restricted to the corpus luteum in **monoparous** (number of fetuses = 1, see Table 12.5) or the corpora lutea in **multiparous** (number of fetuses >1) species, e.g., in humans, horses, and sheep, the placenta will take over progestin production from a certain stage of pregnancy (Table 12.6). Among the different progestin functions, regulation of fetal nourishment and maintenance of myometrial quiescence ("progesterone block") are of utmost importance.

At the end of gestation, luteal or placental progestin production ceases due to signals from the fetus and thus parturition becomes possible. **Parturition** is a complex cascade of events, which is not understood in detail. However, it is well accepted that the fetus triggers this cascade. Maturation of the fetal hypothalamic-pituitary-adrenal axis occurs at the end of pregnancy. Due to limitations in space or nourishment, the fetus starts to increase adrenal cortisol production. Cortisol does not only stimulate tissue maturation but also causes dramatic endocrine changes on the maternal side. These include removal of the progesterone block, a rise in estradiol, and changes in the reproductive tract with special regard to the cervix. Removal of the progesterone block also initiates secretion of the hormone $PGF_{2\alpha}$, which causes luteolysis in species where pregnancy is maintained via luteal progesterone. Further on $PGF_{2\alpha}$ together with oxytocin, which is released from the pituitary, is responsible for the stimulation of **myometrial contractions** (stage I of parturition) which finally result in **expulsion of the fetus** (stage II) and the **fetal membranes** (stage III). The duration of these phases is species specific (Table 12.7).

Further Reading

Johnson MH (2012) Essential reproduction, 7th edn. Wiley-Blackwell, Chichester, West Sussex

Senger PL (2005) Pathways to pregnancy and parturition, 2nd edn. Current Conceptions, Moscow, ID

Schillo KK (2008) Reproductive physiology of mammals: from farm to field and beyond, 1st edn. Delmar Learning, Clifton Park, NY

Benirschke K, Burton GJ, Baergen RN (2012) Pathology of the human placenta, 7th edn. Springer, Berlin

Noakes DE, Parkinson TJ, England GCW (2009) Veterinary reproduction and obstetrics, 9th edn. Saunders, Philadelphia

Common Concepts of Immune Defense 13

Franziska Roth-Walter, Angelika B. Riemer, Erika Jensen-Jarolim, and Hannes Stockinger

Contents

F. Roth-Walter • E. Jensen-Jarolim (✉)
Comparative Medicine, Messerli Research Institute, University of Veterinary Medicine Vienna, Medical University Vienna and University Vienna, Vienna, Austria

Institute of Pathophysiology and Allergy Research, Center of Pathophysiology, Infectiology and Immunology, Medical University Vienna, Vienna, Austria
e-mail: franziska.roth-walter@vetmeduni.ac.at; erika.jensen-jarolim@meduniwien.ac.at

A.B. Riemer
Immunotherapy and -prevention, German Cancer Research Center (DKFZ), Heidelberg, Germany
e-mail: a.riemer@dkfz.de

H. Stockinger (✉)
Institute of Hygiene and Applied Immunology, Center of Pathophysiology, Infectiology and Immunology, Medical University Vienna
e-mail: hannes.stockinger@meduniwien.ac.at

E. Jensen-Jarolim (ed.), *Comparative Medicine*,
DOI 10.1007/978-3-7091-1559-6_13,

13.1 Innate Immunity, the Oldest Type of Defense

Franziska Roth-Walter, Angelika B. Riemer, Erika Jensen-Jarolim

Adapted with permission from "Natürliche Abwehr" in: Krankheit, Krankheitsursachen und -bilder, MCW Block8 (Ed. Marian B) 2012, 6th Ed., of Facultas.wuv Universitätsverlag & Maudrich

13.1.1 Abstract

The immune system has the challenging tasks to discriminate self from non-self, but also to discern harmless from harmful foreign antigens or entities and to attack and eliminate foreign threats.

In the classical view the immune system can be divided into an innate and adaptive branch, where the innate immune system represents a quick first-line defense against pathogens, whereas the adaptive immune system is slower, but more diverse and sophisticated, able to memorize pathogens, and confer long-lasting immunity to the host. The innate immune system relies on recognition of evolutionarily conserved pathogen-associated molecular patterns (PAMPs) by innate pattern recognition molecules and receptors (PRMs and PRRs), whereas the adaptive immune system is principally trained to recognize foreign molecules and to memorize them by highly adapted, specific receptor molecules.

13.1.2 Introduction

The previous chapters focused on comparative anatomy and physiology in complete health. Health is only possible when the immune defense protects the organism from harm. Immune strategies involve immediate, short-term defense called **innate immunity** discussed here, and the slightly delayed but highly specific **adaptive immunity** with memory function (Sect. 13.2). The evolution of innate immunity first occurred in primitive unicellular organisms, but due to its great overall success is highly conserved in all species. Hence, the innate immune system is the evolutionary oldest defense strategy, which is basically found in every living organism, from plants, fungi, insects, and up to vertebrates (Janeway et al. 2001). With the movement from water to earth of vertebrates during the Cambrian explosion, and in parallel rapid development of parasitic and infective life forms, the adaptive immune mechanisms were urgently required and initiated then, some 500 million years ago (Kaufman 2010). This was not only confined to jawed vertebrates, but also invertebrates developed specific adaptive immune mechanisms (Adema et al. 1997; Zhang et al. 2004) (Fig. 13.1). Thus from the beginning, disease-eliciting **pathogens** were a threat for higher life forms. Generally, an agent being recognized by the immune mechanisms is called **antigen**; an agent that induces an immune response is an **immunogen**. An antigen harbors **epitopes**—specific sites that are recognized preferentially by soluble (humoral) or cellular **immune receptors**. The organism being invaded is termed **host** in case of infectious antigens.

The major functions of the innate immune system in vertebrates are to

1. Be a physical and chemical barrier to infectious agents and toxins
2. Recruit immune cells to the sites of infection through the production of humoral factors called chemokines and cytokines
3. Activate the complement system to mark, e.g., bacteria and promote clearance of dead cells or antibody complexes
4. Identify and remove foreign substances in the host by specialized white blood cells
5. Activate the adaptive immune system through a process known as antigen presentation

13.1.3 Body Barriers

The epithelial barrier in the different organs represents a very effective first-line defense against the majority of invading pathogens (see Chap. 7). Desquamation of

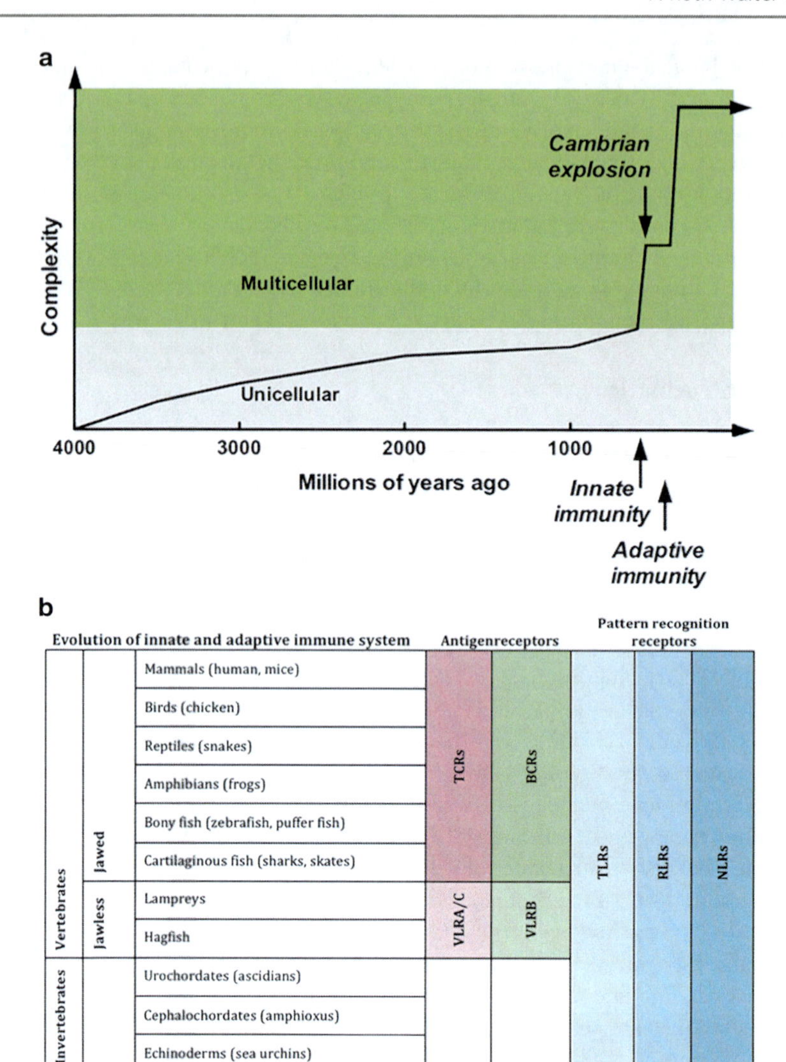

Fig. 13.1 The evolution of immunity. (**a**) The Cambrian explosion around 542 million years ago is characterized by the relatively rapid appearance of most major animal phyla which was accompanied by major diversification of other organisms. With the Cambrian explosion also the innate immune system and later the adaptive immune system evolved. Before the Cambrian explosion organisms were simple and composed of individual cells, occasionally they were organized into colonies (adapted with kind permission from a lecture of Dr. Alexander McLellan, Dept. Microbiology & Immunology, University of Otago, New Zealand). (**b**) Both jawless and jawed vertebrates possess acquired immune systems. However, jawless vertebrates have a unique antigen receptor repertoire VLRA/C and VLRB using LRR-based variable segments, which seem to exert similar function like the TCRs and BCRs in jawed vertebrates which are generated by RAG recombinase. The acquired immune system is activated by the evolutionarily older innate immune system that comprises pathogen recognition receptors and in nearly all living organisms

Table 13.1 Most important examples for barrier defense mechanisms in vertebrates

Anatomical barriers in animals	Additional defense mechanisms
Skin, skin derivatives, and appendices	Sweat, desquamation, camouflage, toxin release
Gastrointestinal tract	
– Stomach	– Peristalsis
	– Gastric acid, digestive enzymes
– Intestine	– Bile acids, mucus, lysozyme, lactoferrin, thiocyanate, defensins, tight junctions
	– Gut flora: *E. coli* produce "colicins" in mammalians
Respiratory tract	
– Nasopharynx	– Mucus, saliva, lysozyme
– Bronchi	– Surfactant, defensins
	– Mucociliary apparatus
	– Smooth muscle contraction and expulsion
Secretions, such as tears from eyes	Contain soluble defense molecules of innate and adaptive immunity

skin epithelium removes bacteria and other infectious agents, the hostile environment in the gut with digestive enzymes, low pH, and mucus helps in degrading and trapping infectious agents. Movement due to peristalsis or cilia in the respiratory system as well as the flushing action of tears and saliva helps to remove agents (Table 13.1). The gut flora can prevent colonization of pathogenic bacteria by competing for nutrients and attachment to cell surfaces as well as by secreting toxic substances (Mayer and Shao 2004a).

13.1.4 Inflammation

Any physiological response to exogenous stimuli such as infections and injuries, as well as to endogenous aberrant cell development such as in cancer and autoimmunity, involves local inflammation first. At the second level of escalation, the local events may lead to a systemic response, also called acute phase reaction (APR).

Stimuli, which affect the integrity of barriers, will lead to activation of the innate immune response constituted by soluble molecules followed by cellular infiltration (Table 13.2). In the following, inflammation is discussed which is a comparable, multicellular process in vertebrates. Interestingly, key molecules and principles can be found in non-vertebrates and monocellular organisms, too, and seem thus to be important for survival of a species.

13.1.4.1 Acute Inflammation

Acute inflammation is a short-term process, which usually appears within minutes or hours and ceases upon the removal of the injurious stimulus (Chandrasoma and Taylor 1998). It is characterized by five cardinal signs:

Table 13.2 Types of stimuli leading to inflammation

Classification of stimuli	Specification
Exogenous triggers	
– Mechanical	Pressure, cuts
– Physical	Radiation, heat/cold
– Chemical	Acidic or alkaline solutions, toxins, enzymes
– Infectious microorganisms	Viruses, bacteria, fungi, protozoa, and derived toxins
– Parasites	Worms, insects
– Innocuous antigens eliciting hypersensitivity	Allergens
Endogenous triggers	
	Benign or malignant neoplasms

(a) *Rubor* (*redness*)
(b) *Calor* (*heat*)
(c) *Tumor* (*swelling*)
(d) *Dolor* (*pain*)
(e) *Functio laesa* (*loss of function*)

At the very beginning of a pathogenic stimulus, a short constriction of the capillaries due to catecholamine release may be visible (paleness at the injury site); however, this is quickly followed by a pronounced dilation of the capillaries initiated by the release of histamine from mostly mast cells. This leads to an increased blood flow into the injury site perceptible as **redness** and **heat** (*rubor*, *calor*), in order to bring in soluble and cellular defense tools. Postcapillary vasoconstriction is maintained to keep the pathogen at the site of entry. All this lowers the blood flow rate already in the beginning of the reaction. Optimally, the agent will be removed, leading to *restitutio ad integrum* (complete healing).

However, if the innate defense mechanisms are not able to remove the stimuli, the lowered blood flow rate results in a postcapillary lack of oxygen. This leads to postcapillary dilation, endothelial leakage, and to a loss of fluid into the interstitial tissue—**swelling** (*tumor*) occurs. When fluid also contains cells, it is called inflammatory **exudate**. Compression of lymphatic vessels hinders draining of the tissue for a prolonged time. Simultaneously, remaining blood cells get concentrated and slower, or even get stopped at inflamed endothelial cells expressing adhesion molecules (*stasis*). As a result a blood clot may be formed, in the worst case leading to necrosis around the injury due to hypoxia. During the whole process, white blood cells (**leukocytes**) slow down, roll along the endothelium, and enter the interstitium (**diapedesis**) (Fig. 13.2). In this acute response, the most important leukocytes are neutrophil granulocytes, which eradicate particulate antigens by phagocytosis (described in more detail below). Other cells with phagocytic function are macrophages and dendritic cells, which are key elements for innate *and* adaptive immunity. Like neutrophils they are able to recognize pathogens via their surface receptors and eliminate them via phagocytosis in an innate response. They may escalate the reaction by transporting these antigens to the next immunocompetent site, e.g., lymph nodes, and initiating a specific immune response. Similar functions

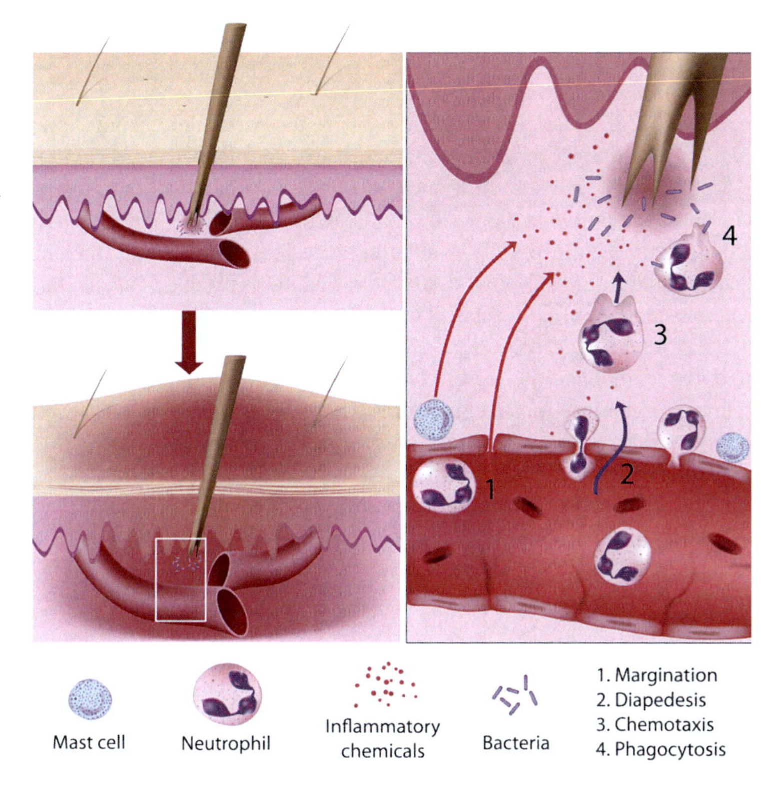

Fig. 13.2 The initial response to a harmful stimulus (injury, pathogen) by acute inflammation in mammals. Upon entrance of a foreign body or pathogen, resident macrophages, dendritic cells, and mast cells initiate inflammation by releasing inflammatory mediators. Vasodilation causes accumulation, followed by migration of leukocytes, mainly neutrophils to the injured/infected tissue. Leukocytes extravasate the blood vessel by margination and diapedesis, before migrating to the site of infection by chemotaxis and removing of the stimulus by phagocytosis. The depicted inflammation is caused by innate immune mechanisms. © [Alila Medical Images]—Fotolia.com

can be found in insects, where the so-called hemocytes are capable of phagocytosis (Bergin et al. 2005). In fact, the famous immunologist **Metchnikoff** detected phagocytosis in the invertebrate starfish; it might be the oldest form of cellular defense and can be tracked down to protozoans like amoeba. It is a "combination of food-getting and defense" (Cooper 2001).

Pain (*dolor*) is the result of liberation of potassium ions from intracellular compartments of destroyed tissue cells, and a release of histamine, substance P, kinines, and prostaglandins into the injured tissue capable of stimulating sensitive nerve endings. **Itch** is a minor form of pain and leads to scratching as a defense reaction.

13.1.4.2 Chronic Inflammation

When repair cannot be achieved, prolonged, chronic inflammation results. It is accompanied by a progressive shift in the type of cells present and is characterized by simultaneous destruction and healing of tissue caused by the inflammatory process.

Several diseases like asthma, periodontitis, atherosclerosis, rheumatoid arthritis, and even cancer (e.g., gallbladder carcinoma) are the results of chronic inflammation. Implants in modern regenerative medicine are complicated by chronic inflammation. Also cancer and autoimmune disease is characterized by chronic inflammation, as the antigens cannot be eradicated.

Macrophages play an important role here, since they are typically the cells that capture particulate antigens. After their activation they produce interleukin IL-1, which activates collagen synthesis in **fibroblasts** for repairing the tissue defect. Capture and storage of antigens that cannot be broken up and killed, such as a foreign body or some bacteria, lead to chronic macrophage activation. The typical result of chronic inflammation is a **granuloma**, containing a center with the stimuli, surrounded by macrophages, which may phenotypically be changed to flat **epitheloid cells**, or multinuclear **giant cells**, as well as **lymphocytes**. A collagen capsule around the site of chronic inflammation may aim at limiting the damage. Typically granulomas are found in diseases like tuberculosis, sarcoidosis, and Crohn's disease.

13.1.4.3 Acute Phase Reaction

When damage or noxious stimuli overwhelm local defense mechanisms and become systemic, the organism initiates the so-called APR, which is characterized by (1) leukocytosis, (2) fever, (3) generation of acute phase proteins, and (4) an increase of the erythrocyte sedimentation rate.

Leukocytosis

During leukocytosis increased numbers of leukocytes (white blood cells) are found in the peripheral blood, which can be determined by a white blood count. In a Gaussian curve, immature cells are found on the left side, normal cells in the middle, and aberrant cells on the right. 70 % of peripheral leucocytes are neutrophil granulocytes characterized by a segmented nucleus.

Leukocytosis is frequently a sign of inflammation, but may also occur after strenuous exercise, e.g., convulsions, labor, or crying of infants. In these cases, the increase of leukocytes is due to recruitment of mature cells (leukocytosis by **re-distribution**) from other tissues into the blood.

Leukocytosis may also be due to the generation of new immune cells (leukocytosis by **production**), induced in the bone marrow by granulocyte-macrophage-derived colony-stimulating factor, GM-CSF, from macrophages during inflammation. Newly produced immature granulocytes are characterized by an unsegmented nucleus, rendering a "left shift" in the Gaussian curve of the hemogram. Granulozytes appear in the blood within 6–12 h, monocytes upon 48 h, whereas lymphocytes appear at much later time points as a sign of involvement of the adaptive immune response.

Fever

Warm-blooded animals, mammals, and birds have a constant body temperature between, e.g., 31 °C (echidna), 35 °C (kangaroo), 37 °C (human, dog), and up to 40 °C in birds, with some variations caused by circadian rhythm or during hibernation. Most other animals show body temperatures adapted to the surroundings. John Hunter described this phenomenon first (see Chap. 1). Some animals like lizards may actively move into the sun during infections to collect warmth and elevate the body temperature passively.

However, active fever production is a phenomenon in warm-blooded animals only. Fever is due to an elevation of the temperature regulatory set-point directly in the hypothalamus by **pyrogens** (fever-producing agents). An **exogenous** pyrogen may be, e.g., bacterial products, causing the release of **endogenous pyrogens** like the cytokines IL-1, IL-6, TNFα, and prostaglandin E2, mostly from activated macrophages.

In warm-blooded animals and humans, an increase of body temperature is achieved by heat production through enhanced muscle tone and **shivering** (release of energy by exothermic reactions), and hormones like epinephrine (adrenaline), as well as in preventing heat loss by **vasoconstriction** (paleness). Further, in vertebrates, small **hair-erecting muscles** are contracted rendering an increase of an isolating layer of air within the animals' coat. In humans, a rudimentary phenomenon can be observed called **gooseskin**. In fact, the replication rate of some bacteria is inhibited by fever. Upon a temperature of 41.5 °C fever becomes life-threatening due to a denaturation of body proteins. The type of fever may be characteristic for a specific disease (e.g., continuous type in pneumonia or typhus, intermittent in malaria).

Acute Phase Proteins (APPs)

There are over 400 APPs known, which are produced in response to inflammation during the acute phase predominantly in the liver. APPs belong to the innate immune defense and are often highly conserved molecules suited for recognition of foreign (or altered self-) antigens (see below) (Manley et al. 2006). The most prominent ones are **C-reactive protein (CRP)** which is used in diagnostics as an inflammation marker, **Mannan-binding lectin (MBL)** able to activate complement by the lectin-way, **fibrinogen** an important coagulation factor, as well as **serum amyloid protein A (SAA)**, which is an apolipoprotein, often found in chronic inflammation in the inflamed tissue as well as in tumors. SAA deposits in kidney, spleen, and gastrointestinal tract can lead to organ failure (secondary amyloidosis).

Erythrocyte Sedimentation Rate

The rate, in which red blood cells sediment in a period of 1 h, is a marker of inflammation, but can also be elevated in physiological processes (older age, pregnancy). During inflammation, the elevated fibrinogen in the blood causes that negatively charged red blood cells form rouleaux and as a consequence sediment faster. In some animals like dogs and cats rouleaux formation is seen in healthy condition, too. The basal ESR is slightly higher in female humans (Feldman et al. 2013). The physiology behind may be to support the traffic of erythrocytes in small capillaries; however, there is a tight balance between enhanced flow rate and coagulation.

13.1.5 Host Defense by Phagocytosis

Extracellular antigens (toxins, viruses, bacteria, other noxious agents) can be cleared from the system by phagocytosis.

Immune cells capable of phagocytosis are leukocytes comprising neutrophilic, basophilic, and eosinophilic granulocytes as well as macrophages and dendritic cells. Phagocytes are able to recognize **pathogen-associated molecular patterns** (PAMPs) of pathogens by their **pattern recognition receptors** (PRRs).

PAMPs are composed by repetitive, uniform membrane antigens, typically expressed by bacteria, viruses, protozoa, and parasites, e.g., lipoproteins on gram-negative bacteria, proteoglycan on gram-positive bacteria, or S-layers on bacteria, and archea.

PRRs are germ line-encoded and hence inherited. Due to a much shorter replication time of primitive organisms, they can adapt quickly to new pathogens by usage of the innate receptors. Moreover, PRRs are directed against essential pathogen structures, which the invading pathogen usually cannot change fast. However, for hosts with longer life cycles like mammals, the innate system was no more sufficiently flexible, prompting the development of the more advanced adaptive immune system (see Sect. 13.2).

The phagocyte by its PRRs is prepared to meet both extracellular and intracellular pathogens (Table 13.3) (Mogensen 2009). The most prominent surface-expressed PRRs are **Toll-like receptors (TLRs);** intracellular PRRs recognizing predominantly viral pathogens are (a) retinoic acid-inducible gene 1 **(RIG-1)-like receptors (RLRs)**, melanoma differentiation-associated gene 5 (MDA5), and LGP2 proteins, (b) nucleotide oligomerization domain **(NOD)-like receptors** (NLRs), e.g., NOD1 and NOD2 sensing peptidoglycan moieties, and (c) cytosolic DNA sensors like DAI (DNA-dependent activator of IFN-regulatory factors) or AIM2 (absent in melanoma 2) (Hornung et al. 2009; Mogensen 2009).

For improved recognition of infectious agents, phagocytes get help through soluble innate (secreted PRRs) or adaptive molecules (immunoglobulins) that are able to mark the pathogen and hence support phagocytosis. This process is called **opsonization**. Opsonin molecules include the following:

- Components of the complement system: C3b, C4b, and iC3b (see below)
- Acute phase proteins (CRP, MBL, Fibrin)
- Immunoglobulins (IgG and IgM)

13.1.5.1 Neutrophils

Neutrophilic granulocytes are the most abundant type of white blood cells in mammals and are an essential part of the innate immune system. They are derived from the myeloid lineage of hematopoietic stem cells (Fig. 13.3). Approximately 10^{11} are produced daily in humans and they account for approximately 50–70 % of

Table 13.3 Pathogen recognition receptors and their ligands (Mogensen 2009)

Receptor	Cellular localization	microbial component(s)	Origin(s)
TLRs			
TLR1/TLR2	Cell surface	Triacyl lipopeptides	Bacteria
TLR2/TLR6	Cell surface	Diacyl lipopeptides	*Mycoplasma*
		Lipoteichoic acid	Gram-positive bacteria
TLR2	Cell surface	Lipoproteins	Various pathogens
		Peptidoglycan	Gram-positive and -negative bacteria
		Lipoarabinomannan	Mycobacteria
		Porins	*Neisseria*
		Envelope glycoproteins	Viruses (e.g., measles virus, HSV, cytomegalovirus)
		GPI-Mucin	Protozoa
		Phospholipomannan	*Candida*
		Zymosan	Fungi
		β-Glycans	Fungi
TLR3	Cell surface/endosomes	dsRNA	Viruses
TLR4	Cell surface	LPS	Gram-negative bacteria
		Envelope glycoproteins	Viruses (e.g., RSV)
		Glycoinositolphospholipids	Protozoa
		Mannan	*Candida*
		HSP70	Host
TLR5	Cell surface	Flagellin	Flagellated bacteria
TLR7/8	Endosome	ssRNA	RNA viruses
TLR9	Endosome	CpG DNA	Viruses, bacteria, protozoa
RLRs			
RIG-I	Cytoplasm	dsRNA (short), 5′-triphosphate RNA	Viruses (e.g., influenza A virus, HCV, RSV)
MDA5	Cytoplasm	dsRNA (long)	Viruses (picorna- and noroviruses)
NLRs			
NOD1	Cytoplasm	Diaminopimelic acid	Gram-negative bacteria
NOD2	Cytoplasm	MDP	Gram-positive and -negative bacteria
NALP1	Cytoplasm	MDP	Gram-positive and -negative bacteria
NALP3	Cytoplasm	ATP, uric acid crystals, RNA, DNA, MDP	Viruses, bacteria, and host
Miscellaneous			
DAI	Cytoplasm	DNA	DNA viruses, intracellular bacteria
AIM2	Cytoplasm	DNA	DNA viruses
PKR	Cytoplasm	dsRNA, 5′-triphosphate RNA	Viruses
FPR	Plasma membrane	*N*-formyl-methionyl peptides	Bacteria, mitochondria

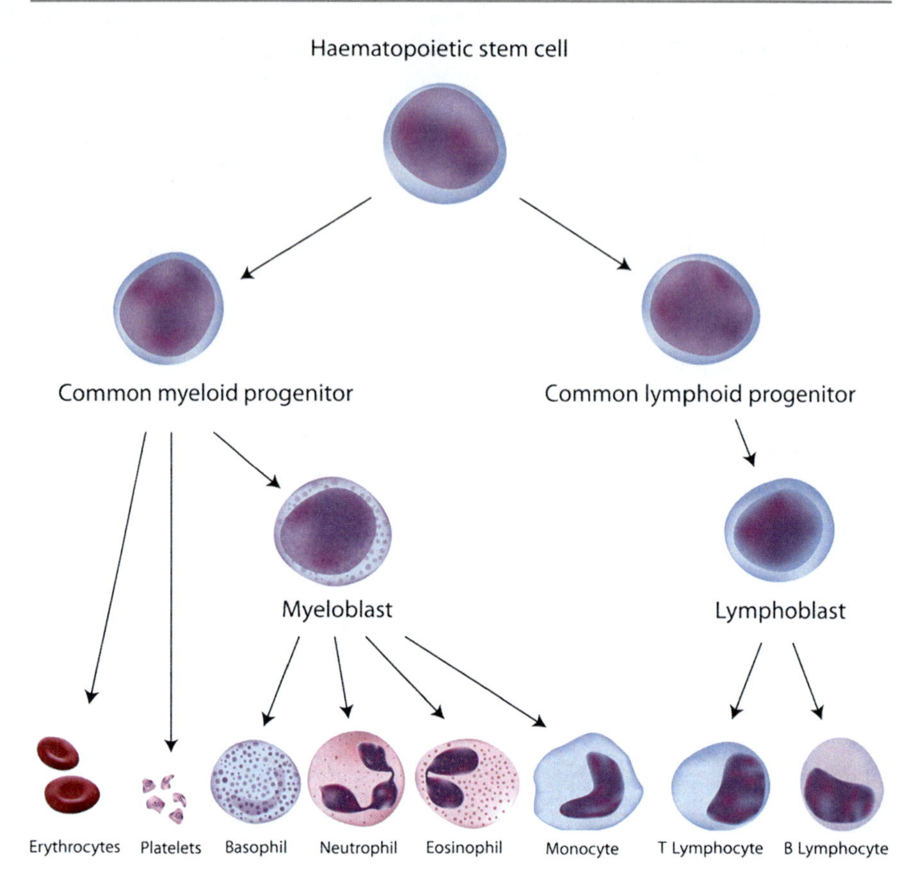

Fig. 13.3 All cellular blood components derive from hematopoietic stem cells. Hematopoietic stem cells residing in the bone marrow give rise to the myeloid and lymphoid lineage. Myelocytes, which include granulocytes (neutro-, baso-, and eosinophils), monocytes, but also platelets and erythrocytes, originate from the common myeloid progenitor. In contrast, lymphocytes (B, T, and NK cells) derive from the common lymphoid progenitor. For simplicity, NK cells and other innate lymphocytes are not depicted. © [Alila Medical Images]—Fotolia.com

all white blood cells (leukocytes). Their average life span in the circulation is relatively short with approximately 5 days (Pillay et al. 2010). Upon activation, they migrate to the tissue, where they die within 2 days. Neutrophils are one of the first responders during inflammation, particularly during bacterial infections, but also in some cancers (Waugh and Wilson 2008). They are usually found in the blood, but are attracted by chemokines and anaphylatoxins (see below) released from the site of inflammation. Neutrophils dominantly contribute to leukocytosis during acute inflammation (see above).

Neutrophils can migrate towards site of infection or inflammation by a process called **chemotaxis**. Their cell surface receptors allow neutrophils to detect chemical gradients of molecules such as interleukin-8 (IL-8), which are released from epithelial cells upon damage; further, these cells use interferon gamma (IFN-gamma), C5a, and Leukotriene B4 to direct the path of neutrophil migration.

Neutrophils have a variety of specific receptors, including complement receptors, cytokine receptors for interleukins and IFN-gamma, receptors for chemokines, receptors to detect and adhere to endothelium, receptors for leptin and proteins, and Fc receptors for opsonizing immunoglobulins like other cells (Pillay et al. 2010).

Extravasation (diapedesis) is the process in which leukocytes leave the circulatory system towards the site of infection or inflammation and is mediated by the following steps: Chemoattraction, rolling adhesion, tight adhesion, and transmigration (Fig. 13.2).

- **Chemoattraction:** Upon recognition of pathogens resident macrophages secrete cytokines like IL-1, and TNFα, which cause the endothelial cells of the blood vessels to express cellular adhesion molecules like selectins.
- **Rolling margination (adhesion):** Circulating leukocytes bind via their carbohydrate residues (sialyl-lewis-unit) to selectin molecules with low affinity (binding strength), causing them to slow down and begin rolling along the inner surface of the vessel wall.
- **Tight adhesion:** During this time, chemokines released by macrophages also induce the expression of integrins (intercellular adhesion molecule-1; ICAM-1), which will bind to the integrins (e.g., lymphocyte function antigen-1, LFA-1; complement receptor 3, CR3) of leukocytes with high affinity. This causes the immobilization of leukocytes, despite the shear forces of the blood flow.
- **Transmigration/Diapedesis:** The cytoskeleton of the leukocytes is reorganized allowing them to extend pseudopodia and passing through the gaps between endothelial cells. Leukocytes penetrate the basal membrane by proteolytic digestion of the membrane and/or by mechanical force (Sorokin 2010). Once in the interstitial fluid leukocytes migrate along a chemotactic gradient towards the site of inflammation or infection.

Function of Neutrophils

Neutrophils exert their antimicrobial function by phagocytosis, degranulation, and extracellular traps.

(a) **Phagocytosis**

Neutrophils can recognize and phagocyte microorganisms that are "naked" or opsonized. When opsonized by an immunoglobulin, the process is called antibody-dependent cellular phagocytosis (ADCP). Phagocytosis involves the following steps:

- **Attachment**
- **Engulfment** via the formation of pseudopodia—a phagosome (eating body) is formed by invagination of the cell membrane around the antigen.
- **Fusion with a lysosome** (see Chap. 2) creating a phagolysosome; like in gastrointestinal digestion (see Chap. 9), pH shifts direct a series of hydrolytic enzymes being secreted into the phagosome. The first set of enzymes works in acidic pH (Table 13.4), followed by an increase in pH, which ensues the activation of the remaining enzymes.
- **Lysis** of the outer lipid bilayer of bacteria

Table 13.4 The granula contents of neutrophils

Granula	Containing
Primary (azurophile)	– Microbial enzymes: myeloperoxidase, lysozyme, neuraminidase
	– Acid hydrolyases
	– Proteases: elastase, cathepsin G, protease 3, esterase N, collagenase
	– Bactericidal/permeability-increasing protein (BPI), defensins,
	– Serine proteases
	– Other proteins like ubiquitin
Secondary (specific)	– Microbial enzymes: lysozyme, neuraminidase
	– Proteinases: collagenase, gelatinase
	– Inhibitors like apolactoferrin, VitB12-binding protein, histaminase
	– Other proteins like β2-microglobulin, lipocalin, plasminogen activator
Tertiary (Gelatinase)	– Proteinases: gelatinase
	– Inhibitors like lactoferrin
	– Other proteins like acetyltransferase

Further, reactive oxygen molecules are generated in the phagolysosomes by a process called ***respiratory burst*** due to the consumption of oxygen. It involves the activation of the enzyme NAPDH oxidase, which produces large amounts of superoxide. Superoxide, a reactive oxygen species (ROS), spontaneously decays to hydrogen peroxide (H_2O_2), which is then converted by the enzyme **myeloperoxidase** to hypochlorous acid (HClO), being uttermost potent to kill phagocytosed microbes. ROS help in depolymerization of collagens, proteoglycans, and hyaluronic acids, as well as oxidize lipids on the cell membrane, denature enzymes, and deactivate serin protease inhibitors (Freitas et al. 2009). Neutrophils are characterized by extensive phagocytosis until disrupture. Then, granula contents damage surrounding tissue leading to necrosis and visible pus.

(b) **Degranulation** (Moraes et al. 2006)

Neutrophils have three types of granules containing an assortment of products with antimicrobial properties. The same contents digest pathogens that have entered a phagosome after phagocytosis (see above). Expulsion of the granulas also helps to combat infections and to kill extracellular pathogens (Table 13.4).

(c) **Extracellular traps (ETs)**

In 2004 Brinkmann et al. (Brinkmann et al. 2004) first described the release of web-like structures, the so-called extracellular traps by neutrophils after stimulation with microbial products. ETs have been found not only in humans and mice, but also in other animals including ox, horses, fish, cats, and even in invertebrates (Goldmann and Medina 2012; Gupta et al. 2005). Extracellullar nucleic acid released by oenocytoid cells has been reported to be an important defense mechanism towards pathogenic microorganisms in insects. ETs are also apparent in plants where they have been demonstrated to play an important role in the defense against fungal infections of the root tip.

ETs are composed of nuclear DNA decorated with antimicrobial molecules such as neutrophil elastase (enzyme digesting elastic fibers) and histones that are able to trap and kill not only bacteria, but also fungi and yeast. ETs provide for a high local concentration of antimicrobial components able to bind and kill microbes independent of phagocytic uptake. Moreover, ETs serve as a physical barrier to prevent further spreading of the pathogen. Other immune cells including mast cells, eosinophils, and macrophages can also form ETs (Goldmann and Medina 2012).

Neutropenia (low neutrophil counts) renders an individual highly susceptible to infections and to colonization with intracellular parasites (Ferencik et al. 2004). It can have congenital reasons (genetic disorders), but also can develop later in life due to aplastic anemia, some types of leukemia or as a negative side-effect of medication, most prominently chemotherapy.

13.1.5.2 Eosinophils

Only about 1–6 % of all leukocytes in the blood are eosinophil granulocytes, since they are primarily tissue-resident cells. In healthy individuals they are found in the gut, mammary glands, uterus, thymus, lymph nodes, bone marrow, and adipose tissues, but not in the lung or the skin. They are related to the neutrophils and basophils (Fig. 13.3).

Eosinophils have an affinity for "acidic dyes" and appear brick-red after staining with eosin using the so-called Romanowsky method. The distinctive feature of eosinophils is their bright pink cytoplasmic granules containing the following defense molecules: major basic protein (MBP), eosinophil peroxidase (EPO), histamines, ribonuclease (RNase), deoxyribonucleases (DNase), lipase, and plasminogen. These mediators are released by degranulation following activation of the eosinophil and are toxic to both parasite and host tissues.

The nucleus is bilobed to multilobed and has coarse, dark, clumped chromatin. The morphology of these granules varies slightly with species. In the dog, granules are pink-red and variably sized in most breeds. Greyhounds and some other breeds may have clear granules due to altered stain uptake, so eosinophils may be more difficult to distinguish (called "gray" or "vacuolated" eosinophils). In cats, the granules are small and rod-shaped, in horses they are very large and bright pink, and in cattle they are small and uniform-sized (French et al.).

Function

Eosinophils persist in the circulation for 8–12 h and can survive in the tissue for an additional 8–12 days in the absence of stimulation. Similarly to neutrophils, eosinophils can kill pathogens by

(a) Phagocytosis
(b) Degranulation
(c) Extracellular traps

(a) **Phagocytosis**

Eosinophils can phagocyte microorganisms or particles, when they are coated with opsonins, since they have receptors for antibodies of the IgG, IgA, and IgE classes (FcγR, FcαRI—CD89, FcεRI) and for complement components. Moreover, they express PPRs enabling them to pathogen binding and phagocytosis (see above Sect. 13.1.5.1). Also here the generation of reactive oxygenic molecules ensures effective killing.

(b) **Degranulation**

Traditionally, eosinophils are considered to be effector cells (Janeway et al. 2001)

- First, by releasing highly **toxic granule proteins** and free radicals upon activation, which can kill microorganisms and parasites but can also cause significant tissue damage in allergic reactions.
- Second, by inducing the synthesis of **chemical mediators** such as prostaglandins, leukotrienes, and cytokines like IL-3, IL-5, and GM-CSF, which amplify the inflammatory response by activating epithelial cells, and recruiting and activating more eosinophils and leukocytes.

Eosinophilic granules contain a wide assortment of cytotoxic granule cationic proteins, which include MBP, eosinophil cationic protein (ECP), EPO, and eosinophil-derived neurotoxin (EDN). MBP, ECP, and EPO are toxic to many tissues by creating toxic pores in the membranes of target cells and allowing potential entry of other cytotoxic molecules into the cell (Young et al. 1986). Moreover, they induce degranulation of mast cells and stimulate fibroblast cells to secrete mucus and glycosaminoglycan (Venge et al. 1999). EPO forms ROS and reactive nitrogen intermediates that promote oxidative stress in the target, causing cell death by apoptosis and necrosis (see below).

Recently, eosinophils have been associated with tissue morphogenesis and homeostasis, e.g., being important in maintenance and repair of epithelial barrier integrity and postnatal mammary gland development (Kita 2013). In the bone marrow, eosinophils colocalize with plasma cells and support their survival (Chu et al. 2011).

13.1.5.3 Macrophages/Monocytes

Circulating monocytes give rise to a variety of tissue-resident macrophages (MΦ) throughout the body, in particular in the kidney and the lung, as well as to specialized cells such as dendritic cells (DCs) and osteoclasts. However, newer studies suggest that macrophages of several tissues that closely associate with epithelial structures, such as Kupffer cells in the liver, epidermal Langerhans cells, and microglia in the brain, can also originate from a different source, namely the yolk sac (see Chap. 12) and not from hematopoietic stem cells in the bone marrow (Schulz et al. 2012). Dependent on the site of differentiation, they have been attributed with different names (Table 13.5). In general, human macrophages are large cells with a diameter of about 20 μm. In contrast to neutrophils with a life span of a few days, macrophages are able to survive in the body for several months.

Function of Macrophages

Macrophages are phagocytes and are pivotal in innate as well as adaptive immunity. Their role is to engulf and digest cellular debris and pathogens. By cytokine secretion and antigen presentation, they are also able to stimulate not only innate immune cells, but also the lymphocytes of the adaptive immune response (see Sect. 13.2). Macrophages play an essential part in the immune-regulatory machinery and host defense. They also seem to have some important function during tissue regeneration (Novak and Koh 2013).

Table 13.5 The different phenotypes of macrophages

Name	Location
Alveolar macrophages	Pulmonary alveolus of lungs
Adipose tissue macrophages	Adipose tissue
Kupffer cells	Liver
Microglia	Neural tissue
Osteoclasts	Bone
Sinusoidal lining cells	Spleen
Histiocytes	Connective tissue
Giant cells	Connective tissue
Peritoneal macrophages	Peritoneal cavity
Macrophage	Serosa and lymphoid organs
Hofbauer cells	Placenta

Main tasks of macrophages

- **Removal of dying cells**: usually performed by tissue-residing macrophages. Upon infection they are also responsible to eliminate aged neutrophils.
- **Antigen presentation**: along with dendritic cells, they are the major contributors to antigen presentation and activation of an immune response (see Sect. 13.2).
- **Tissue repair**
- **Phagocytosis** of pathogens:
  Similarly to granulocytes, they are able to recognize opsonized pathogens due to receptors for IgG and complement (CR3), but also via PRRs.
  Macrophages express
- LPS-receptor CD14 (Mac-2)
- TLR2 and TLR4 (PRRs)
- Scavenger receptor (important in lipid uptake)
- Glycan receptor (recognizes carbohydrate-patterns on pathogen surfaces)
- Mannose receptor (recognizes carbohydrate-patterns on pathogen surfaces)
- CD11b/CD18 (=CR3, Mac1)
- CD11c/CD18 (=CR4, Mac1) (Novak and Koh 2013)

Macrophages secrete among other substances

- Endogenous pyrogens: IL-1, IL-6, and TNFα to induce fever
- All complement components
- Coagulation factors
- Prostaglandins, leukotrienes, and platelet-activating factor (PAF)

Macrophages are thus pivotal for host defense. However, some pathogens have evolved **escape strategies** to survive in their phagosomes. For example, *Mycobacterium tuberculosis* can survive in the phagosome of an unstimulated macrophage due to a robust membrane of long-chain mycolic acids, or the *leishmania* parasite is able to inhibit fusion of the phagosome with the lysosome and is thus able to replicate. These microorganisms can only be eliminated by the host with the help of T lymphocytes. When T lymphocytes secrete IFN-γ and other macrophage-activating factors (MAFs), macrophages are activated and can kill the intracellular phagocytic particle.

13.1.6 Host Defense by Mediator Secretion

13.1.6.1 Mast Cells and Basophils

Mast cells are tissue-resident cells containing granules rich in histamine and heparin. Mature mast cells normally reside close to epithelia, blood vessels, nerves, and near smooth muscle cells of the airways and gut as well as mucus-producing glands. In some species, including murine rodents, mast cells also occur within mesothelium-lined cavities, such as the peritoneal cavity (Galli et al. 2005). In contrast, basophils represent a rare population predominantly circulating in the peripheral blood.

Even though mast cells and basophils share some phenotypic and functional properties, they differ:

- Basophils are circulating granulocytes and can be recruited to inflamed tissue, whereas mast cells are predominantly resident in tissues
- Mast cells are long and can proliferate locally, whereas mature basophils lack this ability and undergo apoptosis
- Basophils leave the bone marrow already mature, whereas mast cells mature when entering the tissue

Both cell types however have

- Granules containing histamine, anticoagulant heparin as well as leukotrienes
- The high-affinity receptor for IgE, FcεRI
- Complement receptors CR1 and CR3, and receptors for the so-called anaphylatoxins C3a and C5a

Granules of mast cells also contain acidic and alkaline phosphatase and 30× more histamine than the granules of basophils. Mast cells have an average diameter of 10 μm. They have more, but also smaller granules (1,000 granules per cell) than basophils (80 granules which are 6× larger than mast cell granules).

Mast cells are heterogeneous. In this respect, mucosal mast cells differ from the mast cells from connective tissue by their dependence on T cells activation (Galli et al. 2005).

Function of Mast Cells and Basophils

Mast cells and basophils are important during inflammation. Once activated, mast cells and basophils rapidly release their granules. Mast cells can degranulate by

- Physical (injury) or chemical stimuli (neuropeptide, opioids, alcohols)
- Cross-linking of IgE by antigens (or in an aberrant immune response by allergens) or lectins (pseudo-allergy) or by anti-FcεRI autoantibodies
- Activation by complement proteins (Ferencik et al. 2004)

Histamine is an important mediator; it dilates postcapillary venules, activates the endothelium, and increases blood vessel permeability (Fig. 13.2). The consequences are the cardinal symptoms of acute inflammation: local edema (swelling), warmth, redness, and the attraction of other inflammatory cells to the site of release. Moreover, the depolarization of nerve endings leads to itching or pain. In concert with IgE receptors and elevated IgE levels in parasitic infestations, mast cells may be physiologically important in the defense against parasites.

Mast cells seem also to be important in tissue regeneration (Wulff and Wilgus 2013), but they are best known for their key role in allergic inflammation.

13.1.7 Host Defense by Cytotoxicity

Intracellular antigens, e.g., viruses, can only be eliminated by killing the whole infected cells. Lymphocyte-mediated cytotoxicity does not necessarily have to be mediated by receptors of the adaptive system. There are certain cells of the lymphocytic lineage that are regarded as innate cells:

- Natural killer (NK) cells
- NK T cells

The immune system has additionally evolved forms of cytotoxicity which depend on opsonization of the antigen with soluble molecules of the innate or adaptive immunity: complement-dependent (CDC) or antibody-dependent cellular cytotoxicity (ADCC). (The cytotoxic T lymphocytes/T cells belong to the adaptive immunity, see Sect. 13.2). Still, all cytotoxic mechanisms are using the same eradication principles resulting in apoptosis.

13.1.7.1 Natural Killer (NK) Cells

NK cells are mostly constituted by **large granular lymphocytes** (LGL) and have a common lymphoid progenitor with B and T lymphocytes. NK cells are known to differentiate and mature in the bone marrow, lymph node, spleen, tonsils, and thymus where they then enter into the circulation.

Similarly to cytotoxic T cells in the vertebrate adaptive immune response, NK cells provide rapid responses to **virally infected** cells, and act also against aberrant cells (e.g., **malignant cells**). In contrast to cytotoxic T cells, which are dependent on antigen presentation by specific presentation molecules (major histocompatibility complex, MHC, or human leukocyte antigen, HLA), NK cells have the unique ability to directly recognize stressed cells, allowing for a much faster immune reaction. They were named "natural killers" because of the initial notion that they do not require activation in order to kill cells. Thereby, they may also attack cells that are *missing* "self" markers (MHC class I molecules).

Functional MHC molecules are present in cartilaginous fish, but not in more primitive species. In this respect, NK cells have not been identified in species lower than fish (Lanier 2005).

NK cells differ from phagocytes (macrophages and granulocytes) in that they solely rely on conserved PRRs, e.g., toll-like receptors. However, today they are considered to be at the interface between innate and adaptive immunity. Despite a lack of receptor diversity generated by DNA rearrangement, as found in B and T cells, NK cells share some properties with cells of the adaptive system, which are

- The capacity to distinguish infected from healthy cells
- To maintain a pool of long-lived cells that expands during a response
- The ability of antigen-specific adaptive recall responses ("immunological memory") (O'Leary et al. 2006; Paust et al. 2010)

NK cells appear to work by the integration of numerous signals from activating receptors and are strictly controlled by **inhibitory receptors**.

NK cell recognition involves

- The initial binding to potential target cells via pattern recognition.

- Interactions between activating and inhibitory receptors with ligands available on the target cell.
- The integration of signals transmitted by these receptors, which determines the fate of the target cell (Lanier 2005).

An important **activating receptor is NKR-P1**, whereas a likewise important **inhibitory receptor is KIR** (killing inhibitory receptor), which recognizes MHC class I molecules on target cells. Target cells expressing MHC class I are therefore protected from the cytotoxic activity of NK cells. However, a low expression of MHC class I molecules is suspicious for NK cells and they attack. Some viruses downregulate protein synthesis and MHC class I expression of the infected cell to escape an attack of cytotoxic lymphocytes of the adaptive response. Thereby, they get detectable by NK cells.

NK cells express the IgG antibody receptor FcγRIII (CD16) and hence are able to mediate apoptosis in infected cells opsonized with antibodies by antibody-dependent cell-mediated cytotoxicity (ADCC).

NK cells are "ready-to-go" effector cells containing **toxic enzymes, granzymes, and perforin** in their granules. NK cells also secrete high levels of **cytokines** like IFNγ, TNFα, IL10, and TGFβ. Their lytic response can be triggered within minutes, without requiring transcription, translation, or cell proliferation.

Release of the granule contents in close proximity to the target cell (**kiss of death**), leads to formation of pores in the cell membrane of the target cell by perforin, creating an aqueous channel allowing the entrance of granzymes and associated molecules, which induce **apoptosis**.

The distinction between **apoptosis** and **necrosis** is important in immunology: Necrosis of a virus-infected cell could potentially release the virions, whereas apoptosis leads to destruction of the virus including its protein, RNA or DNA inside the infected cell. Moreover, during necrosis the release of mediators not only affects the target cell, but also harms the surrounding tissue (pus). Whereas neutrophilic granulocytes usually provoke necrosis in target cells, NK and cytotoxic T cells induce apoptosis without causing further collateral damage.

Apoptosis can be easily discriminated morphologically from necrosis. The biochemical events during apoptosis lead to characteristic changes which include blebbing, cell shrinkage, nuclear fragmentation, chromatin condensation, and chromosomal DNA fragmentation. **Apoptotic bodies** are contracted from the apoptotic cell which are then engulfed by phagocytes. In contrast, necrosis results from an acute cellular injury leading to a disintegration of the cell and release of cytosolic and inflammatory components.

Function of NK Cells

- Control of infections: NK cells are recruited to the site of infection by chemokines and perform their effector function through perforin-dependent cytotoxicity of infected cells and IFNγ secretion. NK cells cannot entirely clear viral infections but are essential for controlling virus titers until an adaptive T cell response eliminates infected cells (Long et al. 2013).

- Control of tumor establishment: They have antitumor activity through upregulation of ligands for activation receptors and/or loss of MCH-I on the side of the tumor cells.
- Control of inflammation: NK cells can promote inflammation by secretion of IFNγ and TNFα, but also control inflammation by killing APCs and activated T cells (Long et al. 2013).
- Role in reproduction: NK are the predominant lymphocyte population in the uterus and have been implicated to be important in uterine vasculature remodeling and in immune suppression. In pregnancy, impaired NK cell activation in humans is associated with pathological elevated blood pressure and proteinuria (preeclampsia) (Erlebacher 2013).

13.1.7.2 NK T Cells

NK T cells are a heterogeneous group of T cells that share properties of both T cells and natural killer (NK) cells. The majority of these cells recognize **CD1** which (in contrast to MHC molecules which present peptides) is a **nonclassical antigen-presenting molecule** able to bind and present self and foreign lipids and glycolipids. Approximately 0.1 % of all peripheral blood T cells are NK T cells.

NK T cells are able to secrete large amounts of IFNγ and other cytokines. Dysfunction or deficiencies of NK T cells are associated with the development of autoimmune diseases, i.e., immune reactions towards self-antigens.

13.1.8 Host Defense by Innate Humoral Factors

Soluble plasma proteins and mediators are released to the surrounding tissue to mediate chemotaxis, opsonize, exert toxic properties on pathogens, or to directly activate cells.

Soluble mediators of inflammation are

- Lipid mediators (prostaglandines, leukotrienes, thromboxane, PAF).
- Cytokines, like interleukins IL-1, IL-6, IL-8, and interferons IFNγ and TNFα.
- Complement components.
- Furthermore, reactive oxygen metabolites, biogenic amines (histamine, serotonin), and the kinin–kallikrein, coagulation- and fibrinolytic system.

13.1.8.1 Lipid Mediators

Immune cells produce lipid mediators upon activation. Lipid mediators are derived from phospholipids of the cell membrane. The enzyme phospholipase A2 (PLA2) releases arachidonic acid that is further enzymatically edited, by cyclooxygenase to produce prostaglandins, or by lipoxygenase to produce. The most well-known lipid mediators are prostaglandins and leukotrienes, with pro- or anti-inflammatory characteristics, and many other diverse functions such as smooth muscle contraction. They are not considered hormones, since they are not produced at a discrete site but in many places throughout the body.

13.1.8.2 Cytokines

Cytokines are small, soluble polypeptides or glycoproteins of less than 30 kDa that act as signaling molecules and local mediators within the immune system. The act on target cells via high affinity receptors, but their effects can be pleiotropic, i.e., inducing different responses in different targets.

Their constitutive production is usually low and synthesis must usually be triggered. Cytokines often act in a synergistic manner, but may also have antagonistic effects.

Cytokines alter the biological response of the target cell, by influencing the expression of other proteins including cytokines.

Cytokines can be categorized in different families like interferons (IFN) including type I interferons, such as IFNα and IFNβ, type II interferons (tumor necrosis factor; TNF) interleukins (IL) chemokines colony-stimulating factor (CSF), etc. Using structural criteria several hormones show clear similarity with cytokines, making discrimination between these two difficult (Secombes et al. 2001).

Some key facts about cytokines are as follows:

- IL-1, IL-6, and TNFα the endogenous pyrogens.
- Macrophages are the predominant cell type secreting TNFα upon activation with, e.g., LPS (endotoxins). TNFα induces local coagulation in capillaries and increases endothelial permeability in acute inflammation. Upon release of high TNFα amounts, systemic toxicity can be observed, characterized by loss of volume to the interstitial tissue, resulting in generalized edema accompanied by disseminated intravascular coagulation (DIC). TNFα has been implicated in septic shock, cachexia, but also tumor regression (Kriegler et al. 1988).
- IFNα and IFNβ are mainly involved in innate immune response against viral infection. Whereas IFNα is mainly released by leukocytes, the main source for IFNβ is fibroblasts. IFN activate immune cells such as NK cells and macrophages and induce upregulation of antigen presentation in T cells. Further, they induce the antiviral state in infected cells, resulting in the transcription of various cellular antiviral genes coding for host defense proteins. Viral protein prevent the activation of the antiviral state leading to HLAI/MHC class I downregulation as a mechanism to escape the specific defense by cytotoxic T-lymphocytes. However, NK cells are specialized to sense cells with aberrant HLAI expression and may the killing then (see above).
- Homologous molecules of IFN have been found in many species, including mammals, birds, reptiles, amphibians, and fish species.

13.1.8.3 Complement

The complement system is evolutionary highly conserved and can not only be found in all vertebrates, but also in protostomes (e.g., nematodes, molluscs) and deuterostomes (Krem and Di Cera 2002).

The complement system consists of over 25 small proteins in the blood, which help ("complement") antibodies and phagocytic cells to clear pathogens from an organism. Complement proteins are generally synthesized in the liver and are found as inactive precursors in the circulation. When stimulated by one or several stimuli, proteases in the system trigger a cascade leading to **inflammation**, support of phagocytes by **opsonization** and at the end to the activation of the cell killing **membrane attack complex (MAC)**.

Three biochemical pathways exist that activate the complement system: the classical, the alternative, and the lectin complement pathway (Janeway et al. 2001).

The **classical complement** pathway starts at the so-called C1 component and typically requires antigen-antibody complexes for activation (specific adaptive immunity). It is thus the evolutionary youngest part of the system and expressed in vertebrates only. Much older are the innate pathways: the **alternative pathway** (starting at the C3 component) and the **lectin pathway** (using mannose-binding lectin, MBL, as an innate starter molecule). Independent of the different starting points, all three pathways use the enzyme **C3-convertase**, which cleaves and activates the component C3, creating a smaller fragment (C3a) and a larger fragment (C3b). Starting of the cascade results in a series of further cleavage and activation events.

C3b is an important opsonin of pathogens for phagocytosis. **C5a** is an important chemotactic protein and helps in the recruitment of immune cells. C3a, C4a, and C5a are called **anaphylatoxins**, as they increase vascular permeability and smooth muscle contraction by directly triggering degranulation of mast cells through their anaphylatoxin-receptors.

The formation of the **MAC** is initiated by C5b and consists further of C6, C7, C8, and polymeric C9 (Ferencik et al. 2004). MAC is the cytolytic end product of the complement cascade leading to the formation of a transmembrane channel and ensuing osmotic lysis of the target cell. The pore insertion is a primitive, conserved process and reminds of the mechanism of cellular cytotoxicity.

The Classical Pathway

The classical pathway is activated when C1q binds antibodies (IgG or IgM) complexed with antigen, which subsequently leads to cleavage of C4 and C2. C4b and C2b together form the C3-convertase, which promotes cleavage of C3 into C3a and C3b. C3b later joins the C3-convertase to make C5-convertase. Different immunoglobulin subclasses have different capabilities to fix complement (humans IgG1=IgG3, IgG2>>IgG4; mouse IgG2a=IgG2b, IgG1>IgG3) (Scott et al. 1990).

The Alternative Pathway

The alternative pathway is activated when C3b accumulates on cell surfaces. The alternative pathway is continuously activated at a low level as a result of spontaneous hydrolysis of the thioester bond of C3. However, accumulation of C3b does not occur in the healthy host due to the expression of complement regulatory proteins on the cell membranes like CD35, CD46, CD55, and CD59, which rapidly degrade any attached C3b. In contrast, pathogenic and foreign surfaces do not have regulatory proteins and get tagged by C3b. Once the alternative C3-convertase enzyme is formed on a pathogen or cell surface, it may bind another C3b, and initiate the cascade.

The Lectin Pathway

The lectin pathway is homologous to the classical pathway, but is activated by acute-phase proteins like MBL, and carbohydrate moieties of bacteria and viruses. This pathway uses, except C1, the same complement components (C2, C3, C4) for the start as in the classical pathway. MBL is a pentamer and typically a pattern recognition molecule (PRM) (Degn and Thiel 2013).

Functions of the Complement System

The main tasks of the complement system are

- Opsonization: labeling pathogens for phagocytosis
- Chemotaxis: attracting macrophages and neutrophils
- Cell lysis: killing of pathogens
- Aggregation and clumping of pathogenic structures

Some by-products of the complement cascade are called anaphylatoxins like C3a and C5a. They enhance the permeability of endothelia in acute inflammation and may also serve as opsonins labeling pathogens and surfaces for phagocytosis.

The complement system also helps to remove immune complexes from the circulation, since many soluble antigens form antibody complexes with too few IgGs to facilitate binding via Fcγ receptors. Such immune complexes can be toxins or remnants of dead microorganisms fixed by neutralizing antibodies (IgG, IgM in humans). Conventionally, immune complexes can be found after infections and antibody responses. The formed immune complexes can activate the classical complement pathway. This leads to C3b, which is recognized by erythrocytes expressing complement receptor 1 (CR1). Like decay shuttles, erythrocytes bind the opsonized complexes and transport them to the spleen or the liver, where macrophages and Kupffer cells remove the immune complex from the erythrocytes.

Complement and Disease

The complement system has to be tightly regulated due to its damaging potential. In this respect several diseases and deficiencies are associated with defects of single or more components of the complement system (reduced opsonization leading to more infections, overload of immune complexes, overload of anaphylatoxins), or defects in the complement regulatory proteins (attack of self). Deficiencies of the terminal pathway predispose to both autoimmune disease and infections (Das et al. 2013; Nilsson and Ekdahl 2012).

13.1.9 Synopsis

The innate immune is important for the survival of all multicellular organisms. It comprises evolutionary old soluble molecules, and various innate immune cells. The major principle of innate antigen identification is based on pattern recognition. Resulting defense functions range from phagocytosis, expulsion of toxic metabolites, enzymes, extracellular traps, cytokines, and mediators, to sophisticated channeling of these substances into the foreign/aberrant cell by a "kiss of death." Some cells of the innate immunity are essential for the initiation of adaptive immune responses in vertebrates.

13.2 Principles and Comparative Aspects of Adaptive Immunity

Franziska Roth-Walter, Erika Jensen-Jarolim, and Hannes Stockinger

Adapted from "Spezifische Abwehr" in: Krankheit, Krankheitsursachen und –bilder, MCWBlock8 (Ed. Marian B) 2012, 6th Ed., with permission of Facultas.wuv Universitätsverlag & Maudrich

13.2.1 Abstract

All organisms evolved mechanisms to protect themselves against pathogen invasion and hence possess an innate immune system. However, vertebrates have also evolved an additional strategy, the so-called adaptive immune system to further protect themselves against pathogens. In the classical view the immune system can be divided into an innate and adaptive branch of immunity with distinct function (Table 13.6) (Zinkernagel et al. 1996).

In contrast to innate defense mechanisms, which occur within minutes, the adaptive immunity is **slower** and usually needs 4–5 days from the primary encounter with the pathogen to be activated. One hallmark of adaptive immunity is the generation of **immunological memory** enabling the host to "remember" pathogens for years. As a consequence, the host can respond to a subsequent encounter stronger, very specifically and immediately. Another hallmark of the adaptive immune system is its high antigen-diversity. This is reached by the random a priori generation of a seemingly unlimited **repertoire of antigen receptors** with different specificities, in contrast to the innate immune cells with a limited number of pathogen recognition receptors.

13.2.2 Introduction

The cells of the adaptive immune system are special types of leukocytes, called lymphocytes. **B cells** and **T cells** are the major types of lymphocytes and are derived from hematopoietic stem cells in the bone marrow (Fig. 13.3). From single progenitors of the bone marrow a huge number of T- and B cells derive, which all carrying a different antigen receptor. These different antigen receptors are generated by chance. As a consequence, the body theoretically a priori must have an antigen receptor for every antigen, pathogen, or toxin.

B cells secrete immunoglobulins and thus are involved in the soluble (humoral) immune response, whereas T cells are involved in cell-mediated immune response. Basically, in adult mammals lymphocytes are produced in the **bone marrow**, B cells do also mature here (equivalent in birds is maturation in *Bursa fabricii*, a lymphatic organ); T cells mature in the **thymus**. Having never met their antigen, the still naïve lymphocytes leave these **primary lymph organs** and start patrolling in

Table 13.6 Classical view of the immune system

Innate immunity	Hall marks
• Physical/chemical barriers	• Immediate, within minutes
• Complement system	• Conserved
• Phagocytes	• No memory
• NK-cells	
• Dendritic cells	
Adaptive immunity	**Hall marks**
• T-Lymphocytes	• Slower, primary response 4–5 days, recall- response immediate
• B-Lymphocytes	• Specific due to antigen receptors, can mature due to receptor editing
	• Immunological memory for years

the periphery. In order to enhance the likelihood to rise an adaptive immune response to an antigen, the so-called antigen-presenting cells (APC) (macrophages, dendritic cells, and B cells themselves) trap antigens ("antigen trapping" and "antigen focusing") and transport them via the lymphatic system (shown for the human in Fig. 13.4a, b) to the **secondary lymph organs**. Hematogenic antigens (transported via the blood) reach the **spleen**; antigens that entered the tissue via **afferent lymphatic vessels** reach the **lymph nodes** (Fig. 13.4b) or related organs such as **Peyers' patches** in the intestine, which all have a strictly organized microanatomy: B cells found in the B cell zone and/or T cells in the T cell zone encounter the antigens and upon recognition get activated and clonally expanded (only the relevant B or T cell clone is expanded), forming a **germinal center**. Matured T- and B cells leave the secondary lymph organs via efferent lymphatic vessels or via the blood. Enlargement of the spleen or lymph nodes indicates stimulation of the adaptive immunity, e.g., in a patient suffering from an infection.

This classical view on adaptive immunity is today challenged by facts such as that NK cells harbor immunological memory (see Sect. 13.1), or that subpopulation of lymphocytes, like B1 and γδ T cells also have characteristics of innate immunity.

13.2.3 B Cells

B cells are lymphocytes, which can be distinguished from other lymphocytes, such as T cells and NK cells by the expression of the **Y-shaped B cell receptor (BCR).** The BCR consists of a **constant domain**, two paired heavy chains fixed in the B cell membrane, and a **variable domain** for antigen recognition composed by the heavy and additionally two light chains. Upon B cell activation and a functional and phenotypic change to a **plasma cell**, the BCR can be secreted into the circulation as immunoglobulin (antibody) (Fig. 13.5a, b), see Sect. 13.2.10 below. The antibody's variable domain determines **specificity** for an antigen, whereas the constant domain determines its **antibody class** and thereby its effector function.

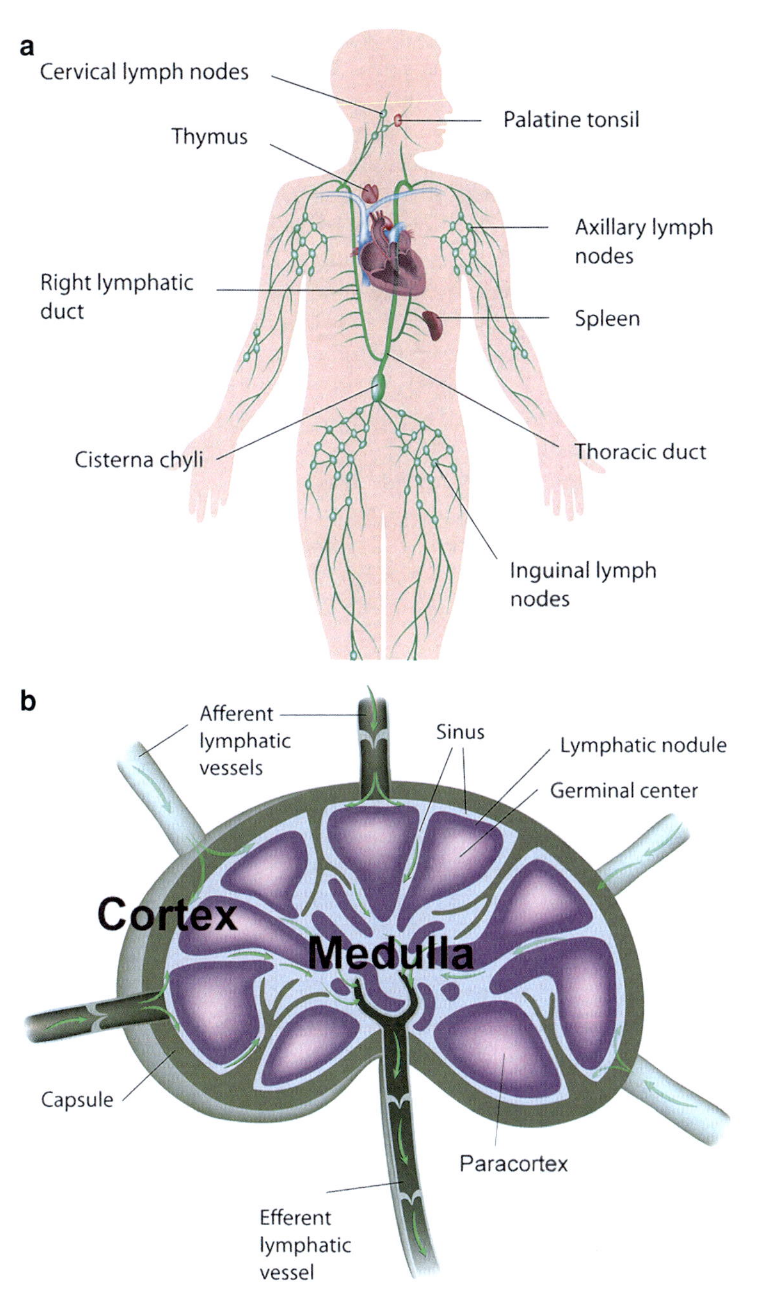

Fig. 13.4 The architecture of lymphoid tissues in mammalians. (**a**) The lymphoid tissue consists of mesh-like areas of connective tissues within the body containing immune cells and fluid. Primary lymphoid tissues are the bone marrow and thymus, in which immune cells develop. In the secondary lymphoid tissues, immune cells come into contact with foreign antigens. They include the lymph nodes, tonsils, spleen, and lymphoid tissues of the gut and respiratory tract. ©

Millions of different B cells each with a unique BCR are generated each day by recombination of the immunoglobulin genes. They circulate in the blood and do not produce antibodies until they become fully activated by a multivalent antigen. Once a B cell encounters its cognate antigen and receives an additional stimulatory signal from a **T helper cell**, it can further differentiate into **plasma B cells** or **memory B cells**. The B cell may either directly differentiate to one of these cell types or it may undergo an intermediate differentiation step: In the germinal center within secondary lymph organs the B cell induces mutations into the variable region of its immunoglobulin gene ("somatic hypermutation") by an enzyme, the **activation-induced cytidin deaminase (AID).** This evolutionary old AID enzyme is also involved in creation of the antibody **repertoire** (from birds on), in antibody class switching (occurring from amphibians upwards), during somatic mutation as well as in the recombination of genes of variable lymphocyte receptors (VLRs) in fish (Kaufman 2010; Conticello et al. 2007).

Together, during the adaptive immune response B cells (not the T cells) are instructed in respect to better antigen recognition and binding strength (affinity). Simultaneous cytokine stimulation may also improve the effector function by the usage of other heavy chains (class switching), with different characteristics.

Other functions of B cells include antigen presentation, cytokine production, and lymphoid tissue organization.

13.2.3.1 B Cell Types

- Plasma cells secreting large amounts of antibodies.
- Memory B cells are long-living antigen-specific activated B cells, which can respond quickly following secondary exposure with the antigens.
- B1 cells: is a subtype of B cells with more innate characteristics; they do not have memory, have polyspecific receptors, and secrete high levels of IgM or IgG. B1 cells are produced in the fetus and undergo self-renewal in the periphery (Montecino-Rodriguez et al. 2006).
- B2 cells are the commonly called B cells.
- Marginal zone B cells which are similar to B1 cells, can act in a T cell-independent manner, express high levels of IgM, are sessile in the marginal zone in the spleen, and are long-living cells (Pillai et al. 2005).

Fig. 13.4 (continued) [Alila Medical Images]—Fotolia.com. (**b**) Lymph nodes are generally ovoid, encapsulated aggregates with precise architecture. Its primary function is to enable interactions of immune cells with foreign antigens for an efficient and effective immune response. Many immune cells enter via the afferent and leave via the efferent lymphatic vessels the lymph node, T- and B-lymphocytes enter via the so-called high endothelial venules (HEVs). The lymph node contains germinal centers with predominantly B cells, but also T cells, macrophages, and follicular dendritic cells are found. The paracortex, the zone between follicles and the medulla, is rich of T cells and HEVs. The sinus often contains large numbers of macrophages. The medulla is the central area of the lymph node and contains plasma cells, macrophages, and B cells. On demand lymphoid follicles can be formed at almost any mucosal site, harboring a similar architecture like a lymph node, but devoid of a capsule. © [Alila Medical Images]—Fotolia.com

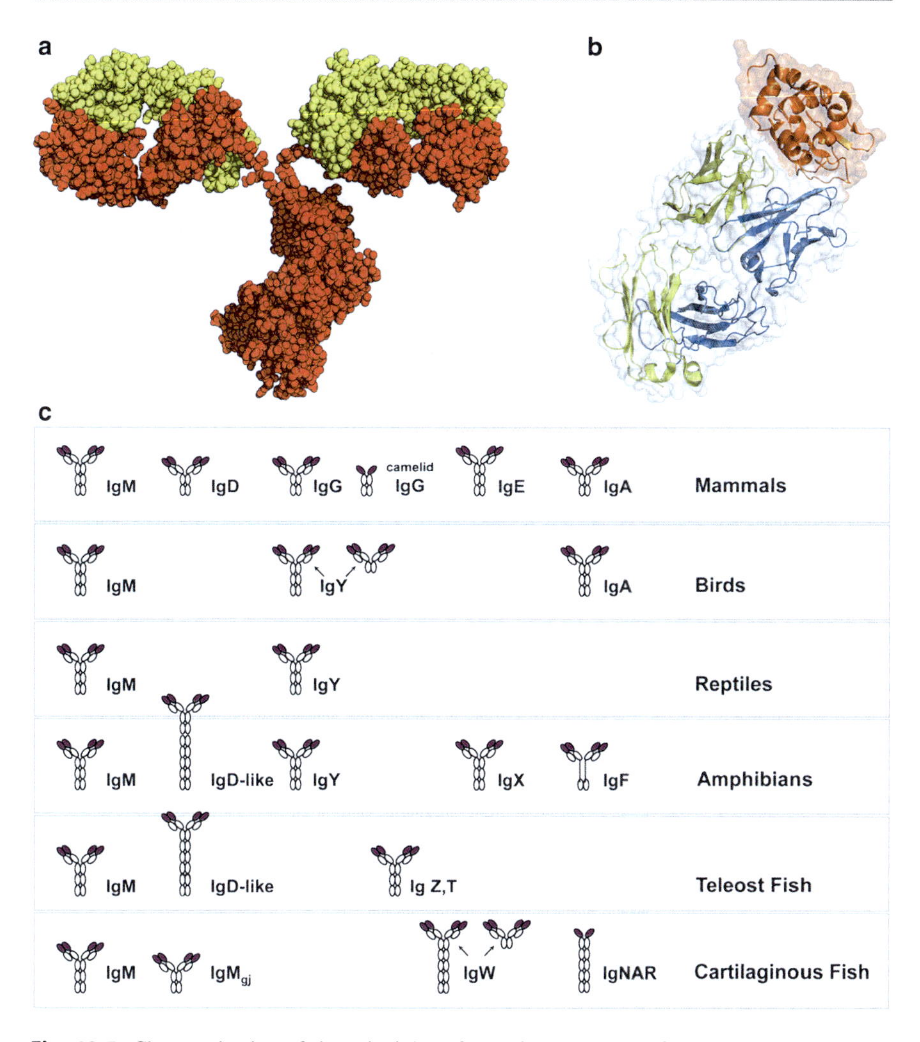

Fig. 13.5 Characterization of the principles of adaptive humoral defense: Antibodies. (**a**) An antibody is a large (140–180 kDa) Y-shaped protein, which is composed of two heavy chains (*red*) and two light chains (*yellow*). © [petarg]—Fotolia.com. (**b**) A single Fab arm of an immunoglobulin binding to its cognate antigen. The variable domains of heavy and light chains (*yellow* and *blue*) are the most important regions for binding to an antigen (*red*). The contact zone representing the ultimate tip of the Y of an antibody is called hypervariable region or complementary determining region (CDR). Exactly here antibodies make contact to a specific antigen. The CDRs can be improved for binding to an antigen by a process called affinity maturation. © [petarg]—Fotolia.com. (**c**) Overview on immunoglobulin classes in animals. Cartilaginous fishes (e.g., shark) have immunoglobulins IgM and IgW, which are closely related to mammalian/human IgD, as well as the heavy chain immunoglobulin new antigen receptor (IgNAR). Bony fish possess IgTs. Amphibians, reptiles, and birds have functional equivalents to IgGs, namely IgY antibodies, whereas IgX seem to be functional equivalent with mammalian IgA antibodies. (Adapted with kind permission of Prof. Martin Flajnik, University of Maryland, Baltimore, and Prof. Louis du Pasquier, Evolutionary Biology, Zoological Institute, University of Basel, Switzerland)

- Follicular B cells are B2 cells that mature in the primary follicles of the spleen/lymph nodes and hence are called "follicular" B cells.
- Regulator B cells have also been described having regulatory function and secreting cytokines like IL-10 and TGFβ (Vadasz et al. 2013).

13.2.4 T Cells

T cells are lymphocytes, which can be distinguished from other lymphocytes, such as B cells and NK cells by the expression of the T cell receptor (TCR), classically composed by an α and a β chain, more seldom by a γ and a δ chain. T cells have their name from the thymus, where they mature. Similarly to B cells several subtypes have been described.

13.2.4.1 T Cell Types

- CD4+ T helper cells (Fig. 13.6) assist other leukocytes in their immune response (activation of cytotoxic T cells, maturation of B cells, class switch) once they are activated by APC via MHC class II molecules. Dependent on their cytokine profile, they differentiate into Th subtypes including Th1, Th2, Th3, Th9 Th17, Tregs, or T_{FH}.
- CD8+ cytotoxic T cells (Fig. 13.7) are able to induce apoptosis in infected cells presenting antigens via MHC class I molecule.
- Memory T cells confer long-lasting antigen-specific protection after an infection has resolved.
- NK T cells recognize glycolipid antigens presented by CD1 molecule.
- γδ T cells are a small subset of T cells that possess a distinct TCR on their surface and usually reside in the gut mucosa and are not MHC restricted and have been suggested to recognize lipid antigens (Hayday 2000).

13.2.5 Antigen Receptor Diversity by V(D)J Recombination

The generation of the high diversity number of antigen receptors in B and T cells takes place in the primary lymphoid tissues: the bone marrow for B cells/Bursa fabricii in birds and in the thymus for T cells. The diversity is reached by three molecular mechanisms (Kato et al. 2012):

(a) Gene rearrangement: V(D)J recombination randomly combines **V**ariable, **D**iverse, and **J**oining gene segments in vertebrate lymphocytes. Each lymphocyte has multiple sets of V, D, and J gene segments. A combination of a random selection of V, D, and J already increases the diversity.

As an example, the TCR is composed of an α- and a β-chain. For the combination of the α-chain humans have a set of 70 V-α segments, 61 J-segments, and 1 constant segment for disposition. The number of these segments largely depends

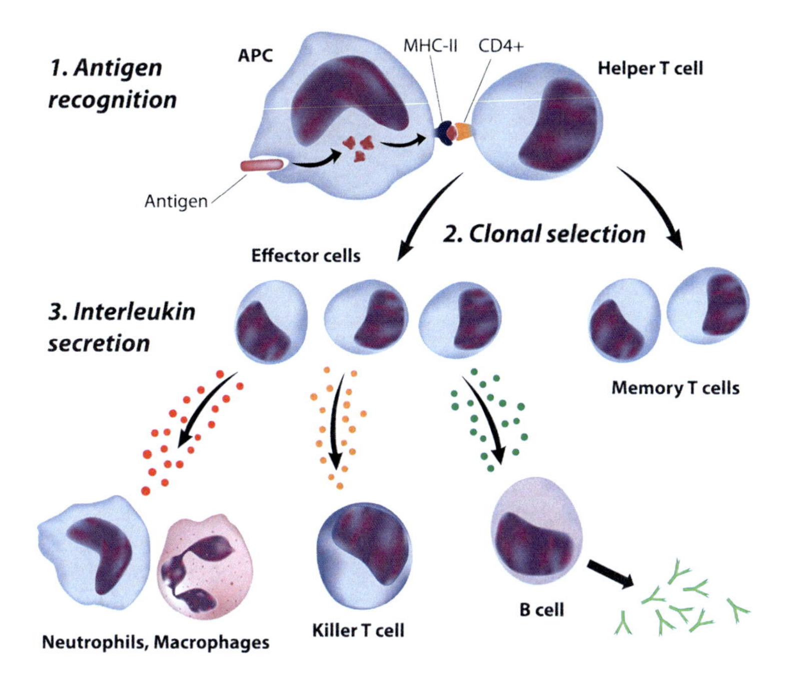

Fig. 13.6 Helper T cell activation and action. T helper cells recognize antigens presented via MHC class II by antigen-presenting cells. Upon further activation with costimulatory molecules and secreted cytokines, T helper cells begin to proliferate into effector and memory T helper cells. Effector T helper cells are important in the host defense against intra- and extracellular pathogens by secreting mediators to further activate innate immune cells, as well as cellular immune cells and stimulating B cells to secrete antibodies. © [Alila Medical Images]—Fotolia.com

on the species and determines combinatorial diversity. For construction of the β-chain, it can choose a set of 52 V-segments, 13 J-segments, 2 D-segments, and 2 (constant) C-segments.

(b) Combination of chains: each antigen receptor consists of two different chains. The TCR consists of an α- and a β-chain, the BCR (and hence its soluble form, the immunoglobin; Fig. 13.5a, b) are composed of two identical heavy and light chains. The constant (C) and variable (V) regions of both, TCR and BCR, are built symmetrically facing each other. As a consequence, e.g., out of two different α- and β-chains four different antigen receptors can be built.

(c) Junctional diversity: during recombination joining of the gene segments can be additionally diversified by insertion of P- and N-nucleotides. The inaccuracies of joining provided by junctional diversity is estimated to triple the diversity initially generated by these V (D) J recombination, e.g., if two cells A and B combine the same gene segments, the antigen receptors would have different specificities due to differences in joining the segments.

Through the mechanisms for antigen receptor diversity, a vast number of different antigen receptors (10^{18}) determining the **immunological repertoire** can be generated (see Table 13.7). Among the total number of cells of an individual

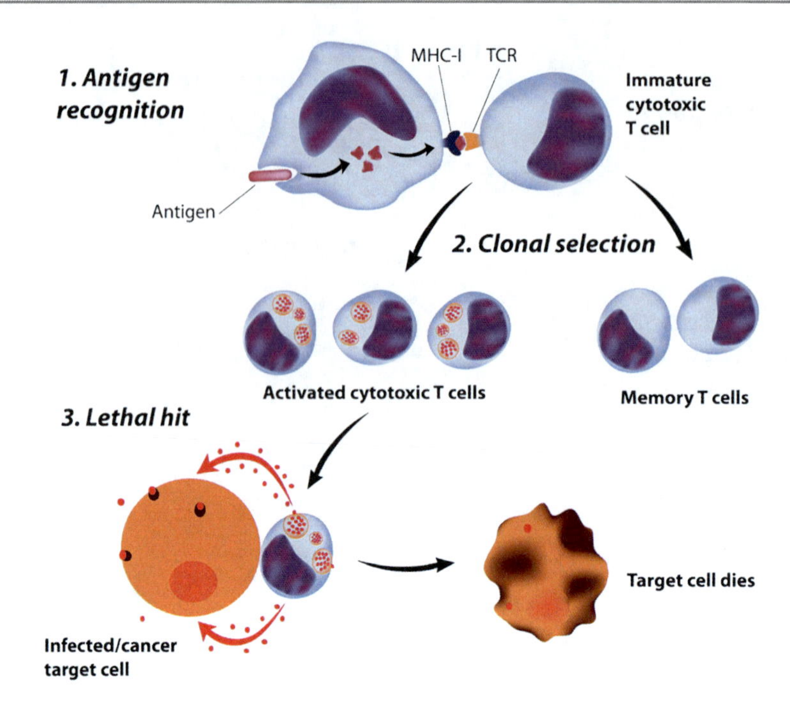

Fig. 13.7 Cytotoxic T cell activation and action. Intracellular antigens are presented via MHC class I molecules to cytotoxic T cells. Upon further binding with costimulatory molecules (CD28/CD80, CD86), cytotoxic T cells proliferate into effector cytotoxic T cells and memory T cells. Cytotoxic T cells release perforin and granzymes which eventually lead the target cell to apoptosis. © [Alila Medical Images]—Fotolia.com

(10^{14}), an average human normally has around 1×10^{12} lymphocytes. With the assumption that 10^2–10^3 cells exist per antigen-specificity, the human body harbors around 10^9 specificities. Each individual is composed with a different set of specificities, which are changing during live. Hence, during an average life of about 70 years, up to 10^{13} antigen receptors with different specificities are generated.

13.2.5.1 Structure of Antigen Receptors

TCR- (α- and β-recombinated chains) as well as BCR- (recombinated heavy and light chains) complexes are composed of two parts: the ligand binding receptor and a signal transduction moiety. The signal transduction moiety of T cells is called CD3. CD3 is built up of four polypeptide chains called γ-, δ-, ε-, and ζ-chains. On the BCR the signal transduction moiety is called CD79, consisting of an α- and a β-chain.

By its first encounter with an antigen, T- or B cells are activated and CD3 and CD79, respectively, becomes phosphorylated, the cell proliferates and differentiates, e.g., to generate a population of antibody-secreting plasma B cells and memory B cells.

Table 13.7 The mechanisms for antigen receptor diversity create a vast number of different antigen receptors

Possible numbers of variation in Lymphocytes	
B cells	1×10^{18}
T cells	1×10^{18}
Total	2×10^{18}
Cells in humans	
Total	1.1×10^{15}
Bacteria	1×10^{15}
Human cells	0.1×10^{15}
Lymphocytes	1×10^{12}
Specificities	10^2–10^3

13.2.5.2 Adaptive Immune Responses in Jawless Vertebrates

Vertebrates are classified as jawed and jawless, with the later being the most primitive vertebrates and which diverged approximately 500 million years ago (Figs. 13.1a, b, and 13.5c) (Kasamatsu 2013). Jawless vertebrates also called mouthless fish (hagfish, lampreys) do have lymphocyte-like cells with morphological similarities to B and T cells of mammals (Fig. 13.1b). Moreover, they are able to produce antigen-specific immunoglobulin analogues and can form immunological memory. However, they have developed unique antigen receptors, the so-called **VLRs.** Jawless vertebrates do not harbor the antigen receptors of jawed vertebrates such as TCR, BCR, MHCs, and RAGs. Similarly to mammals having T- and B cells, jawless vertebrates evolved two lymphocyte-lineages (**VLRA+** and **VLRB+**, respectively) (Kasamatsu 2013).

13.2.6 Tolerance

13.2.6.1 Advantage and Disadvantage of the Adaptive Immunity

The adaptive immune system has to keep the right balance between tolerance to harmless external antigens (dietary antigens, commensal bacteria) as well as self-antigens, but responding to harmful exogenous antigens (viruses, bacteria, helminthes) and aberrant endogenous antigens (tumors).

Due to the ability to generate *a priori* receptors able to bind to potential exogenous antigens (like bacteria, viruses, and helminthes) as well as endogenous antigens (tumors), the adaptive immunity has a risk to generate antigen receptors that recognize either innocuous exogenous antigens (dietary antigens) or self-antigens (Table 13.8). As a consequence of a break in immune tolerance autoimmunological disorders, inflammatory bowel disease or allergies may arise. Many pathogens (external antigens) have learned to exploit immune tolerance. They actively induce tolerance and escape immune recognition to invade the host. By very similar mechanisms malignant cells escape immune surveillance due to the inability of the immune system to recognize the tumor as harmful.

Table 13.8 Advantages and disadvantages of the great receptor repertoire of T and B cells

Advantage against	Disadvantage against
Harmful exogenous antigens	Harmless exogenous antigens
Viruses, bacteria, helminthes	Dietary antigens, commensal bacteria → allergy, inflammatory bowel disease
Harmful endogenous antigens: Malignant cells	Harmless self → autoimmunity

The adaptive immune system has evolved different tolerogenic strategies to cope with hyper-activated immune cells. Dependent on site and reactivity to endogenous or exogenous antigens, tolerance is subdivided into

I. **central tolerance,** which is induced at the level of the bone marrow (for B cells) or the thymus (for T cells) against cells reacting to self-antigens
II. **peripheral tolerance**, which happens in the periphery; lymphocytes reacting towards self become anergic (paralyzed), or regulatory immune cells dampen their immune response by cytokine secretion
III. **acquired tolerance,** where the major site for tolerance induction to exogenous harmless antigens happens in the periphery and in the liver, respectively, and leads to generation of regulatory and anergic immune cells (important forms: tolerance to commensal bacteria, dietary antigens, tolerance in gestation)

The consequence of tolerance mechanisms may involve **elimination** of the auto-aggressive lymphocytes (induction of cell death by apoptosis—recessive tolerance), or **anergy** (induction of unreactivity, anergy), and of **dominant tolerance** by **suppressor T cells**.

13.2.6.2 Recessive Central Tolerance in the Thymus

It is estimated that over 95 % of all generated lymphocytes will recognize self-antigens due to the randomly occurring V(D)J recombination. Hence, the system has evolved a strategy to eliminate these self-reacting immune cells. For B cells the site of elimination is the bone marrow, whereas for T cells this occurs in the thymus. The process of elimination is called **clonal deletion**. Only T cells not recognizing self are allowed to proceed to the periphery. Once this naïve T cell encounters an exogenous antigen from a pathogen, it will mature and proliferate (**T cell clone**) and generate multiple daughter cells with the same specificity (T cells undergo clonal selection). The generation of T cell clones capable of combating pathogens takes 4–5 days. The functional mature T cells are the so-called effector T cells. Simultaneously, also memory cells are produced.

Characteristics for Central Tolerance in the Thymus

In mammals, the thymus is pivotal for tolerance induction for self-antigens by clonal deletion. As this process is dependent on encountering the self-antigens of the whole body, the thymus had to generate an additional source of all possible antigens through the action of the transcription factor **AIRE (autoimmune regulator),** which allows the expression of organ-specific antigens, e.g., of the pancreas or prostate, in the thymus. Medullary thymic epithelial cells (mTEC) then act as APC and activate AIRE gene (aire) to present self-antigens via MHC class I and MHC class II molecules. A comparative analysis between human and several species including opossum, chicken, zebrafish, and pufferfish revealed that an aire-dependent T cell tolerance mechanism dates back as a minimum to bony fish (Saltis et al. 2008).

For induction of central tolerance in T cells, immature T cells have to undergo positive and negative selection in the thymus.

Positive selection for T cells is induced in the cortex, of the thymus. Developing thymocytes (i.e., T cells in thymus) are exposed to cortical thymic epithelial cells presenting self-antigens via MHC class I or II. Only those thymocytes that bind to the MHC class I- or MHC class II-peptide complex receive a "survival signal." T cells not binding to the MHC-peptide complex die "by neglect," which means that these thymocytes do not receive growth stimulatory signals. Also a thymocyte's function is determined during positive selection. Thymocytes binding to MHC class I-peptide complex will mature to **cytotoxic T** cell precursors expressing CD8 (Fig. 13.7), whereas those binding to MHC class II-peptides will differentiate to **T helper (Th)** precursors expressing CD4 (Fig. 13.6) (Ma et al. 2013).

Subsequently the thymocytes migrate towards the medulla in the thymus, where they undergo **negative selection**. There mTEC cells present self-antigens to them. Thymocytes that bind too strongly to "self" MHC-peptide complex undergo apoptosis. Thymocytes surviving the selection exit the thymus as naïve T cells. Some of these cells are selected to become natural Tregs. One key element for regulatory T cells (Tregs) is the transcription factor **Foxp3**. Mice missing this gene, the so-called Scurfy mice, suffer severely from autoimmune disorders. Similarly, also humans having a defect on this gene suffer from the very severe autoimmune disorder IPEX (immune dysregulation, polyendocrinopathy, enteropathy, X-linked).

However, despite the stringent selection criteria, a small number of cells will still retain the capability to react to self. Autoreactive cells that escape negative selection in the thymus but then later encounter an abundance of self-antigen in the periphery may be either anergized or deleted in a process known as peripheral tolerance (Murphy 2011), fight against cancer or cause autoimmunity later.

13.2.6.3 Tolerance vs. Immune Activation

The immune system is poised to react to insults, but should remain tolerant to harmless or beneficial antigens. Introduction of antigens in the peripheral tissues, whether by subcutaneous or intramuscular injections or injury, usually leads to local infiltration of inflammatory cells including T cells, B cells, and specific immunoglobulin production. By contrast, antigens where the portal of entry is through mucosal sites (oral, nasal, and respiratory) usually elicit active inhibition of the immune response to those antigens, systemically. Also antigens that are directly injected in the blood and hence flow into the liver will induce tolerance (overview in Table 13.9).

Table 13.9 Routes of antigen entry determines immune outcome

Route of antigen	Immune outcome
Subcutaneous	Immunity
Intramuscular	Immunity
Intravenous	Tolerance
Mucosal	Tolerance

The decision to initiate immunity or tolerance usually depends on the site of antigen entry and integrates signals from several sources. For example, **danger signals** can be transmitted by pattern recognition receptors (PRRs) recognizing PAMPs (see Sect. 13.1). These receptors, including the TLRs and the NOD (nucleotide-binding oligomerization domain) family of proteins, detect common motifs that are present on pathogens and can initiate an inflammatory cascade by activating **costimulatory molecules**.

To activate T cells, antigens have to be presented by APC using antigen presentation molecules such as MHCs, see below.

13.2.6.4 Antigen Presentation

Antigen presentation can occur via professional APC-like dendritic cells or by tissue cells. Professional APC can present antigens via MHC class I and MHC class II, whereas cells of tissue usually only express MHC class I and are less efficient to activate the immune system. Moreover, professional APCs have the capacity to costimulate, whereas nonprofessional APCs most often rather induce tolerance due to a lack of costimulation. The most important professional APCs are the dendritic cells (DCs), but also macrophages, certain B cells and certain activated epithelial cells can present antigens via MHC class I and MHC class II. The subsequent differentiation and activation of T cells depend on

Signal 1: interaction of TCR with MHC-peptide complex

Signal 2: binding of costimulatory molecules like CD28/CD80 or CD86 or inhibitory molecules like the cytotoxic T-lymphocyte antigen 4 (CTLA4), or programmed death 1 (PD1) with CD86, PD-L1 (PD-ligand 1), or PD-L2

Signal 3: soluble cytokines secreted by immune cells or other cell types

When only Signal 1 occurs, without additional Signal 2 and Signal 3, cells become anergic. With an additional inhibitory Signal 2, cells will be usually deleted by apoptosis.

However, an additional activating Signal 2 in combination with Signal 1 activates T cells. The type of generated T cell (Th1, Th2, Th17, Treg) however, is dependent on Signal 3. All soluble factors secreted by surrounding cells are here decisive.

13.2.6.5 Acquired Immune Tolerance

Tolerance to foreign antigens also can be acquired. Since the 1960s it is known that systemic tolerance can be induced by either administration of a single high dose of antigen (>20 mg) or repeated exposure to lower doses (100 ng to 1 mg) in mice. These two forms of tolerance, now termed high- and low-dose tolerance are mediated by distinct mechanisms. Similarly to *peripheral tolerance*, high doses of oral antigen can induce lymphocyte anergy and/or deletion. **High-dose tolerance** occurs by induction of apoptosis, whereas anergy occurs through TCR ligation without adequate costimulation (by cognate interactions between CD80 and CD86 on APCs with CD28 on T cells, or by soluble cytokines such as IL-2).

Low-dose tolerance is mediated by active suppression of the immune response by Tregs. Tregs are mainly CD4+ T cells and can—at least in humans and mice—be divided into three subgroups: CD4+CD25+ Tregs, TH3 cells, and TR1 cells. Noteworthy, even though each of these subtypes of Tregs is activated in an antigen-specific manner, they can suppress immune responses in the immediate surrounding area in an antigen nonspecific manner by secretion of suppressive cytokines and expression of inhibitory cell-surface ligands. This phenomenon is known as **bystander suppression** (Shao and Mayer 2004; Mayer and Shao 2004a, b).

13.2.7 Host Defense

There are five main types of pathogens: viruses, bacteria, fungi, protozoa, and worms. Virtually all pathogens have an extracellular phase where they are vulnerable to antibody-mediated effector mechanisms. However, during intracellular phases pathogens are not accessible to antibodies. Therefore, cytotoxicity is needed for completing the host defense (Janeway et al. 2001).

Dependent on their replication procedure, pathogens can be classified (Goodpasture 1936) as

(a) ***Extracellular pathogens***, which replicate and live in extracellular spaces and are resistant to phagocytosis, e.g., *E. coli*, Streptococci. They are recognized by B cells or dendritic cells and their antigens internalized via vesicles. Antigens that are internalized via vesicles are presented via MHC class II molecules to immune effector cells.

(b) ***Facultative (vesicular) intracellular pathogens***, which are able to survive inside host cells, e.g., salmonella, listeria, mycobacterium tuberculosis, legionella, fungi, and protozoa. They are generally resistant to complement and antibodies. They survive inside the tissue macrophages where they usually reside and replicate in the vesicular compartment (endosomes, phagosomes). Since their antigens also reside in vesicles, also in this setting antigens are presented by MHC class II. As a consequence, in (a) and (b) T helper cells are activated which contribute by cytokines to killing and antibody production.

(c) ***Obligate (cytoplasmic) intracellular pathogens***: they must reside in host cells to survive, e.g., viruses, chlamydia, and rickettsia. They usually replicate in the cytosol. Cytosolic antigens are presented by MHC class I molecules (Silva 2012). This makes sense as these antigens only can be eradicated by cytotoxic T cells.

13.2.7.1 Cytotoxic T Cells

Cytotoxic T cells expressing CD8 in close association with their TCR complex recognize cytosolic antigens presented by MHC class I molecules in virus-infected cells (Fig. 13.7). Cytotoxic T cells kill these infected cells by induction of apoptosis similarly to NK cells (see Sect. 13.1). Through the action of perforin, which forms holes in the plasma membrane of the target cells, granzymes enter the cytoplasm of the target cell and their serine protease function triggers the caspase cascade, which eventually lead to apoptosis (programmed cell death).

13.2.7.2 T Helper Cells

Cells presenting pathogenic antigens by MHC class II molecules are helped by **T helper** cells expressing CD4 (Fig. 13.6). For instance, macrophages infected with mycobacteria are attacked by T helper cells. Th cells activate B cells, which produce antibodies against the extracellular pathogenic antigens leading to destruction and expulsion of the pathogen. Depending on the nature of the antigen and the costimulatory signal, several subsets of T helper cells may be activated with diverse functions. Th1 cells produce IFNγ, are able to activate macrophages and induce an isotype switch to IgG in B cells. Th2 cells preferentially stimulate B cells to Th2-type antibody production (IgE and IgG4 in humans; IgE and IgG1 in rodents); Th17 cells secrete IL-17, which mobilizes neutrophils able to phagocytose extracellular bacteria and fungi.

Taken together, presentation of exogenous antigens by MHC class I molecules leads to cytotoxic T cell activation and death of the infected cell on the one hand, whereas presentation of antigens by MHC class II proteins rather leads to immune activation of macrophages and B cells.

MHC Molecules

Major histocompatibility complex (MHC) molecules cannot present the entire pathogen. Thus cells have to dissect the proteins of the pathogens into small peptide fragments of 8–10 amino acids (aa) for the cavity of the MHC class I protein and 20–30 aa for the MHC class II protein. The process is termed antigen processing. The antigens are then exhibited on the cell surface of the cell, a process termed **antigen presentation**.

13.2.8 Antigen-Processing and Presentation

13.2.8.1 Antigen-Processing and Presentation by MHC class I Molecules

Antigen processing starts when the pathogen enters the cytoplasm of a host cell. For example, a virus infects the cell and integrates its DNA or RNA into the nucleus of the host cell. Subsequently, the replication machinery starts with the production of viral proteins in the cytosol. These **intracellular proteins** are then degraded into small peptides by a highly intricate protease system and transported through the transporter associated with antigen processing (TAP) into the endoplasmic reticulum (see Chap. 2). There MHC class I molecules bind to the TAP complex transporting cytosolic 8–10 aa long peptides and, after binding of peptide, the peptide/MHC complex is transported through the Golgi apparatus to the cell surface. Every mammalian cell is capable (exception red blood cells without nucleus) of presenting peptides by MHC class I molecules and hence able to be killed by CD8 cytotoxic T cells (Andersen et al. 2006).

13.2.8.2 Antigen-Processing and Presentation by MHC class II Molecules

The ability to process and present antigens via MHC class II molecules is restricted to APCs, which have acquired **extracellular proteins** by endocytosis/phagocytosis such as B cells, macrophages, dendritic cells, and activated epithelial cells. B cells capture antigens via their BCRs, which subsequently is endocytosed. In the endosomal pathway, the pH constantly decreases leading to activation of proteases, which further digest the proteins into peptides. These peptide-carrying vesicles fuse with MHC class II containing vesicles. MHC class II is produced in the endoplasmic reticulum and its peptide-binding cleft blocked with a special polypeptide known as the invariant chain. In the MHCII containing endosomes the invariant chain is dissected into CLIP peptide functioning as a place holder for the MHC class II peptide slot. During fusion, the HLA-DM protein catalyzes loading of the extracellular peptides with MHC class II. The MHC class II-peptide complex is then transported via vesicles to the cell membrane for further presentation to CD4 T helper cells (Vascotto et al. 2007).

13.2.9 Lymphocyte Activation

13.2.9.1 T Cell Activation

T cell activation occurs via binding to MHC-peptide complex. T cells can be categorized into cytotoxic and helper T cells. Cytotoxic T cells express the molecule CD8, whereas helper T cells express CD4 (see Figs. 13.7 and 13.6). CD8 can bind the constant region of the MHC class I protein, whereas CD4 is able to bind to the constant region of the MHC class II. Since CD4 and CD8 recognize the MHC molecules, they are termed co-receptors of the TCRs.

CD4 drew special attention due to the fact that the HIV virus infects T helper cells via this receptor. As a consequence CD4-infected cells are not only attacked by the virus itself, but they are also destroyed by the cytotoxic T cells, which thereby not only eliminates the virus, but also cells playing a central role in the adaptive immune response.

In conclusion, for protection of the cellular compartments cytotoxic CD8+ T cells as well as NK cells are responsible by inducing programmed cell death in the infected cells. Specific protection of the extracellular compartments is primarily performed by antibody-producing B cells, especially when activated by T helper cells.

13.2.9.2 B Cell Activation

A critical difference between B cells and T cells is how each lymphocyte recognizes its antigen. B cells recognize their antigen in their native form, whereas T cells only recognize their cognate antigen in context with MHC molecules, as a processed peptide. B cell activation leads to clonal proliferation and terminal differentiation into plasma cells secreting antibodies. B cell can be activated in a T cell-dependent or -independent manner.

T Cell-Dependent B Cell Activation

Most antigens are T cell-dependent, meaning that B cells require further stimulus by T helper cells for antibody production (Noelle and Snow 1991). For **T cell-dependent activation** of B cells, the first signal comes from antigen cross-linking at least two BCRs, and the second signal comes from costimulation provided by a T cell. T-dependent antigens hence comprise proteins that are presented by a B cell via MHC class II to a T helper cell. Subsequently, the primed Th cell expresses the costimulatory molecule **CD40L** that binds to CD40 on the B cell. Moreover, the T cell secretes cytokines that activate the B cell leading to its proliferation and differentiation into plasma cells. Plasma cells secrete then the soluble form of the BCR: antibodies/immunoglobulins.

The chance that an antigen-specific B cells will meet an antigen-specific T cells for further activation is with a probability of 1×10^{18} rather slim. Hence, nature has provided other APC-like dendritic cells that are able to take up extracellular proteins independent of any receptors by macropinocytosis and further process and present them via MHC class II. In contrast to tissue cells, dendritic cells are mobile and once activated by engulfing extracellular proteins and after secondary activation by PRRs, they travel to the lymph node, where immune cells are concentrated and the probability for binding to its cognate partner increases. Alternatively, a dendritic cell may also recognize antigens via surface bound immunoglobulins. This process of antigen recognition is termed **antigen trapping** and **antigen focusing**.

Antigen Presentation in the Lymph Node

Immune cells can only enter the lymph node by passing the T cell zones via the high endothelial venules (Fig. 13.4b). In the T cell zones, B cells become fully activated and move to the primary follicle consisting of a network of follicular dendritic cells. Follicular dendritic cells are specialized non-hematopoietic stromal cells that reside in the lymph nodes. They possess long dendrites and carry intact antigen on their surface. Here B cells start to proliferate rapidly and after a few days of vigorous proliferation, the characteristic structure of the germinal center becomes apparent: a dark zone consisting almost exclusively of proliferating B cells and a light zone containing T cells, follicular dendritic cells and macrophages, and B cells which undergo immunoglobulin class switching and improved affinity (see also Sect. 13.2.3) (Klein and Dalla-Favera 2008; Liu 1997). These processes together are pivotal for T cell-dependent antigens.

T Cell-Independent Activation

T cell-independent activation occurs when a B cell binds to an antigen and only receives secondary activation by PRRs, e.g., TLRs, but not by a T cell. Alternatively, the antigen itself may be a molecule with multiple repetitive segments able to simultaneously cross-link enough BCRs to fully activate the B cell (Cerutti et al. 2013).

13.2.10 Structure and Function of Antibodies

An antibody is a large **Y-shaped** protein consisting of two identical heavy and two light chains (Milstein 1985). Both chains have constant (C) and variable (V) regions and are built symmetrically facing each other. They are also called immunoglobulins.

The amino acid sequence in the tips of the "Y" varies greatly among different antibodies. This variable region, composed of 110–130 amino acids, gives the antibody its specificity for binding antigen. The variable region includes the ends of the light and heavy chains (antigen-binding fragment, Fab). The constant region (constant Fragment, Fc) determines the mechanism used to destroy antigen. Mammalians have developed diverse classes and subclasses of antibodies and receptors, depending on their environmental behalf (Clark 1997). Generally, antibodies are divided into five major classes **IgM, IgG, IgA, IgD,** and **IgE**, based on their constant region. The differentiation of antibodies in vertebrates is depicted in Figs. 13.1b and 13.5c. The different classes are termed isotype, further specialization within classes are termed subclasses. For instance, in humans there are 4 IgG subclasses (IgG1, 2, 3, 4), and 2 IgA (IgA1, 2) subclasses; in mouse there are 3 IgG subclasses (IgG1, IgG2a, b, IgG3) and only 1 IgA; in the rabbit there are 13 IgA subclasses (Schneiderman et al. 1989). They differ in their biological properties, functional locations, and the ability to deal with different antigens (Janeway et al. 2001). A major difference is given by their ability to bind to different immunoglobulin Fc receptors: neonatal receptor FcRn, receptor for polymeric immunoglobulins: polyIgR, for IgG: FcγR, for IgE: FcεR, for IgA: FcαR and their subtypes.

Isotype or class switching occurs during differentiation of B cells in secondary lymph organs by changing the constant region. That means, that an antigen-specific B cells can differentiate into daughter cells of different isotypes, while maintaining their antigen specificity (Milstein 1985). A switched B cell cannot switch back to the previous class, because during the isotype switch the respective DNA segment is looped and cut out.

Pathogens usually enter the host by breaking the epithelial barrier of mucosal surfaces of the gastrointestinal, urogenital, and respiratory tract, or by injury. They usually cause inflammation in the invaded tissue. Very rarely toxins of insects, or pathogens by injection needles or by wounds invade the body directly via the blood. However, all different compartments are protected from infections by antibodies with their specialized isotypes.

The first isotype generated during primary infection is the **IgM** molecule **(primary antibody response).** IgM mostly assembles to a polymeric complex consisting of 5 IgM molecules (pentamer). It eliminates pathogens in the early stages of B cell-mediated (humoral) immunity before there is sufficient IgG. Because of its size, it cannot diffuse across blood vessels and hence is mainly found in the blood. IgM is very efficient in activating the complement by the classical pathway (see Sect. 13.1). The primary response is slow and usually has a lag phase of 5 days.

Further antigen contact may lead to a **secondary immune response**, also called **memory response**. The quality of this response depends on the costimulatory signal, antigen amount, number, and interval of antigen encounters (see vaccination below). By a subsequent encounter of the antigen/pathogen, the antibody-response immediately accelerates and vastly exceeds the levels obtained during primary infection. IgG, IgA, and IgE are the typical mammalian immunoglobulin classes during a secondary response.

Half of the produced **IgA** is found as a dimer in peripheral blood and can as such be secreted to the mucosal sites, such as the gut, respiratory and urogenital tract. Its main function is to prevent colonization by pathogens by neutralization and thus only has weak opsonization and complement activation capacity. IgA is also found in saliva, tears, and breast milk. IgA in the breast milk protects small infants until they are able to produce antibodies by themselves.

But also **IgG** is able to cross the human placenta and be transferred from the mother to the newborn. As a result, newborns have similar IgG levels than their mothers. However, this is not the case in all mammalians; for instance, the porcine placenta in impermeable for immunoglobulins. Generally, IgG is the primary isotype in the blood, highly conserved in mouse, rat, and humans (Clark 1997) and can be further divided in human into four subclasses. IgG opsonize pathogens and can activate the complement system efficiently.

Only low concentrations of **IgE** are detected in the blood. This is due to the outstanding affinity of (the cytophilic) IgE to its FcεRs, e.g., on mast cells and eosinophils. Mast cells are scattered under the skin and the mucosa and along blood vessels. When antigens are captured and cross-link minimally two IgEs on mast cells, they trigger cell activation and release of mediators like histamine. The results are wheezing, coughing, and vomiting to expel the pathogen. An overreaction is usually seen in aberrant allergic reaction to harmless antigens. **IgE** has been implicated to be important against parasitic infections by promoting eosinophilic attack and recruiting phagocytes and macrophages by the release of chemokines and cytokines.

The precise function of **IgD** as an ancestral membrane-bound antigen-receptor between fish and humans and is still under investigation; it may especially interact with innate cells such as basophils (Chen and Cerutti 2011).

13.2.10.1 Antibody Function

Antibodies protect the extracellular compartments from pathogens and their products by three mechanisms.

Table 13.10 The Immunoglobulin classes of vertebrates

Vertebrates	IgG classes	Class switch	Class switch
Jawless—(e.g., lampreys, hagfish)	VLR-producing lymphocytes	–	–
Cartilaginous fish (sharks)	IgM, IgW, IgNAR	–	–
Teleost fish (zebra fish, trout)	IgM, IgD, IgT	–	–
Amphibian	IgM, IgY, IgX, IgD	+	+
Reptiles	IgM, IgY, IgA?	+	+
Birds	IgM, IgY, IgA	+	+
Mammals	IgM, IgG, IgA, IgD, IgE	+	+

1. Neutralization: Here the antibodies bind via the hypervariable part of its Fabs to the epitopes of pathogenic structures and molecules and can thereby prevent entry and binding (Fig. 13.5b).
2. Opsonization: Antibodies bind to pathogens causing them to agglutinate. By coating the pathogen, antibodies stimulate effector functions against the pathogen in cells that recognize their Fc region. Subsequently, phagocytes will eliminate the pathogen by phagocytosis. Due to the antibody diversity, nearly an unlimited amount of specificities can be generated.
3. Complement activation: Antibodies bind to pathogenic structures, which enables complement to bind via the antibodies indirectly to the pathogen and activate the complement cascade. Moreover, phagocytes are able to better recognize the pathogen via their complement receptors (see Sect. 13.1).

13.2.10.2 Antibodies from Fish to Mammals

Cartilaginous fish (e.g., shark, rays) produce three types of immunoglobulins IgM, IgW, and IgNAR (immunoglobulin new antigen receptor) (see Fig. 13.5c and Table 13.10). Phylogenetic analysis suggests that IgW is closely related to IgD. Whereas IgNAR is similar to camelid heavy-chain IgG antibodies lacking light chains.

Bony fish contain tetrameric IgM (as opposed to the pentameric mammalian IgM), IgD, and IgT/Z. They have poor affinity maturation of their IgM responses and require much longer time periods to generate an antigen-specific immune response (3–4 weeks as opposed to 5 days in mammals). IgY is the functional equivalent of IgG in birds, reptiles, and amphibians. IgX seem to be the analogues of mammalian IgA and can be found in amphibians. Some species have developed unique isotypes of Ig such as IgF in Xenopus and IgO in the platypus (Sun et al. 2013).

13.2.11 Immunological Memory

One of the hallmarks of adaptive immunity is its ability to memorize pathogens that were previously encountered (Zinkernagel 2012). As a consequence, it is able to eliminate the pathogen fast and very specific upon consecutive encounter.

Since it is still unclear, how the immune system is able to memorize these pathogens, there are several hypothesis and models available.

It seems that specific populations of B and T cells, the so-called memory cells are responsible for memory. Memory cells are long-living, antigen-specific lymphocytes that are generated during primary infection. During primary infection the numbers of antigen-specific T cells are increasing. After elimination of the pathogens, most of these cells die by apoptosis. However, a small quantity of these highly specific T cells, which is around 1,000 times bigger than the original population, is maintained. In contrast to naïve T cells, they express receptors similarly like effector T cells on their plasma membrane. Similarly, also memory B cells can be discerned from naïve B cells by their numbers and surface expression pattern.

13.2.11.1 Vaccination

The ability of the immune system to remember specific antigens is exploited during vaccination. In the twentieth century, typically not the disease-causing organism is introduced to the immune system, but dead or inactivated organisms, or purified or synthetic products derived from them. The agent stimulates the body's immune system to recognize the agent as foreign, destroy it, and "remember" it, so that the immune system can more easily recognize and destroy any of the same microorganisms that it later encounters. The addition of adjuvants may enhance a danger signal and improve the protective effect. The route of antigen entry may direct the type of immune response (mucosal or systemic) (Table 13.9). Effective vaccination programs in countries have eliminated or reduced mortality and infection rates considerably, e.g., for diphtheria, polio, and small pox. Personalities involved in the development of vaccines have often been interested in comparative medicine, pathology, and microbiology (see Chap. 1). Since vaccination does not eliminate the pathogens but only protects an individual by inducing memory, vaccination programs in which a high percentage of the population is immunized, confer the best "herd" protection against these pathogens.

Passive vaccination has emerged as a newer form for immune therapy since the invention of monoclonal antibodies (mAb) and recombinant engineering of immunoglobulins. In contrast, to active immunization, pre-synthesized antibodies are transferred to a person so that the body does not have to generate the antibodies itself. Passive vaccination is used as specific therapy for cancer, allergy, and inflammatory disorders such as autoimmunity. Since 2000, the human market for therapeutic monoclonal antibodies has grown exponentially with antibodies like bevacizumab, trastuzumab (both oncology), adalimumab, infliximab (both autoimmune and inflammatory disorders, "AIID"), and rituximab (oncology and AIID) accounting for 80 % of revenues in 2006. In 2007, 8 of the 20 best-selling biotechnology drugs in the USA were therapeutic monoclonal antibodies (Scolnik 2009; Kelley 2009). This rapid growth in demand for monoclonal antibody production has been well accommodated by the industrialization of mAb manufacturing. Due to the low probability of generic threats, mAbs are now the largest class of biological therapies under development. The high cost of these drugs, which usually engross 40 % of the therapy cost in cancer, and the lack of generic competition conflict with a financially stressed health system, setting reimbursement by payers as the major limiting factor to growth.

Passive vaccination with monoclonal antibodies has the advantage that it works immediately; however, it is short lasting due to the fact that the antibodies are naturally broken down and not synthesized by the body itself.

So far, only a single canine antibody has been developed for veterinarian passive anticancer immunotherapy, i.e., an IgG antibody directed against the highly conserved cancer antigen EGFR (Singer et al. 2012, 2013).

Acknowledgements The authors would like to thank the Austrian Science Funds FWF (grants SFB F4606-B19, P 23398-B11, and W1205-B09) for supporting the project. Thanks to crossip communications, Vienna for kind support in preparing the figures.

References

Adema CM, Hertel LA, Miller RD, Loker ES (1997) A family of fibrinogen-related proteins that precipitates parasite-derived molecules is produced by an invertebrate after infection. Proc Natl Acad Sci U S A 94(16):8691–8696

Andersen MH, Schrama D, Thor Straten P, Becker JC (2006) Cytotoxic T cells. J Investig Dermatol 126(1):32–41. doi:10.1038/sj.jid.5700001

Bergin D, Reeves EP, Renwick J, Wientjes FB, Kavanagh K (2005) Superoxide production in Galleria mellonella hemocytes: identification of proteins homologous to the NADPH oxidase complex of human neutrophils. Infect Immun 73(7):4161–4170. doi:10.1128/IAI.73.7.4161-4170.2005

Brinkmann V, Reichard U, Goosmann C, Fauler B, Uhlemann Y, Weiss DS, Weinrauch Y, Zychlinsky A (2004) Neutrophil extracellular traps kill bacteria. Science 303(5663): 1532–1535. doi:10.1126/science.1092385

Cerutti A, Cols M, Puga I (2013) Marginal zone B cells: virtues of innate-like antibody-producing lymphocytes. Nat Rev Immunol 13(2):118–132. doi:10.1038/nri3383

Chandrasoma P, Taylor CR (1998) Concise pathology, 3rd edn. McGraw-Hill, New York

Chen K, Cerutti A (2011) The function and regulation of immunoglobulin D. Curr Opin Immunol 23(3):345–352. doi:10.1016/j.coi.2011.01.006

Chu VT, Frohlich A, Steinhauser G, Scheel T, Roch T, Fillatreau S, Lee JJ, Lohning M, Berek C (2011) Eosinophils are required for the maintenance of plasma cells in the bone marrow. Nat Immunol 12(2):151–159. doi:10.1038/ni.1981

Clark MR (1997) IgG effector mechanisms. Chem Immunol 65:88–110

Conticello SG, Langlois MA, Yang Z, Neuberger MS (2007) DNA deamination in immunity: AID in the context of its APOBEC relatives. Adv Immunol 94:37–73. doi:10.1016/S0065-2776(06)94002-4

Cooper E (2001) Immune response: evolution. Enzyclopedia Life Sci 1–8. http://immuneweb.xxmu.edu.cn/reading/adative/5.pdf

Das N, Biswas B, Khera R (2013) Membrane-bound complement regulatory proteins as biomarkers and potential therapeutic targets for SLE. Adv Exp Med Biol 735:55–81

Degn SE, Thiel S (2013) Humoral pattern recognition and the complement system. Scand J Immunol 78(2):181–193. doi:10.1111/sji.12070

Erlebacher A (2013) Immunology of the maternal-fetal interface. Annu Rev Immunol 31:387–411. doi:10.1146/annurev-immunol-032712-100003

Feldman M, Aziz B, Kang GN, Opondo MA, Belz RK, Sellers C (2013) C-reactive protein and erythrocyte sedimentation rate discordance: frequency and causes in adults. Transl Res 161 (1):37–43

Ferencik M, Rovensky J, Matha V, Jensen-Jarolim E (2004) Wörterbuch Allergologie und Immunologie: Fachbegriffe, Personen und klinische Daten von A-Z. Springer, New York

Freitas M, Lima JL, Fernandes E (2009) Optical probes for detection and quantification of neutrophils' oxidative burst. A review. Anal Chim Acta 649(1):8–23. doi:10.1016/j.aca.2009.06.063

French T, Blue J, Stokol T (eds) Hematology atlas. Cornell University College of Veterinary Medicine. https://ahdc.vet.cornell.edu/clinpath/modules/hemogram/eos.htm

Galli SJ, Kalesnikoff J, Grimbaldeston MA, Piliponsky AM, Williams CM, Tsai M (2005) Mast cells as "tunable" effector and immunoregulatory cells: recent advances. Annu Rev Immunol 23:749–786. doi:10.1146/annurev.immunol.21.120601.141025

Goldmann O, Medina E (2012) The expanding world of extracellular traps: not only neutrophils but much more. Front Immunol 3:420. doi:10.3389/fimmu.2012.00420

Goodpasture EW (1936) Immunity to virus diseases. Am J Public Health Nations Health 26(12): 1163–1167

Gupta AK, Hasler P, Holzgreve W, Gebhardt S, Hahn S (2005) Induction of neutrophil extracellular DNA lattices by placental microparticles and IL-8 and their presence in preeclampsia. Hum Immunol 66(11):1146–1154. doi:10.1016/j.humimm.2005.11.003

Hayday AC (2000) [gamma][delta] cells: a right time and a right place for a conserved third way of protection. Annu Rev Immunol 18:975–1026. doi:10.1146/annurev.immunol.18.1.975

Hornung V, Ablasser A, Charrel-Dennis M, Bauernfeind F, Horvath G, Caffrey DR, Latz E, Fitzgerald KA (2009) AIM2 recognizes cytosolic dsDNA and forms a caspase-1-activating inflammasome with ASC. Nature 458(7237):514–518. doi:10.1038/nature07725

Janeway C, Travers P, Walport M, Shlomchik M (2001) Immunobiology, 5th edn. Garland Science, New York

Kasamatsu J (2013) Evolution of innate and adaptive immune systems in jawless vertebrates. Microbiol Immunol 57(1):1–12. doi:10.1111/j.1348-0421.2012.00500.x

Kato L, Stanlie A, Begum NA, Kobayashi M, Aida M, Honjo T (2012) An evolutionary view of the mechanism for immune and genome diversity. J Immunol 188(8):3559–3566. doi:10.4049/jimmunol.1102397

Kaufman J (2010) Evolution and immunity. Immunology 130(4):459–462. doi:10.1111/j.1365-2567.2010.03294.x

Kelley B (2009) Industrialization of mAb production technology: the bioprocessing industry at a crossroads. mAbs 1(5):443–452

Kita H (2013) Eosinophils: multifunctional and distinctive properties. Int Arch Allergy Immunol 161(suppl 2):3–9. doi:10.1159/000350662

Klein U, Dalla-Favera R (2008) Germinal centres: role in B-cell physiology and malignancy. Nat Rev Immunol 8(1):22–33. doi:10.1038/nri2217

Krem MM, Di Cera E (2002) Evolution of enzyme cascades from embryonic development to blood coagulation. Trends Biochem Sci 27(2):67–74

Kriegler M, Perez C, DeFay K, Albert I, Lu SD (1988) A novel form of TNF/cachectin is a cell surface cytotoxic transmembrane protein: ramifications for the complex physiology of TNF. Cell 53(1):45–53

Lanier LL (2005) NK cell recognition. Annu Rev Immunol 23:225–274. doi:10.1146/annurev.immunol.23.021704.115526

Liu YJ (1997) Sites of B lymphocyte selection, activation, and tolerance in spleen. J Exp Med 186 (5):625–629

Long EO, Sik Kim H, Liu D, Peterson ME, Rajagopalan S (2013) Controlling natural killer cell responses: integration of signals for activation and inhibition. Annu Rev Immunol 31:227–258. doi:10.1146/annurev-immunol-020711-075005

Ma D, Wei Y, Liu F (2013) Regulatory mechanisms of thymus and T cell development. Dev Comp Immunol 39(1–2):91–102. doi:10.1016/j.dci.2011.12.013

Manley PN, Ancsin JB, Kisilevsky R (2006) Rapid recycling of cholesterol: the joint biologic role of C-reactive protein and serum amyloid A. Med Hypotheses 66(4):784–792. doi:10.1016/j.mehy.2005.10.018

Mayer L, Shao L (2004a) Therapeutic potential of oral tolerance. Nat Rev Immunol 4(6):407–419. doi:10.1038/nri1370

Mayer L, Shao L (2004b) The use of oral tolerance in the therapy of chronic inflammatory/autoimmune diseases. J Pediatr Gastroenterol Nutr 39(suppl 3):S746–S747

Milstein C (1985) From the structure of antibodies to the diversification of the immune response. Nobel lecture, 8 December 1984. Biosci Rep 5(4):275–297

Mogensen TH (2009) Pathogen recognition and inflammatory signaling in innate immune defenses. Clin Microbiol Rev 22(2):240–273. Table of Contents. doi:10.1128/CMR.00046-08

Montecino-Rodriguez E, Leathers H, Dorshkind K (2006) Identification of a B-1 B cell-specified progenitor. Nat Immunol 7(3):293–301. doi:10.1038/ni1301

Moraes TJ, Zurawska JH, Downey GP (2006) Neutrophil granule contents in the pathogenesis of lung injury. Curr Opin Hematol 13(1):21–27

Murphy K (2011) Janeway's immunobiology, vol 8. Garland Science, New York

Nilsson B, Ekdahl KN (2012) Complement diagnostics: concepts, indications, and practical guidelines. Clin Dev Immunol 2012:962702. doi:10.1155/2012/962702

Noelle RJ, Snow EC (1991) T helper cell-dependent B cell activation. FASEB J 5(13):2770–2776

Novak ML, Koh TJ (2013) Macrophage phenotypes during tissue repair. J Leukoc Biol 93(6):875–881. doi:10.1189/jlb.1012512

O'Leary JG, Goodarzi M, Drayton DL, von Andrian UH (2006) T cell- and B cell-independent adaptive immunity mediated by natural killer cells. Nat Immunol 7(5):507–516. doi:10.1038/ni1332

Paust S, Gill HS, Wang BZ, Flynn MP, Moseman EA, Senman B, Szczepanik M, Telenti A, Askenase PW, Compans RW, von Andrian UH (2010) Critical role for the chemokine receptor CXCR6 in NK cell-mediated antigen-specific memory of haptens and viruses. Nat Immunol 11(12):1127–1135. doi:10.1038/ni.1953

Pillai S, Cariappa A, Moran ST (2005) Marginal zone B cells. Annu Rev Immunol 23:161–196. doi:10.1146/annurev.immunol.23.021704.115728

Pillay J, den Braber I, Vrisekoop N, Kwast LM, de Boer RJ, Borghans JA, Tesselaar K, Koenderman L (2010) In vivo labeling with 2H2O reveals a human neutrophil lifespan of 5.4 days. Blood 116(4):625–627. doi:10.1182/blood-2010-01-259028

Saltis M, Criscitiello MF, Ohta Y, Keefe M, Trede NS, Goitsuka R, Flajnik MF (2008) Evolutionarily conserved and divergent regions of the autoimmune regulator (Aire) gene: a comparative analysis. Immunogenetics 60(2):105–114. doi:10.1007/s00251-007-0268-9

Schneiderman RD, Hanly WC, Knight KL (1989) Expression of 12 rabbit IgA C alpha genes as chimeric rabbit-mouse IgA antibodies. Proc Natl Acad Sci U S A 86(19):7561–7565

Schulz C, Gomez Perdiguero E, Chorro L, Szabo-Rogers H, Cagnard N, Kierdorf K, Prinz M, Wu B, Jacobsen SE, Pollard JW, Frampton J, Liu KJ, Geissmann F (2012) A lineage of myeloid cells independent of Myb and hematopoietic stem cells. Science 336(6077):86–90. doi:10.1126/science.1219179

Scolnik PA (2009) mAbs: a business perspective. mAbs 1(2):179–184

Scott MG, Briles DE, Nahm MH (1990) Selective IgG subclass expression: biologic, clinical and functional aspects. In: Shakib F (ed) The human IgG subclasses: molecular analysis of structure and function. Pergamon Press, Oxford

Secombes CJ, Wang T, Hong S, Peddie S, Crampe M, Laing KJ, Cunningham C, Zou J (2001) Cytokines and innate immunity of fish. Dev Comp Immunol 25(8–9):713–723

Shao L, Mayer L (2004) Treatment of disease through oral tolerance. Discov Med 4(23):338–343

Silva MT (2012) Classical labeling of bacterial pathogens according to their lifestyle in the host: inconsistencies and alternatives. Front Microbiol 3:71. doi:10.3389/fmicb.2012.00071

Singer J, Weichselbaumer M, Stockner T, Mechtcheriakova D, Sobanov Y, Bajna E, Wrba F, Horvat R, Thalhammer JG, Willmann M, Jensen-Jarolim E (2012) Comparative oncology: ErbB-1 and ErbB-2 homologues in canine cancer are susceptible to cetuximab and trastuzumab targeting. Mol Immunol 50(4):200–209. doi:10.1016/j.molimm.2012.01.002

Singer J, Fazekas J, Wang W, Weichselbaumer M, Matz M, Mader A, Steinfellner W, Meitz S, Mechtcheriakova D, Sobanov Y, Stockner T, Spillner E, R K, Jensen-Jarolim E (2013) Generation of a canine anti-EGFR (ErbB-1) antibody for passive immunotherapy in dog cancer patients (manuscript in review)

Sorokin L (2010) The impact of the extracellular matrix on inflammation. Nat Rev Immunol 10 (10):712–723. doi:10.1038/nri2852

Sun Y, Wei Z, Li N, Zhao Y (2013) A comparative overview of immunoglobulin genes and the generation of their diversity in tetrapods. Dev Comp Immunol 39(1–2):103–109. doi:10.1016/j.dci.2012.02.008

Vadasz Z, Haj T, Kessel A, Toubi E (2013) B-regulatory cells in autoimmunity and immune mediated inflammation. FEBS Lett 587(13):2074–2078. doi:10.1016/j.febslet.2013.05.023

Vascotto F, Le Roux D, Lankar D, Faure-Andre G, Vargas P, Guermonprez P, Lennon-Dumenil AM (2007) Antigen presentation by B lymphocytes: how receptor signaling directs membrane trafficking. Curr Opin Immunol 19(1):93–98. doi:10.1016/j.coi.2006.11.011

Venge P, Bystrom J, Carlson M, Hakansson L, Karawacjzyk M, Peterson C, Seveus L, Trulson A (1999) Eosinophil cationic protein (ECP): molecular and biological properties and the use of ECP as a marker of eosinophil activation in disease. Clin Exp Allergy 29(9):1172–1186

Waugh DJ, Wilson C (2008) The interleukin-8 pathway in cancer. Clin Cancer Res 14(21): 6735–6741. doi:10.1158/1078-0432.CCR-07-4843

Wulff BC, Wilgus TA (2013) Mast cell activity in the healing wound: more than meets the eye? Exp Dermatol 22(8):507–510. doi:10.1111/exd.12169

Young JD, Peterson CG, Venge P, Cohn ZA (1986) Mechanism of membrane damage mediated by human eosinophil cationic protein. Nature 321(6070):613–616. doi:10.1038/321613a0

Zhang SM, Adema CM, Kepler TB, Loker ES (2004) Diversification of Ig superfamily genes in an invertebrate. Science 305(5681):251–254. doi:10.1126/science.1088069

Zinkernagel RM (2012) Immunological memory not equal protective immunity. Cell Mol Life Sci 69(10):1635–1640. doi:10.1007/s00018-012-0972-y

Zinkernagel RM, Bachmann MF, Kundig TM, Oehen S, Pirchet H, Hengartner H (1996) On immunological memory. Annu Rev Immunol 14:333–367. doi:10.1146/annurev.immunol.14.1.333

Laboratory Animal Law: An Introduction to Its History and Principles

14

Regina Binder

Contents

R. Binder (✉)
Division for Animal Law, Institute of Animal Husbandry and Animal Welfare, University of Veterinary Medicine, Vienna, Austria
e-mail: Regina.binder@vetmeduni.ac.at

E. Jensen-Jarolim (ed.), *Comparative Medicine*, 267
DOI 10.1007/978-3-7091-1559-6_14, © Springer-Verlag Wien 2014

Abstract

To place the topic into an appropriate context, it proves to be a rewarding endeavour to trace back the development of lab animal law to its very beginning. The consideration of when and why experiments on animals were legally regulated for the first time conveys interesting insights into the social impact on legislation. At the same time it proves worthwhile to notice that many of the principles characterising modern lab animal law were already formulated in the nineteenth century (e.g. the demand to reduce pain and suffering to the necessary extent, the obligation to anaesthetise animals and general requirements for persons in charge of the experimental procedures as well as for institutions carrying out animal experiments). Thus, the following chapter provides a short introduction into the history of laboratory animal law and an outline of the basic principles of European laboratory animal law as enacted in Directive 2010/63/EU on the protection of animals used for scientific purposes (A more detailed outline of the Directive and of Austrian lab animal law will be provided in the 3rd semester of the IMHAI programme).

14.1 Introduction: Worldview, Religious Belief and the Experimental Use of Non-human Animals

The topics of comparative medicine documented in the preceding chapters have explored numerous analogous as well as homologous anatomical structures and physiological functions in the human body and the bodies of non-human animal species. Interest in these more or less obvious similarities goes back to the ancient world, when Hippocrates, Aristotle and others succeeded in making the first systematic anatomical observations by dissecting and vivisecting animals. Throughout history, however, man's attitude towards animals in general and their experimental use in particular was strongly influenced by the prevailing worldview or, more precisely speaking, by the predominant religious convictions. Animal experiments were regarded as methodically useful as long as the relationship between human beings and other animals was intuitively realised or taken for granted and as long as man was interested in gaining knowledge of the natural world (Greek antiquity, Humanism and Renaissance, modern times). They were, on the other hand, considered as useless, when man's interest was focused on the hereafter and a fundamental difference was supposed to separate man from any other species (e.g. Middle Ages). Paradoxically, religious belief not only inhibited but at the same time indirectly encouraged the use of animals in early medical science by strictly prohibiting the dissection of dead humans, whose corpses had to be left intact to await the resurrection of the body. Similarly, the status of medical observations concerned with aetiology, diagnosis and possibilities to treat diverse symptoms significantly depended on the question, if sickness was considered to be a natural phenomenon or regarded as a divine punishment.

14.2 Historical Development of Animal Experiments and Lab Animal Law

14.2.1 The Rise of Animal Experiments in the Sixteenth and Seventeenth Centuries

The evolution of lab animal law was, of course, closely connected with the history and development of animal experiments, which reached its first climax in the sixteenth and seventeenth centuries. During Humanism and Renaissance, medieval thinking was replaced by an intense interest in nature, leading to new insights and an enormous increase in knowledge, but at the same time laying the foundation to instrumentalise and even exploit nature.

In his novel *Nova Atlantis* (1624), famous English philosopher Francis Bacon expressed the spirit of the age, describing a utopian state equipped with facilities for animal experimentation:

> We have [. . .] parks and enclosures of all sorts of beasts and birds which we use not only for view or rareness, but likewise for dissections and trials; that thereby we may take light what may be wrought upon the body of man. [. . .] We try also all poisons and other medicines upon them, as well of chirurgery, as physic. By art likewise, we make them greater or taller than their kind is; and contrariwise dwarf them, and stay their growth: we make them more fruitful and bearing than their kind is; and contrariwise barren and not generative. Also we make them differ in colour, shape, activity, many ways. We find means to make commixtures and copulations of different kinds; which have produced many new kinds, and them not barren, as the general opinion is.[1]

In the course of the sixteenth and seventeenth centuries, many important discoveries were made by using animals; Gaspare Aselli (1581–1626), for example, described the lymphatic vessels, William Harvey (1578–1657) discovered the circulatory system, and Richard Lower (1631–1691) experimented with blood transfusions between dogs. The frontispiece of Reinier de Graaf's *Tractatus anatomico-medicus de succi pancreatici natura & usu* (1671)[2] illustrates that the use of dead as well as of live animals, the dissection of the human corpse and the treatment of patients were regarded as equal parts of early medical science.

Many animal experiments were carried out in public, to demonstrate to layman anatomical structures and the functioning of physiological systems. These public gatherings continued to be very popular in the nineteenth century, but at the same time were fiercely criticised. In England René Descartes's famous hypothesis, maintaining that animals are unable to experience pain, was increasingly challenged and the justification to use animals for experimental purposes started to be questioned. Jeremy Bentham, an English philosopher and legal scholar, maintained the view that animals should be considered morally because of their

---

[1] F. Bacon: The New Atlantis (1624). E-Book #2434, produced by M. Pullen and W. Fishburne. Posting date: October 23, 2008, 19.

[2] http://digital.lib.uiowa.edu/cdm/compoundobject/collection/jmrbr/id/2564.

capacity for suffering. In his treatise *An Introduction to the Principles of Morals and Legislation* (1789), Bentham formulated the widely known credo of pathocentrism by stating: "The question is not, can they reason? Nor can they talk? But: Can they suffer?"[3]

14.2.2 The Need for Regulation: Climax of "Vivisection Controversy"

In the nineteenth century the animal welfare movement originated in England[4] and the public debate on animal experiments culminated in the so-called vivisection controversy. It was this social debate which eventually was to lead to the first legal regulations on animal experiments.

In the middle of the nineteenth century, antivivisectionists started to compile and publish information on animal experiments.[5] Many of these documents show that scientists often disregarded the increasing demand for animal welfare, using animals in a thoughtless and even cruel way. Experiments often were carried out without anaesthetic drugs or painkillers, although morphine, ether and similar substances were already available. Moreover, some experiments did not meet the scientific standards of the time, being criticised even by contemporary scientists.

Finally it was clear that abusive animal experiments had to be stopped. Although it proved to be difficult to define an "abusive" or unjustified experiment, it finally was clear that such experiments had to be banned. It was the merit of Friedrich von Althoff, a Prussian jurist and politician, to determine two criteria characterising unjustified animal experiments. According to Althoff, an experiment was to be regarded as unjustified if it (a) did not serve a "serious scientific purpose" or (b) if the pain inflicted on the animals exceeded the degree which would have been necessary to achieve the legitimate purpose. Remarkably, these two criteria have been enshrined in lab animal legislation up to the present day.

14.2.3 The Beginning of Lab Animal Legislation

The first legal provisions on animal experiments were passed in England,[6] Germany[7] and Austria in the second half of the nineteenth century. The Austrian

[3] J. Bentham (1789): Introduction to the Principles of Morals and Legislation, Chapter XVII (Of the Limits of the Penal Branch of Jurisprudence), footnote 122. Library of Economy and Liberty (http://www.econlib.org/library/Bentham/bnthPML.html).

[4] In 1824 the first animal welfare NGO, the Royal Society for the Prevention of Cruelty against Animals (RSPCA), was founded in England.

[5] Cf. H. Bretschneider (1962): Der Streit um die Vivisektion im 19. Jahrhundert, 12ff.

[6] Cruelty to Animals Act (1876).

[7] *Gossler'sche Verordnung* (1885).

decree on vivisection (*Viviselektionserlass*), enacted in 1885, is a typical example for these early regulations, containing provisions on the following issues:

- Animal experiments were only allowed for purposes of serious research and for indispensable educational purposes.
- Animal experiments had to be carried out in public institutions, i.e. mainly at universities.
- Basically only university teachers were permitted to carry out or supervise animal experiments.
- It was obligatory to choose the "lowest" animal species sufficient to achieve the experimental purpose.
- The animals had to be anaesthetised unless the use of narcotic drugs was incompatible with the experimental purpose.

Thus, in a very general way, early lab animal regulations anticipated modern principles such as the need of the experiment to be justified by a legitimate scientific or educational purpose, requirements for institutions and qualification of staff and specific strategies of refinement (choice of species, obligatory anaesthesia).

14.2.4 Pre- and Post-war Lab Animal Legislation

Unfortunately, in the twentieth century it was the German *Reichstierschutzgesetz* 1933 which has to be regarded as a corner stone in the development of modern lab animal law. Additionally to the requirements of lab animal regulation of the nineteenth century for the first time, it was mandatory under this law (a) to apply for an authorisation to carry out an animal experiment and (b) to keep records of the animals and of the experimental procedures, both requirements being important features of current lab animal law.

14.3 Modern European Lab Animal Law

Modern lab animal law comprises different levels of regulation, international and supranational provisions, on the one hand, and national legislation, on the other hand.

On the level of *international law*, the European Convention for the Protection of Vertebrate Animals used for Experimental and other Scientific Purposes (ETS 123), which was issued by the Council of Europe in 1986, lays down a regulatory framework for the experimental use of animals. The Council of Europe, which was founded in 1949, strives to enhance international cooperation on political as well as on cultural level. Conventions issued by the Council of Europe are treaties, i.e. international agreements, which may be signed and ratified by individual states.[8]

[8] For further information cf. European Council: The Treaty Office in a Nutshell http://www.conventions.coe.int/Treaty/TreatyOffice-Nutshell.pdf.

On *supranational level* lab animal legislation started in 1986, when the European Economic Community (EEC), the forerunner of today's European Union, passed the first directive on the experimental use of animals.[9] In 2010 a new directive on animal experiments was enacted: Directive 2010/63/EU[10] (short: Directive) had to be transformed into the national legislation of the member states by the 1 January 2013, thus constituting the basis of the Austrian *Tierversuchsgesetz* 2012. In 2007 the European Commission had passed Council Recommendation 2007/526/EC, defining guidelines for the accommodation and care of lab animals.

Austrian lab animal legislation[11] currently comprises:

- The *Act on Animal Experiments* (*Tierversuchsgesetz—TVG 2012*), which came into force on the 1 January 2013
- A decree on animal experiments (*Tierversuchsverordnung—TVV 2012*), issued by the Minister of Science and Research and establishing the requirements for accommodation and care of great number of lab animal species
- A decree on the collection of statistical data (*Tierversuchsstatistik-Verordnung*), which will have to be renewed according to Commission Implementing Decision 2012/707/EU[12]

14.3.1 Legal Definition of the Term "Animal Experiment"

The term "animal experiment" is defined by Art. 3 of the Directive, mentioning the following constitutive elements for an animal experiment in the legal sense: (a) use of living animals in the sense of the Directive, (b) a specific experimental or other scientific purpose and (c) adverse effects (pain, suffering, distress or lasting harm) inflicted on the animals.

[9] Council Directive of 24 November 1986 on the approximation of laws, regulations and administrative provisions of the member states regarding the protection of animals used for experimental and other scientific purposes (86/609/EEC).

[10] Directive 2010/63/EU of the European Parliament and of the Council of 22 September 2010 on the protection of animals used for scientific purposes, OJEU 20.10.2010, L 276/33. Directive 2010/63/EU will be referred to as "Directive"; for full text see http://www.vetmeduni.ac.at/de/tierschutzrecht/infoservice/tierversuchsrecht/.

[11] Full text may be downloaded from http://www.vetmeduni.ac.at/de/tierschutzrecht/infoservice/tierversuchsrecht/.

[12] Commission Implementing Decision of 14 November 2012 establishing a common format for the submission of the information pursuant to Directive 2010/63/EU of the European Parliament and of the Council on the protection of animals used for scientific purposes (2012/707/EU), OJEU, 17.11.2012, L 320/33.

14.3.1.1 Live Animals

The Directive applies to all vertebrate animals (including larvae, e.g. tadpoles, and foetuses of mammals in the last trimester of their prenatal development) and to cephalopods (e.g. squids, octopus).[13]

14.3.1.2 Legal Purposes

In accordance with Art. 5 of the Directive, experiments may only be carried out for the purpose of (a) basic research, (b) specified categories of translational and applied research and (c) development, manufacturing and testing of products (e.g. drugs, foodstuff). Further legitimate purposes of experiments are (d) the protection of the natural environment, (e) the preservation of species, (f) higher education and training and (g) forensic inquiries.

Agricultural practices (e.g. the dehorning of calves) and clinical veterinary practices (i.e. diagnosis and treatment of sick or injured animals) do not fall into the scope of the Directive. Furthermore, the marking of an animal—e.g. the implantation of a microchip or the application of a tag—is no animal experiment if it primarily aims at identifying an animal. Finally, the killing of animals for the sole purpose of using their organs or tissues is also not regarded as an experiment under the Directive.

14.3.1.3 Adverse Effects

In order to speak of an animal experiment in the sense of the law, it is necessary that the procedures carried out on the animals cause a certain level of pain, suffering, distress or lasting harm.

Pain is defined as "an unpleasant sensory and emotional experience associated with actual or potential tissue damage, [. . .]".[14] D*istress* means an aversive state in which an animal is unable to adapt to stressors and therefore reacts with maladaptive behaviour, thus being similar to emotional or mental pain.[15] Suffering is a "state of 'mind', which is not identical to, but might be a consequence of pain or distress".[16] The concept of (lasting) *harm* on the other hand refers to a human-induced impairment of an animal's state as compared either to its previous condition or to the species-specific normal condition. While pain, suffering and distress imply a subjective experience on the part of the animal and therefore can only be inflicted on a sentient being, harm, being assessed from the external perspective, can be inflicted on any animal, regardless of what is known about its (degree of)

[13] It is interesting to notice that the US Animal Welfare Act defines "animal" as "warm-blooded animals", excluding birds, mice and rats, thus protecting less than 10% of all lab animals; cf. J. E. Schaffner (2011): An Introduction to Animals and the Law, 90.

[14] FELASA Working Group on Pain and Distress: Pain and distress in laboratory rodents and lagomorphs, Laboratory Animals (1994) 28, 98.

[15] Cf. FELASA Working Group on Pain and Distress (1994), 98.

[16] FELASA Working Group on Pain and Distress (1994), 98.

sentience (e.g. molluscs); thus the concept of harm is independent from the question if the animal is actually able to experience the negative effect of the harmful event (e.g. congenital blindness in various inbred lines of lab mice).

According to lab animal law, an adverse effect has to reach a *minimum level*, which is defined by the intensity "of pain which is at least equivalent to the introduction of a needle in accordance with good veterinary practice" (Art. 3 of the Directive). It is, however, important to note that also non-invasive procedures may cause pain, suffering or distress, which are equivalent to or even (far) more harmful than an injection (e.g. experimental designs preventing animals from expressing natural behaviour, any kind of deprivation, i.e. restrictions on the housing, husbandry and care standards). In accordance with recital nr. 23 of the Directive from an ethical standpoint, there should be an *upper limit* of pain, suffering and distress above which animals should not be subjected in scientific procedures. Thus procedures which are likely to cause "severe pain, suffering or distress which is likely to be long lasting and cannot be ameliorated" are basically forbidden by the Directive, but may be allowed by the member states "for exceptional and scientifically justifiable reasons" (cf. Art. 55/3 of the Directive, the so-called safeguard clause).

14.3.1.4 Severity Categories

Animal experiments may differ significantly with regard to the severity of adverse effects inflicted on the animals. In order to objectify the evaluation of the project applications and to carry out the necessary harm–benefit analysis, Directive 2010/63/EU introduced a system to assess the prospective severity of the experimental impact on the animals. Annex VIII of the Directive establishes the following severity categories, defined by two parameters: (a) the degree and (b) the duration of adverse effects (cf. Table 14.1).

Under Directive 2010/63/EU there are *two instances of severity classification*: on the one hand, scientists are obliged to indicate the expected severity categories in the project applications; on the other hand, specific projects have to be evaluated after having been finished. All projects which have been assigned the category "severe" and each project involving non-human primates have to be evaluated retrospectively. Moreover, a retrospective assessment may be ordained by the competent authority on a case-by-case basis.

14.3.2 The "3Rs" as Key Concept of Lab Animal Law

General animal welfare law aims at protecting animals from the infliction of adverse effects which are regarded as unnecessary. Lab animal legislation constitutes a specialised field of general animal welfare law.

Lab animal welfare is not an end in itself, but usually enhances the quality of experiments by minimising or even eliminating unwanted side effects caused by pain, suffering and distress. The "3Rs" (reduction–replacement–refinement), which

Table 14.1 Severity categories

Severity category		Definition (simplified)
1	Nonrecovery	Procedures performed entirely under general anaesthesia from which the animal shall not recover consciousness
2	Mild	Animals are likely to experience short-term mild pain, suffering or distress
3	Moderate	Animals are likely to experience short-term moderate pain, suffering or distress or long-lasting mild pain, suffering or distress
4	Severe	Animals are likely to experience severe pain, suffering or distress, or long-lasting moderate pain, suffering or distress
[4+]	[Most severe]	Procedure involving severe pain, suffering or distress that is likely to be long-lasting and cannot be ameliorated

as a term were coined in 1959 by Russell and Burch,[17] are the guiding principles of current lab animal regulation. The significance of the "3Rs" has been enhanced by Directive 2010/63/EU; thus Art. 1, defining subject matter and scope, states that the rules of the Directive are meant to contribute (a) to replace animal experiments, (b) to reduce the number of animals used for experimental purposes and (c) to improve living conditions of the animals as well as to minimise the negative impact inflicted on the animals by experimental procedures.

According to the Directive, animals have an *intrinsic value*.[18] Therefore, animals should always be treated as sentient creatures, and their use in procedures should be restricted to areas which may ultimately benefit human or animal health or the environment. Thus the use of animals for scientific or educational purposes should only be considered where nonanimal alternatives are unavailable.

Basically, lab animal law pursues two strategies: on the one hand, lab animal legislation protects animals from being used in experiments, because (a) animal experiments have to be replaced by alternative methods (e.g. in vitro experiments) whenever possible (*replacement*) and (b) every animal experiment has to be carried out with the smallest number of animals, which is necessary to reach the scientific purpose (*reduction*). On the other hand, animals used for experimental purposes have to be protected from any unnecessary adverse effect, caused by the conditions of accommodation and care as well as by experimental procedures (*refinement*) (Table 14.2).

Recently, it has been tried to define *responsibility* as a "4th R". Looking more closely into this matter, it becomes clear, however, that the concept of responsibility is the source of all of the "3Rs" and should therefore be regarded as "meta-R". In a similar way, attitude and qualification of personnel have to be attributed a special status among implementary tools, being crucially important for the execution of each of the "3Rs".

---

[17] W. M. S. Russell and R. L. Burch (1959): The Principles of Humane Experimental Technique (http://altweb.jhsph.edu/pubs/books/humane_exp/het-toc).

[18] Cf. recital nr. 12.

Table 14.2 The "3Rs": replacement–reduction–refinement

<table>
<tr><th colspan="2">Principle</th><th colspan="2">Meaning / aim / scope</th><th colspan="2">Implementation (examples)</th></tr>
<tr><td rowspan="5">Responsibility</td><td>Replacement</td><td colspan="2">Reduction of animal experiments by alternative methods</td><td>• use of methods without live animals (e.g. tissues, isolated organs); in vitro-experiments
• methods without animals (e.g. dummies, computer simulations)</td><td rowspan="5">Qualification of staff & equipment of establishment</td></tr>
<tr><td>Reduction</td><td>Reduction of number of animals used in an experiment</td><td>breeding & use</td><td>• choice of breeding strategy
• choice of complementary methods
• choice of adequate animal model
• choice of adequate statistics</td></tr>
<tr><td>Refinement</td><td colspan="3">• improvement of well-being
• minimisation of pain, suffering, distress and lasting harm with simultaneous optimisation in increase of knowledge</td></tr>
<tr><td>of living conditions</td><td>accommodation & care</td><td rowspan="2">breeding & use</td><td>• environmental enrichment
• group housing
• human-animal-interaction, e.g. competent handling, gentling, training; respectful attitude</td></tr>
<tr><td>of experimental methods</td><td>choice of most humane experimental design and procedures</td><td>• choice of most gentle techniques
• application of adequate anaesthesia & analgesia
• most humane killing method
• humane endpoint</td></tr>
</table>

14.3.3 Examples for Refining Experimental Procedures

Refinement of experimental procedures or techniques comprises all measures contributing to reduce pain, suffering, distress or lasting harm to the necessary minimum or even to eliminate adverse effects on the animals.

14.3.3.1 Adequate Anaesthesia and Analgesia

One of the most important requirements of lab animal law is the obligation to anaesthetise animals appropriately. Therefore in line with Art. 14 of the Directive, national lab animal legislation has to ensure that appropriate (local or general) anaesthesia is basically obligatory, unless anaesthesia would be incompatible with the legitimate experimental purpose. Procedures involving serious injuries which may cause severe pain must never be carried out without anaesthesia.

Member states also have to ensure that pain suffering and distress of the animals are kept to a minimum by adequate analgesia.

14.3.3.2 "Humane" Killing Methods

The use of inappropriate killing methods can cause significant pain, suffering and distress to the animals. Thus it is necessary to ensure that only methods are used, which are appropriate to the particular species.[19] The species–specific killing methods are listed by annex IV of the Directive. Knowledge and competence of persons killing animals are equally important for the protection of lab animals; therefore the persons killing animals have to be properly educated and trained.

14.3.3.3 Humane Endpoint

The so-called "humane endpoint" is one of the most important tools of experimental refinement. It is defined as "the point at which pain or distress in an experimental animal is prevented, terminated, or relieved [...] [providing] an alternative to experimental endpoints that result in unrelieved or severe animal pain and distress, including death".[20] Thus, a humane endpoint not necessarily implies the killing of the animal, but could also result in interventions to alleviate pain, suffering or distress, e.g. by providing analgesics. The humane endpoint should be defined by a set of reliable criteria (e.g. loss of body weight, behavioural changes); its implementation requires a close monitoring of the animals as well as a thorough understanding of the species–specific symptoms of pain, suffering and distress.

14.3.4 Examples for Refining Accommodation and Care

The harmonised requirements for accommodation and care of lab animals are established by annex III of the Directive, comprising a general and a species–specific part. Additionally, the guidelines defined by Commission Recommendation 2007/526/EC should be considered.

Accommodation of lab animals can be refined by all types of environmental enrichment, which should meet the specific ethological needs of the particular species. Generally speaking, environmental enrichment should stimulate the physical and cognitive activities of the animals, thus contributing to their fitness. The most common means to enrich cages are structures to hide and to explore, (nesting) material to manipulate, gnawing material for rodents and lagomorphs, racks and

[19] Cf. recital nr. 15 of the Directive.

[20] Guide for the Care and Use of Laboratory Animals, Committee for the Update of the Guide for the Care and Use of Laboratory Animals Institute for Laboratory Animal Research, 8th ed., 27.

scratching posts for cats, toys for cats and dogs and climbing devices and swings for primates.[21] Besides environmental enrichment, social contact (group housing) of gregarious species is also to be regarded as an important factor to improve lab animal welfare. The proper enrichment of cages for SPF-animals remains an important research issue for lab animal science.

Refinement of care on the other hand especially comprises all means enhancing welfare by providing suitable foodstuff in a manner appropriate to the behavioural needs of the species and by ameliorating interaction with personnel, especially by competent handling and gentling and suitable training programmes (e.g. for cats and dogs, primates and farm animals).

It has already been mentioned that refinement not only benefits animals but usually at the same time serves as a means to enhance the quality of science. In order to implement the principles of the "3Rs" efficiently, the Directive also regulates the operation of users, breeders and suppliers and the obligation to get an authorisation before carrying out an animal experiment.

14.3.5 Authorisations

Under Directive 2010/63/EU users, breeders and suppliers have to be authorised by and registered with the competent authority. Moreover, each individual project needs an authorisation, and staff carrying out specific functions has to be trained appropriately.

14.3.5.1 Users, Breeders and Suppliers

Users, breeders and suppliers are authorised if the relevant requirements established by the Directive are met; these requirements mainly concern (a) adequate equipment and (b) qualified personnel. Installations and *equipment* have to meet the requirements for accommodation and to allow the procedures to be performed efficiently and with the least distress to the animals.

Each user (i.e. any natural or legal person using animals in procedures), breeder and supplier has to employ a sufficient number of qualified *personnel*. Moreover, users, breeders and suppliers have to nominate (a) at least one person responsible for overseeing the welfare and care of the animals in the establishment, (b) a designated veterinarian and (c) establish an animal-welfare body with the primary task of giving advice on animal-welfare issues. The animal-welfare body should also follow the development and outcome of projects, foster a climate of care and provide tools for the timely implementation of recent technical and scientific

[21] For numerous examples cf., e.g. Animal Welfare Institute (http://awionline.org/content/additional-resources-4) or Fund for the Replacement of Animals in Medical Experiments (FRAME; http://www.frame.org.uk/page.php?pg_id=67).

developments in relation to the principles of the "3Rs". The advice given by the animal-welfare body should be properly documented and open to scrutiny during inspections.[22]

14.3.5.2 Persons Designing Procedures and Other Staff

It is generally acknowledged that the welfare of lab animals is highly dependent on the quality and competence of the personnel supervising and performing procedures and of the staff taking care of the animals on a daily basis.[23] Thus a high level of qualification and a professional as well as a caring attitude towards the animals definitely contribute to the implementation of the "3Rs".

Therefore, the staff carrying out procedures on animals, taking care of animals or killing animals has to be adequately educated and trained. Member states are obliged to ensure that persons designing procedures and projects have received instruction in a scientific discipline relevant to their work and have acquired specific knowledge on the animal species they are working with.

14.3.6 Project Licence

A "project" is being defined as "a programme of work having a defined scientific objective and involving one or more procedures [i.e. animal experiments]".[24] Depending on the kind of project there are *two types of administrative procedures*: a *regular* and a *simplified procedure*. The simplified procedure applies to projects classified as nonrecovery, mild or moderate, if (a) the project is ordained by a legal provision ("regulatory projects", e.g. testing of drugs) or (b) animals are used for production or diagnostic purpose with established methods (e.g. vaccines, serums) *and* no non-human primates are involved in the project.

The simplified administrative procedure differs from regular authorisation with regard to the contents of the project application and with regard to the decision deadline. Basically, the competent authorities have to take their decisions within 40 working days; this period may be prolonged by 15 working days in the case of "complex or interdisciplinary" projects subject to the regular administrative procedure.

14.3.7 Further Regulations of Directive 2010/63/EU

Apart from the basic principles dealt with in this chapter, Directive 2010/63/EU contains a series of provisions restricting the experimental use of specific groups of animals (e.g. specimens of endangered species, non-human primates, stray and feral

---

[22] Cf. Recital nr. 31 of Directive 2010/63/EU.

[23] Cf. Recital nr. 28 of Directive 2010/63/EU.

[24] Art. 3 nr. 2 Directive 2010/63/EU.

animals). Furthermore, it comprises regulations on animal identification and records, on documentation and on inspections of users, suppliers and breeders. In order to enhance development and validation of alternative approaches, a Union Reference Laboratory has to be established. Although according to the Directive "the use of live animals continues to be necessary to protect human and animal health and the environment", it considers itself as "an important step towards achieving the final goal of full replacement of procedures on live animals for scientific and educational purposes as soon as it is scientifically possible to do so".[25]

Bibliography

Bretschneider H (1962) Der Streit um die Vivisektion im 19. Jahrhundert: Verlauf, Argumrente, Ergebnisse. Stuttgart: Fischer (Medizin in Geschichte und Kultur, Bd. 2)

Dolan K (2007) Laboratory animal law. Legal control of animals in research, 2nd edn. Blackwell Publishing, Oxford

FELASA Working Group on Pain and Distress; Baumans V, Brain PF, Brugére H, Clausing P, Jeneskog T, Perretta G (1994) Pain and distress in laboratory rodents and Lagomorphs. In: Report of the Federation of European Laboratory Animal Science Associations (FELASA) Working Group on Pain and Distress accepted by the FELASA Board of Management, November 1992, Laboratory Animals, vol 28, pp 97–112

Guide for the Care and Use of Laboratory Animals (2011) Committee for the Update of the Guide for the Care and Use of Laboratory Animals Institute for Laboratory Animal Research, 8th edn. The National Academic Press, Washington, DC

Monamy V (2009) Animal experimentation. A guide to the issues, 2nd edn. Cambridge University Press, Cambridge

Schaffner JE (2011) An introduction to animals and the law, The Palgrave Macmillan animal ethics series. Palgrave Macmillan, Basingstoke

[25] Recital nr. 10 of the Directive.

Ethics in Laboratory Animal Science

15

Herwig Grimm

Contents

Abstract

In the following article, I discuss some core aspects of moral responsibility in the field of laboratory animal science. After a short introduction, I briefly deal with the entanglement of normative and empirical knowledge in animal testing. The argument will be put forward that both kinds of knowledge are needed for a complete account of animal ethics and taking on moral responsibility in the field. This section is followed by an overview of the central theories and approaches in ethics and animal ethics and their relevance for the debate on animal experiments. In the closing chapter, I address some methods and problems of weighing human against animal interests in a plausible and ethically informed way.

H. Grimm (✉)
Ethics and Human-Animal Studies, Messerli Research Institute, University of Veterinary Medicine Vienna, Medical University of Vienna, University of Vienna, Vienna, Austria
e-mail: herwig.grimm@vetmeduni.ac.at

E. Jensen-Jarolim (ed.), *Comparative Medicine*,
DOI 10.1007/978-3-7091-1559-6_15, © Springer-Verlag Wien 2014

15.1 Introduction: Animal Ethics: Scientists' Responsibility in Challenging Situations

An ethical issue is one that challenges us to apply our concepts of right and wrong, good and bad, to new situations (Rollin 2012, p. 23). In general, this is necessary when well-established practices that have been taken for granted become doubtful and questionable (Dewey 1938, p. 164; Grimm 2010, pp. 126–156). Therefore, ethics comes into play when customs, lifestyles, and normative institutions lose their guiding function, and formerly self-evident normative beliefs fail to direct actions (Höffe 2008, p. 10). Human–animal interaction is such a field; formerly unquestioned practices are put into doubt and our dealing with animals in labs, on farms, and so forth has become questionable to many people. This is one of the main reasons why animal ethics became prominent in the last few decades (Grimm 2012). Animal research is only one area among many in this field, which includes the keeping of livestock on farms or animals in zoos or as pets. However, it is a central issue and probably one of the most debated topics in today's practical animal ethics. Certainly, it is one of the most discussed issues related to animals in the public debate.

Questions of ethics in animal testing and its legal regulation have not only been put forward by ethicists. The way of dealing with animals has also been questioned from within by scientists before ethicists started to inquire into animal research intensively in the 1970s. Before Roslind Godlovitch, Stanley Godlovitch, and John Harris published the book *Animals, Men and Morals* (Godlovitch et al. 1971), in which ethical problems in laboratory animal science are addressed, Russell and Burch presented their famous book *The Principles of Humane Experimental Technique* in 1959 (Russell and Burch 1959). As scientists, they suggest the 3Rs (reduction, replacement, refinement) as principles to improve animal testing in terms of animal welfare. The book has been published 50 years after the original publication as an abridged version by Michael Balls (2009) where Russell and Burch state their point clearly: "We [W.M.S. Russell and R.L. Burch] assume throughout that experimental biologists are only too happy to treat their animals as humanely as possible" [Russell/Burch in Balls (2009; p. 9)]. They continue to identify the central problem, which is determining what is and what is not humane, and how humane treatment of animals can be promoted without prejudice to scientific and medical aims [cf. Russell/Burch in Balls (2009, p. 10)]. As Russell and Burch question established practices that have been taken for granted and develop new ways of thinking in animal research, their contribution is ethics at work in the abovementioned sense.

One could trace back the debate on ethics and animal research much longer (e.g., anti-vivisectionists movement), but I am going to focus on the second half of the last century in order to stay within the scope and limits of an article. This article is concerned with issues and concepts that are presently discussed and gives guidance on an active debate. There is no end in sight, neither for animal testing, nor for the related public and ethical debate. Quite the opposite, the debate seems to get more controversial and its impetus, the number of animals used in testing, is increasing. Reasons for the increased use of animals in medical testing can be seen in the rapid development of the life sciences in the last few decades both in its methods and resulting new questions. Second, growing media attention and animal welfare and animal rights organizations raise awareness and fuel a lively debate. Third, against the background of the mentioned development, changes to the legal framework which raise important political questions are necessary (Grimm 2012). For instance, methods of genetic engineering brought about new questions in animal ethics that, in turn, led to sociopolitical questions. However, at the bottom of all these questions, we find the ethical problem that shall be dealt with in the following.

The problematic structure of animal testing can roughly be described as follows: One can either decide to gain knowledge and take the risk of adverse effects on animals or choose to do otherwise and accept not gaining the desired knowledge. This leads to several problems that are intrinsically connected with animal experiments. In general, the desired outcome is related to distressing animals. If we—as is nearly unanimously agreed—consider distressing animals morally problematic, an ethical problem is at hand. Consequently, a plausible way out is to avoid using animal experiments and gain the knowledge by other means. European legislation (directive 2010/63/EU) holds scientists responsible for proving that particular knowledge cannot be achieved without animals: If the relevant knowledge can be gained without animal testing, a proposed experiment is neither legal nor ethically acceptable. Only if we *cannot* gain important knowledge without using animals can animal experiments be justified (which does not necessarily mean that they are justified). But what is important enough to cause adverse effects to animals and to what degree? It is not only a plausible and important ethical responsibility to prove that there is no alternative, but, within the European Union, also a legal one. Obviously, the question of alternative methods is followed by the challenge and responsibility to promote the development of such methods to reduce the total number of animal experiments. Still, even if it can be proved that there is no alternative to animal testing, this does not issue a free pass to scientists. Still the question remains: can gaining specific knowledge justify related animal distress? And more generally, how do we deal with the difficulties and challenges of the moral conflict?

One option is to focus on the scientists' responsibility. In principle, a scientist can decide *for* or *against* an animal experiment. If he chooses to conduct animal research, he consciously accepts animal distress to get the expected results. With this knowledge, justifying reasons need to be provided. These reasons can only be found in the expected results and their significance. Therefore, scientists should be able to pinpoint reasons why they think that the relevance of the expected results justifies doing harm to animals.

The situation is different if a researcher decides *against* an animal experiment. Of course, this is a hypothetical case because scientists are not likely to plan an experiment and reflect afterwards if it leads to relevant insights. Quite the opposite, the research question has to be considered relevant in order to think about carrying out an animal experiment and investing resources. Within the structure of the moral conflict outlined above, choosing not to carry out an experiment means that a researcher considers the level of animal distress too high or the relevance of the expected results too low or both. But is it a moral problem if a researcher does not carry out a promising experiment?

One could argue that it is morally wrong for a researcher not to carry out a particular animal experiment because he renounces relevant potential findings. This argument is not convincing and must not be overstrained. An epistemic reason can be put forward against it; animal testing is a process with an open end, which means that whether the desired results can be accomplished or not is an open question. Therefore, it would not be reasonable to make a scientist accountable for *not* getting a desired result after carefully planning and conducting an experiment. Per definition, experiments are designed to establish the clarity of the hypothesis and not to gain certain results. Uncertainty is an essential component of animal testing.

A strong case could be made to hold scientists responsible for not conducting an experiment, only if scientists had to meet a moral requirement to gain certain knowledge. Scientists cannot be obligated to attain specific knowledge and because there is no moral right to certain scientific knowledge, deciding against an animal experiment is less problematic from a moral point of view.

In brief, it is easier to justify a decision against an animal experiment than a decision for it, as in the latter case no moral claims are violated. Nobody can force a scientist to carry out an animal experiment. However, if a scientist wants to work in a field including animal testing, he will face moral, legal, and scientific claims. This becomes clear when we consider that we have to demonstrate the relevance of the expected knowledge in order to justify animal distress and to responsibly decide for or against an experiment. But where does this moral responsibility stem from and how to deal with it? "What *ought* I do?" is the core question here, which leads back to the field of philosophical ethics.

15.2 Animal Ethics and the Entanglement of Normative and Descriptive Sciences

The crucial question in philosophical ethics is "What *ought* I do?" Therefore, philosophical ethics starts where orientation is needed (Höffe 2008, p. 10). Philosophical ethics is a normative discipline, dealing with the question of what *should* be done. Empirical research disciplines, on the contrary, focus on what *is* the case and consider themselves descriptive science. Despite the fact that we find two different fields of knowledge, "is" and "ought" are of equal importance when it comes to the practical ethical questions in animal research. In the field of applied ethics, normative and descriptive dimensions are always entangled (Grimm 2010, pp. 63–66; Bayertz 1999, p. 74). A simple example illustrates this: If an experimenter knows, for instance, that he or she *should* respect a mouse's welfare, he or she also needs to know what *is* important for a mouse (e.g., because of behavioral requirements) and how many mice are really necessary for an experiment. Looking at the same issue from another angle, a scientist should know what it means if he notices a mouse squatting in a corner raising its back and should react to this behavior in a responsible manner. The basis for responsible action is knowledge of what *is* the case, e.g., that a certain behavior indicates reduced well-being. The question concerning what one *should* do is at the same time a question of what *is* the case, what we know, and what we can do. Simply put: Practical ethics is always entangled with empirical facts (Bayertz 2002, pp. 9–12). Therefore, answering normative questions depends on empirical scientific disciplines as well. As a consequence, animal ethics aims at composite judgments, whose validity is dependent on the validity of judgments that have to be proven in different disciplines (Düwell 2008, pp. 55–59, 100; van den Daele 2008). This has already become clear in the aforementioned example of proving that there is no alternative to animal experiments, in which scientific knowledge is required for dealing with ethical concerns. On the other hand, normative premises that have to be clarified by ethics matter in science. Deciding for or against an animal experiment includes value judgments that proceed empirical, descriptive sciences. Making the methods required to come to such judgments explicit and open to criticism is a part of ethics.

When it comes to the question of what should be done, not only ethics are relevant to the normative considerations of scientists. The fulfillment of legal requirements stemming from their research is also a major concern. As in ethics, living up to the legal standards and norms in animal research requires empirical knowledge and illustrates the connections between descriptive and normative dimensions. However, the basis of legal requirements is often ethical in nature. They express the perception of animals in society and show to what extent animal welfare claims can prevail over other public interests. Although many aspects of scientific responsibility in the field of animal testing are regulated by law, two points show firmly moral claims: (a) the responsibility for seeking the best option within the given context, this responsibility refers mainly to scientists conducting animal experiments in their daily work, and (b) the responsibility for making sure that the conditions within the experiment are reconsidered and refined in a way that the best possibility is also morally acceptable. This responsibility does not only refer to scientists but to all citizens. Both aspects are the cornerstones of scientific responsibility in animal testing. In the following, influential ethical frameworks are illustrated that exemplify important normative sources of our moral responsibility towards animals.

15.3 Ethical Theories in Animal Ethics

In the following section I discuss different corner posts of moral responsibility towards animals. Three important approaches in philosophical ethics that shape the debate on animal testing are introduced (Sandøe et al. 2008; Olsson et al. 2003). These approaches give different answers to the question *if* and *why* we should consider animals *for their own sake* and what this means.

15.3.1 Contractarianism and Anthropocentric Views

In contractarianism moral status can only be attributed to beings capable of making promises, concluding contracts, and understanding and considering the consequential mutual rights and obligations (Wolf 2012, p. 55). Therefore, this is an animal ethical position, which—surprisingly—opposes direct moral respect for animals: Only legally capable beings (persons) have a moral status and have standing in the moral community.

Morals are considered a sort of agreement between rational, independent people with individual interests who benefit from this agreement (Narveson 1977; Carruthers 1994; Cohen 2008). In other words, if I treat other people badly, they will treat me badly too. Therefore, it is in the interest of people to stick to moral principles and norms and not to hurt other people in order not to get hurt themselves. People have an interest to "subscribe" to the "moral contract."

With respect to interests, the contrary is the case in the field of human–animal interaction. Humans do not need to worry that animals will "fight back" if they are used in painful experiments. Therefore, human interests are not threatened, and it would not be reasonable to cut back on satisfying them without good reason. In contractarianism, harming animals is not a moral problem per se. Therefore, we may label this an anthropocentric position (*anthropos* = Greek for *man*). The anthropocentric view holds the opinion that only humans have moral status. In contractarianism, the capability to enter contracts gives humans a special position. As animals cannot conclude contracts, they are not part of the moral community. According to this approach, animal protection can only be justified via *indirect duties*.

Immanuel Kant is probably the most famous representative of the position stating that we have only indirect duties towards animals. He formulates the argument of brutalization also known as the "violence graduation hypothesis" (Arluke et al. 1999): We should not treat animals cruelly because we would desensitize. Treating animals cruelly coarsens our sensitivity to others' suffering, and we become hard also in our dealing with men, according to the most famous eighteenth-century philosopher (Altman 2011, p. 16). Humans do not only have the capacity to moral perfection but also the obligation. Cruelty is contrary to this obligation, which leads to an *indirect* or *derivative* protection of animals. But we do not directly owe moral respect to animals; only rational humans deserve direct moral respect since only they can act in accordance with principles. Our duties towards animals are indirect duties towards mankind.

What contractarianism and pedagogic approaches in animal ethics have in common is that they can justify animal protection only indirectly via obligations towards humans. Even if these positions do not seem to be very plausible at first sight, they are very powerful. Animal protection is most established where it is an advantage for humans. For instance, if experimental research shows that increasing pain and distress has negative influence on the quality of the scientific results, this is of course one of the strongest arguments to save animals from pain and distress. But what happens if this is not the case or if bad husbandry conditions are necessary in order to reach a certain research aim? For contractarianists, this is not a moral problem.

Summary: In contractarianism, animals are not morally relevant themselves. Animals do not have moral status. However, feelings and interests of legally capable persons can justify the obligation to protect animals via indirect duties. This approach can help to define an (accepted) framework for animal experiments, in which research on animals can be conducted in order to benefit humans (Sandøe et al. 2008, p. 105). The abovementioned dilemma would be mitigated. According to contractarianism, there is no dilemma as we do not owe anything directly to animals in terms of morals. But can it be right that animal distress has no significance, even if only trivial interests or no interests are satisfied? Not many would support such a theory. Quite the opposite, most of us seem to consider it unethical to make animals suffer without any justifying reason. Justifying the moral status by characteristics that can *only* be seen in humans is problematic, especially since there are theories that can integrate these strong intuitions. Utilitarianism is a theory in accordance with these fundamental and most important moral intuitions.

15.3.2 Utilitarianism and Pathocentric Positions

According to the utilitarian approach (*utilitas* = Latin for *benefit, utility*), not only human but also animal interests deserve moral respect. Harm and distress in animal research can be justified if the benefits compensate it or prevail. In utilitarianism, the main principle is the *greatest good for the greatest number*. Therefore, acts that lead to the best possible consequences in terms of benefits are morally superior. In this consequentialist conception, pleasure is desirable and pain to be avoided, no matter where and with whom it arises. As opposed to contractarianism, every (comparable) interest has the same value, no matter if it is a human or an animal interest. Moral problems arise if animal interests (e.g., being free from suffering) are frustrated by husbandry conditions or in an animal experiment. But who has interests that count morally?

The English philosopher and legal scholar Jeremy Bentham gave a far-reaching and much-quoted answer to this question. According to Bentham, the ability to suffer is the morally relevant characteristic that makes humans and animals similar in moral terms. He states concisely that ethics does not have its foundations in the ability to think or to speak, but in the ability to suffer: "Is it the faculty of reason, or, perhaps, the faculty of discourse? But a full-grown horse or dog is beyond comparison a more rational, as well as a more conversible animal, than an infant of a day, or a week, or even a month [...] The question is not, Can they *reason*?, nor, Can they *talk*? but can they *suffer*?" (Bentham 1970, chap. XVII, sec. 1). Bentham differs from anthropocentric approaches and shapes the idea that all beings capable of suffering belong to the moral community. These positions have been coined "pathocentric" in ethics (*pathos* = Greek for *passion*).

In the twentieth century, Peter Singer continues this idea and initiates an intensive debate. In his famous essay *All animals are equal* from 1976, he writes:

"If a being suffers, there can be no moral justification for refusing to take that suffering into consideration. No matter what the nature of the being, the principle of equality requires that its suffering be counted equally with the like suffering [...] of any other being" (Singer 1976, p. 154). This argument led to a fundamental turn in animal ethical reflection: If animals and humans are similar concerning morally relevant characteristics—like in terms of the ability to suffer—we should consider them for this reason in our actions. Everything else would be unequal treatment of equals. This opposes our fundamental idea of justice to *consider equals equally* and *unequals unequally*. The assumption of superiority by humans cannot be justified. On the contrary, suggesting that humans are superior to animals because they are human—in the sense of belonging to the human species—can be described as speciesism. Richard Ryder introduced the term "speciesism" (1971), as an analogous concept to racism in the context of the unjustified discounting of animal interests (Ryder 1971, p. 81).

Singer made the argument of equality prominent in the current debate. His reasoning starts from those characteristics that humans consider morally relevant and have in common with animals (McReynolds 2004; Grimm 2013). If it is plausible to hold that these characteristics in animals are comparable to those of humans, a moral status for animals can be inferred (McReynolds 2004; Grimm 2013; Badura 2001; Schmidt 2011, p. 158). In short, if humans acknowledge a moral norm to avoid suffering, it does not matter where and with whom this suffering arises. The norm "avoid suffering!" refers to all beings to whom we can attribute relevant suffering. The relevant group of animals does not only include mammals. For example, an expert opinion written for the Swiss Federal Ethics Committee on Non-Human Biotechnology by philosopher Markus Wild and biologist Helmut Segner on the experience of suffering and pain in fish has been published recently (Wild 2012a; Segner 2012). For a long time it was the tenor in philosophy to emphasize the differences between humans and animals (Wild 2012b). These bounds are becoming increasingly dubious from an empirical and normative point of view (Benz-Schwarzburg and Knight 2011; Benz-Schwarzburg 2012).

However, it is not that easy to decide which beings suffer in a relevant way. We cannot even know for sure if humans are suffering at a certain moment. But as it is not plausible to deny humans the ability to suffer, neither it is plausible to do this in regard to animals. The objection that humans are able to suffer more or more intensely due to their cognitive capacities does not derogate this argument. Maybe humans and animals suffer in different ways, but lacking cognitive capacities such as being aware of one's own future does not prove that the suffering is more or less relevant (Wolf 2012, pp. 115–118).

Even if this problem can be mitigated, some difficulties remain in utilitarianism that stem from its grounding principle (the greatest good for the greatest number) and are central in the field of animal testing: First, the possibility of weighing interests against each other leads to the question of *how* to weigh them. The problem of weighing competing interests arises and the question is if a selective calculation method can be designed to justify animal research in terms of maximizing benefits (Bass 2012). It is questionable if a "calculation method" can be designed in order to justify certain animal experiments to maximize benefits. Nevertheless, the basic structure is clear: Animal distress has to be justified by benefits. This implies that not every weighing of interests has to result in favor of human interests. Utilitarianism and the logic of balancing distress and benefits play a major role in the context of animal testing. Further, the strength of utilitarianism, to consider all interests, leads to the consequent question: How to deal with harm that does not cause suffering or violate claims? What happens if, for instance, a lab mouse is killed without suffering? Does killing without suffering matter in this theory? Singer in fact stated that killing animals without suffering was only a moral evil if interests are violated. He argues that this can only be the case in beings with future interests (Singer 2011/1979, pp. 94–122).

Summary: The focus on the similarities between humans and animals in utilitarian theories has a significant consequence: If animals are similar to us concerning morally relevant characteristics, for example, in terms of their ability to suffer, we should consider them for this reason in our actions. Everything else would be unequal treatment of equals, which opposes our fundamental idea of justice. If we distress animals, this has to be compensated and justified by benefits. Not every weighing of suffering and benefits will be in favor of human goals. Human and animal interests are counted equally. However, according to the animal rights position, this consideration does not go far enough. In this framework, the metaphor of the scales is rejected, and the idea of animal rights that cannot be outweighed is argued for.

15.3.3 Animal Rights Approach: Against Instrumentalizing Animals

In so-called animal rights approaches, it is argued that it is morally wrong to use animals as tools for one's own purposes. Using an animal as a tool or an instrument to reach one's own aims violates moral rights. Such rights cannot be weighed against benefits. The inherent value of animals is not respected when using them as instruments or tools for human purposes. According to this approach, there is no benefit that can justify the violation of animal rights (e.g., the right to physical integrity). Therefore, according to the animal rights approach, we do not need to look for justifying reasons in the context of animal experimentation. Representatives of this approach, such as its most famous representative Tom Regan (2004, p. 393), reject animal testing completely. This general reluctant attitude is called *abolitionism* in the animal ethics debate (Regan 2004; Francione and Garner 2010).

As in the work of Singer, the principle of equal consideration of equals is a central argument for the extension of the moral community for Regan. However, this principle can only become normatively powerful if relevant comparable characteristics can be found in animals. In this case it is to be an *experiencing subject of a life*, whose moral rights should be acknowledged: "And the really crucial, the basic similarity, is simply this: we are each of us the experiencing subject of a life, including our pleasure and pain, our enjoyment and suffering, our satisfaction and frustration, our continued existence or our untimely death [. . .]. As the same is true of those animals that concern us [. . .], they, too, must be viewed as experiencing subjects of a life with inherent value of their own" (Regan 1985, p. 22).

Consequently, Regan pleads for giving up the practices of research involving animal testing. Animal testing has to end because it denies fundamental animal rights, no matter if there is only little distress for the animals or if the desired knowledge is of great significance for mankind. "In the case of the use of animals in science, the rights view is categorically abolitionist. [. . .] Because these animals are treated routinely, systematically as if their value were reducible to their usefulness to others, their rights are routinely, systematically violated" (Regan 1985, p. 24). A less radical interpretation of the animal rights approach states that there should be limits in dealing with animals that are not up for negotiation, but animal experiments can still be carried out (Sandøe et al. 2008, p. 109). In the context of animal research, this could be understood as an upper limit for distress that may never be exceeded or, if it must be exceeded, only in exceptional cases. However, this is diverging from the original idea of animal rights that animals must not be used for human purposes.

The animal rights approach has been faced with the argument that only beings with obligations can have rights and that we cannot reasonably talk about animal obligations. This can be countered by the fact that we acknowledge rights of children, demented, or disabled people without talking about their obligations in a narrow sense.

Summary: The central idea in the animal rights theory is that there are absolute limits in dealing with animals that are not up for negotiation. Treating animals as tools is to be rejected, no matter if there might be benefits. Animal rights representatives like Tom Regan (2004) think that the use of animals as a means to human ends is to be rejected as it violates fundamental animal rights. Other authors like the abovementioned Peter Singer advocate animal rights only for rational and self-conscious animals to whom they attribute the status of a person (Singer 2011/1979). Singer supports the establishment of fundamental rights for great apes in a declaration together with Paola Cavalieri (Singer and Cavalieri 1994). A moderate interpretation of the animal rights approach (Sandøe et al. 2008, p. 109) would, for example, lead to an upper limit for distress in ethically acceptable animal experiments.

15.3.4 Animal Ethics and New Developments in the Life Sciences

Since the 1990s, new concepts have been discussed in animal ethics that go beyond the abovementioned approaches and especially beyond pathocentrism. This is due to the fact that new procedures—such as genetic engineering—appeared that are not inherently connected to negative welfare. Questions like "Is it morally relevant if the genome of a mouse is modified in such a way that the mouse does not suffer, although its abilities and characteristics are affected?" appeared. Three suggestions for dealing with such questions will be sketched in the following: *telos*, *integrity*, and *the dignity of living beings*.

15.3.4.1 Telos: The Nature of Animals

Bernhard Rollin's book *The Frankenstein Syndrome* (Rollin 1995) is a paradigmatic example for authors who analyze current problems in animal ethics against the backdrop of biotechnological innovations and make the close connection between normative and empirical scientific disciplines their starting point (Grimm 2012). Nearly 20 years ago, Rollin identified the main tasks of ethics with the focus on problems of social reality and reasoning for practical solutions (Rollin 1995, p. 6). With regard to questions in the field of genetic modification of animals and traditional approaches in animal ethics, Rollin extends the pathocentric well-being-centered approach significantly: "The well-being that should be protected involves both control of pain and suffering and allowing the animals to live their lives in a way that suits their biological natures" (Rollin 1995, p. 157). The term "biological natures" is central in this quote. Every living animal has characteristic interests, needs, behavioral repertoires, etc. Rollin describes these characteristics, taken together, as *telos*: "Following Aristotle, I call this the *telos* of an animal, the pigness of the pig, the dogness of the dog—"fish gotta swim, birds gotta fly" (Rollin 1995, p. 159; Rollin 1998, p. 146). Only if a being can completely act on its *telos* and satisfy its species-specific interests can its real nature become apparent (Schmidt 2008, p. 263). This criterion, which can also be filled ethologically (Rollin 1998 pp. 92–95) is Rollin's answer to pressing questions in the field of genetic modification of animals: "Thus, *telos* has emerged as a moral norm to guide animal use in the face of technological changes which allow for animal use that does not automatically meet the animals' requirements flowing from their natures. [. . .] *telos* provides the conceptual underpinnings for articulating social moral concern about new forms of animal suffering" (Rollin 1998, p. 161). Rollin responds to technological developments and suggests the "nature" of animals, which has to be respected, as the central criterion in animal ethics. In doing this, he tries to systematize the moral discontent related to genetic modification of animals and to deal with it on the basis of the Aristotelian concept of the *telos*.

15.3.4.2 Integrity: The Wholeness and Completeness of Animals

The following illustration of the concept of integrity is mainly oriented towards the Dutch debate on new concepts in animal ethics. This debate also developed against the backdrop of the need to regulate genetic modification of animals. The concept of animal integrity became a criterion to rationally reflect moral intuitions that are not covered by the pathocentric principle of well-being and avoiding suffering. The standard definition of the term integrity in the Dutch debate has been developed by Bart Rutgers and Robert Heeger. It expresses the relevant aspects in three complementary elements:

> We define animal integrity as follows: the wholeness and completeness of the animal and the species-specific balance of the creature, as well as the animal's capacity to maintain itself independently in an environment suitable to the species [. . .]. This definition is made up of three mutual linked and complimentary elements: (1) the wholeness and completeness, (2) the balance in species specificity, and (3) the capacity to independently maintain itself. It is essential that all three elements are satisfied for there to be a state of integrity.
>
> According to the definition, integrity refers to both the individual animal and the species. One can speak of the integrity of the individual animal because every individual animal is a clearly defined biological entity with its own disposition. (Rutgers and Heeger 1999, p. 4)]

As in the context of *telos*, a morally relevant recourse to biologically given dispositions can be found. It is not necessary that animals experience violation of their integrity. In general, the point is not only well-being, but the psychophysical entity of the animal (Rollin 1998, p. 119): "Animals are psycho-physical wholes expressing a certain species-specific nature which is referred to when we speak about the animal's 'integrity.' We cannot set psychic or conscious experiences apart from the bodily aspects as is done in the receptacle view of consciousness" (Verhoog and Visser 1997, p. 231).

Therefore, this is also an approach in which moral respect for animals goes beyond avoiding pain and suffering. Similarly to the debate about *telos*, the argumentation seems to be motivated by moral discontent caused by the possibility of genetic modification of animals. This moral intuition is the starting point for assessing new technological possibilities in a comprehensible way. Ethics provides rational ways of dealing with it.

15.3.4.3 Dignity of Creatures

The concept of "dignity" within the Swiss debate also aims at regulating our dealing with animals in the age of genetic modification. However, it seems that the concept of dignity involves even more possible interpretations than the concept of integrity. When searching for the "homeland" of the discussion of dignity of animals and of living beings in general (Kunzmann 2007, pp. 13–26), the answer is clear: The Swiss canton Aargau elevated dignity to constitutional status in 1980. Twelve years later, in 1992, dignity was incorporated into the Swiss Federal Constitution. Certainly, this has been a reaction to the technological innovations in the field of genetic engineering, as in the context of the *telos* and integrity concept. It is in this context that reflection on moral respect for animals beyond pathocentrism started. The relevant section in the constitution reads correspondingly:

> Art. 120 Non-human gene technology
>
> (1) Human beings and their environment shall be protected against the misuse of gene technology.
>
> (2) The Confederation shall legislate on the use of reproductive and genetic material from animals, plants and other organisms. In doing so, it shall take account of the dignity of living beings as well as the safety of human beings, animals and the environment, and shall protect the genetic diversity of animal and plant species.

After incorporating the concept of dignity into the Federal Constitution in 1992, a lively debate started on what this paragraph means. Fifteen years after the incorporation of the term into the Swiss Federal Constitution, Peter Kunzmann presents the central arguments in a clear way (Kunzmann 2007). He develops an assistant theory that reflects categories of the term *dignity* (Kunzmann 2007, pp. 110–117). Kunzmann points out that talking about the "dignity of animals" does not only embrace animals but also the human actors. They have to show respect for the animal not only in terms of *action* but also in terms of their *attitude*. Further, both the *inherent value* of the animal and the animal in its *uniqueness* have to be respected.

Summary: As the concepts *telos*, integrity, and dignity show, pathocentric theories are not enough to reasonably reflect our moral intuitions in light of current biotechnological innovations and possibilities. It becomes clear that new questions in animal ethics are implied by scientific developments. Therefore, the development and results of scientific research—e.g., life sciences, ethology, or veterinary medicine–are of great significance as they bring about new problems and provide knowledge for the analysis of problems in animal ethics. As mentioned above, knowledge from different disciplines has to be integrated into the work of ethicists.

15.4 Taking Ethical Considerations into Account: Fairness Towards Lab Animals

Against the background of the outlined positions, the question arises, which approach can help to deal with moral responsibility in animal research. Generally speaking, all of the illustrated positions shed light on morally relevant aspects of animal research (Webster et al. 2010). Despite the fact that animal ethics is far beyond a single "super theory" that only needs to be applied to specific cases, ethicists are increasingly asked to develop feasible tools in order to solve ethical problems (Borchers 2009, p. 21). From at least three angles, this is a challenging situation for ethicists (Borchers 2009): First, there is not a single ethical theory that can be applied without criticism. Second, there are a number of plausible theories that can serve as a basis for problem solving, each bringing about different solutions. So, under practical circumstances, ethicists have the responsibility to illustrate ethical perspectives of relevance, even if they are contradicting. Therefore the problem of which theory best meets theoretical and practical expectations arises. Third, applied ethicists are often working within a given legal framework. It would not be plausible to think of ethics as a discipline that provides a kind of mysterious knowledge that can be used to settle moral conflict and dissent once and for all. In any case, because they are able to relay the possibilities and limitations of moral views, ethicists can bring forth tools and methodologies for a modus vivendi to deal with moral dissent.

From a practical point of view, two important approaches for such a modus vivendi in animal research should be separated. Both approaches reflect aspects of the mentioned theories: First, one can hold that gaining knowledge is a moral obligation and justifies the use and adverse effects on animals in scientific research if certain criteria are met. This position relates to the 3R approach and is anthropocentric in nature. Second, one can hold that intentionally inflicting adverse effects on animals in order to gain knowledge is not justified per se and thereby frame the issue as a matter of justification. This position allows harms and benefits to be weighed openly and is related to a utilitarian, non-anthropocentric theory. In the following, it will be outlined how these approaches deal with the dilemma mentioned above.

15.4.1 The 3R Approach Versus Weighing Harms and Benefits

Within the framework of the 3R approach (replace, reduce, refine), the mentioned dilemma in animal research has a clear solution: Experiments should only be carried out if they are necessary to reach important goals that cannot be reached otherwise and the 3Rs are met. Human benefits are superior to harm and distress of animals. In other words, it is more important to help humans than not to harm animals. Within this logic, the morally relevant interests of humans are prior to animal interests in the field of animal experimentation (Rippe 2012). In brief, if the desired knowledge is of relevance and likely to fulfill the moral obligation to help humans (principle of beneficence), we have the moral obligation to conduct the experiment. Since the interests of animals are secondary, animal interests can only outweigh nonmoral interests of humans. What does this mean in practice?

For instance, take an experiment that provides knowledge for medical treatment of humans and therefore aims at a morally relevant goal. Given that this goal can be reached through a particular experiment and cannot be reached otherwise, the research should be conducted even though unavoidable adverse effects are inflicted. Only the following two questions remain: How to minimize the number of animals used? How to minimize the negative effects and treat and keep the animals well? Whether the animals are harmed severely or not is not relevant; the question is whether *unavoidable harm* is required in order to gain the morally relevant knowledge. The only relevant consideration is to keep adverse effects at a minimum, whether severe or light, long term or short term. This is consistent with inflicting severe pain on a long-term scale, if it is necessary to reach the relevant knowledge and if the criteria of the 3Rs are met (Rippe 2012).

Such a narrow 3R approach reflects a hierarchical order: Human interests override animal interests. Rippe (2009; 2012) mentions a consequent major problem of this approach: We have a strong moral intuition that human interests (even morally relevant ones) should *not* always overrule animals' interests. If an experiment is necessary, it is necessary to reach a *certain goal*. These goals are not of unquestionable value. For instance, take the example of an experiment providing knowledge to lower the risk of a certain disease of a single patient. In the logic of the 3R approach, a necessary experiment to lower this risk is justified even if 10,000 mice or more, perhaps even if all mice in the world, had to be used and killed. This will never be the case under practical circumstances, but this does not matter in the framework of the 3R approach. If it were necessary to use all mice to help a single patient, we would have the moral obligation to act accordingly. This example shows clearly that this approach goes against our moral intuitions. If adverse effects on animals cannot outweigh human benefits under any circumstances, the framework itself is put into doubt. This deficit is reflected in the alternative approach of weighing harms and benefits. And the logic of this alternative approach has been integrated into animal legislation within the EU member states and will be discussed in the following.

15.4.2 Weighing Harms and Benefits: Ethical Tools in Animal Research

On the basis of the directive 2010/63/EU, the EU member states were asked to harmonize their legal requirements in animal research. One point among others was the requirement that authorities have to approve or disapprove research proposals with regard to ethical aspects. Article 38 of the adopted directive (Directive 2010/63/EU) includes a claim to take into account ethical considerations for the harm-benefit analysis:

> (2) The project evaluation shall consist in particular of the following: [. . .] d) a harm-benefit analysis of the project, to assess whether the harm to the animals in terms of suffering, pain and distress is justified by the expected outcome taking into account ethical considerations, and may ultimately benefit human beings, animals or the environment; [. . .]

However, the directive does not provide any specification of how and according to which standards the harm-benefit analysis is to be carried out nor of how "ethical considerations" should be taken into account.

In this context, the question arises whether this analysis should be a "real" weighing of competing interests. If it is agreed that it is meant to be a "real" weighing of competing interests, one has to be aware that the result could also be that it is ethically *not* acceptable to carry out a certain experiment, even if morally relevant goals are strived for. To use a metaphor, the scales are not in favor of carrying out the experiment per se like in the 3R approach. The metaphor of scales has been used since 1987 (Gärtner 1987) to allegorize the weighing of competing interests and to make it more tangible. In this framework, human and animal interests can conflict in a real sense. Still, there is the question of how to weigh these interests and what should be put on the scale pans. Obviously, the concept of "weighing of competing interests" faces several problems (Alzmann 2010):

(a) The problem of *incomparable interests* is inherent in weighing competing values or goods. What kind of benefit in terms of gaining knowledge can justify how much animal distress? Here we enter the problem that there is no common standard or common "currency." In philosophy, this problem has been coined *incommensurability* of values. In everyday language, this problem is known as comparing apples and oranges. How can human benefits be weighed against animal harms? Weighing competing interests would be easy if we put something of the same currency on both scale pans—a certain number of apples on the one pan and another number of apples on the other pan. However, in the context of animal experiments, relevant aspects are comparable only in part.

(b) The *number and weight of interests to be weighed* is a question in itself. Two aspects are essential in order to come to an adequate result when weighing competing interests: First, *whose* interests and *which* of them are put on the scale pans and therefore taken into account? Second, *how* are the interests taken into account? In other words, *how* are the interests *weighted*? Are human and animal interests of the same weight or not?

(c) Further, we are faced with the question of what it means to take into account *all relevant aspects*. The weighing of competing interests is often referred to by using the following criteria: pain, suffering, and harm for the lab animals' scale pan and the benefits for the other scale pan. Can the weighing be reasonably *reduced* to the mentioned criteria? As argued by Alzmann (2009), reducing the procedure to the criteria mentioned above leads to a bias against the "animal side." Whenever it comes to practical terms, the number of the aspects on the scale pans significantly influences the result of the weighing and all relevant aspects should be included; to decide on what is relevant is of course a normative question itself.

(d) *Ethical acceptability as a condition for approval* is a further open question. There have been fierce debates on the evaluation of ethical acceptability of animal experiments for decades. With good reason, as in many countries, ethical acceptability is one of the three essential aspects besides indispensability and the lack of alternatives that have to be fulfilled—besides applying the 3Rs—in order to approve a proposal including animal experiments. It turns out that one of the basic problems is the fact that the people involved in approval procedures (applicants, animal welfare committees, representatives of animal welfare groups, representatives from approving authorities, members of advisory boards that support the authorities, etc.) are often experts in natural sciences, but they *lack ethical expertise*. Only rarely are ethicists members of approving bodies. Weighing competing interests, however, is much more than counting and calculating scientific data. Therefore, there is a need for feasible methodologies to carry out approval procedures.

To tackle the mentioned problems, different systems have been developed in several countries that aim to structure and objectify the weighing of competing interests to evaluate the ethical acceptability of animal experiments. They are in general checklists or catalogues of criteria. These *ethical tools* are opposed to "impact assessments" that are used to prospectively or retrospectively estimate and classify the severity of the adverse effect on lab animals only. When using ethical tools such as "catalogues of criteria," the severity of the adverse effects on animals is only one of many aspects—although an important one—that is evaluated. Further, beneficial aspects—the pros of the experiment—are included as well in a "weighting procedure."

The Canadian biomedical researcher David G. Porter is considered a "pioneer" of ethical tools in the context of animal research (Porter 1992). He published on *Ethical scores for animal experiments* in 1992 in *Nature*. This publication triggered a lively debate and led to improved and more detailed methodologies (Alzmann 2010; De Cock and Theune 1994; Scharmann and Teutsch 1994; Mand 1995a; b; Stafleu et al. 1999; Hirt et al. 2007; SAMW/SCNAT 2007). To give one example, the Swiss Internet program for "self-evaluation" of research proposals with respect to competing interests can provide a good insight into the way such a methodology can work and support decision-making. This ethical tool has been developed by the Ethics Committee of the Swiss Academy of Medical Sciences (SAMW) and the Swiss Academy of Sciences (SCNAT). It works with six categories (some of which contain subcategories) and merges them into a visualized and explicitly formulated overall result. As an evaluation tool, it indicates in which respects a research proposal is weak and needs to be improved.

If the weighing process is understood as "real" weighing, human benefits are not necessarily overriding. Therefore, such weighing procedures reflect a non-anthropocentric approach.

15.4.2.1 Synopsis

Applying our concepts of right and wrong to the field of animal research fuels an ongoing and highly controversial debate. The role of ethics within this debate is not to provide clear-cut solutions to fuzzy problems since there is no unquestionable basis to be found in ethics. Quite the opposite, different ethical theories provide important insights into the topic of animal research. Therefore, a plausible contribution of ethicists is to illustrate possibilities and limitations of moral positions in light of ethical theories. Therewith, they can assist decision-making processes and even develop ethical tools to integrate ethical aspects in the evaluation procedures of experimental designs. These are all ways of dealing with the dilemma outlined at the beginning of this article. It should be acknowledged that the function of ethics is not to *judge* whether a particular research proposal is to be carried out or not. Within the existing legal framework, this is the responsibility of the legal authorities. But ethics can bring forth criteria and methods and describe a modus vivendi that leads to better decision-making. Ethicists can give reasons why scientists should take care of animals and what it means for scientists to take on moral responsibility in their field. But only if the normative and descriptive sciences are working together and take into account the public debate can feasible strategies to answer to question of moral responsibility in the field of laboratory animal science be developed.

References

Altman MC (2011) Kant and applied ethics. The uses and limits of Kant's practical philosophy. Blackwell Publishing, Malden

Alzmann N (2009) Zur Notwendigkeit einer umfassenden Kriterienauswahl für die Ermittlung der ethischen Vertretbarkeit von Tierversuchsvorhaben. In: Borchers D, Luy J (eds) Der ethisch vertretbare Tierversuch—Kriterien und Grenzen. Mentis, Paderborn, pp 141–170

Alzmann N (2010) Zur Beurteilung der ethischen Vertretbarkeit von Tierversuchen. Dissertation at the Faculty of Biology. Eberhard Karl University Tübingen, Tübingen

Arluke A, Levin J, Luke C, Ascione F (1999) The relationship of animal abuse to violence and other forms of antisocial behavior. J Interpers Violence 14(9):963–975

Badura J (2001) Leidensfähigkeit als Kriterium? Überlegungen zu einer pathozentrischen Tierschutzethik. In: Schneider M (ed) Den Tieren gerecht werden. Zur Ethik und Kultur der Mensch-Tier-Beziehung. GhK, Kassel, pp 195–210

Balls M (2009) The three Rs and the humanity criterion. FRAME, Nottingham

Bass R (2012) Lives in the balance: utilitarianism and animal research. In: Garrett JR (ed) The ethics of animal research. Exploring the controversy. The MIT Press, Cambridge, pp 81–105

Bayertz K (1999) Moral als Konstruktion. Zur Selbstaufklärung der angewandten Ethik. In: Kampits P, Weinberg A (eds) Angewandte Ethik. Beiträge des 21. Internationalen Wittgenstein Symposiums, Wien, pp 73–89

Bayertz K (2002) Warum "Selbstaufklärung der Bioethik"?. In: Ach JS, Runtenberg C (eds) Bioethik. Disziplin und Diskurs. Zur Selbstaufklärung angewandter Ethik. Campus, Frankfurt am Main, pp 9–12

Bentham J (1970/1789) An introduction to the principles of morals and legislation, Bd. 2. In: Burns JH, Rosen, Schofield P (eds) The collected works of Jeremy Bentham. University of London/The Athlone Press, London

Benz-Schwarzburg J (2012) Verwandte im Geiste—Fremde im Recht. Sozio-kognitive Fähigkeiten bei Tieren und ihre Relevanz für Tierethik und Tierschutz. Harald Fischer Verlag, Erlangen

Benz-Schwarzburg J, Knight A (2011) Cognitive relatives yet moral strangers? J Anim Ethics 1:9–36

Borchers D (2009) Ethiktools für die Güterabwägung: Wie pragmatisch dürfen Ethiker sein? In: Borchers D, Luy J (eds) Der ethisch vertretbare Tierversuch. Kriterien und Grenzen. Mentis, Paderborn, pp 15–51

Carruthers P (1992/1994) The animals issue: moral theory in practice. Cambridge University Press, Cambridge

Cohen C (2001/2008) Warum Tiere keine Rechte haben. In: Wolf U (ed) Texte zur Tierethik (orig.: Why animals do not have rights. In: Regan T (ed) The animal rights debate. Rowman & Liittlefield, New York/Oxford). Reclam, Stuttgart, pp 51–55

De Cock BT, Theune EP (1994) A comparison of three models for ethical evaluation of proposed animal experiments. Anim Welf 3:107–128

Dewey J (1938) Ethics. The collected works of John Dewey. The later works, vol 7. Southern Illinois University Press, Carbondale

Directive 2010/63/EU of the European Parliament and the European Council September 22, 2010, on the protection of animals used for scientific purposes. Official Journal of the European Union, October 20, 2010 L 276/33–79

Düwell M (2008) Bioethik. Methoden, Theorien und Bereiche. Metzler, Stuttgart

Francione GL, Garner R (2010) The animal rights debate. Abolition or regulation? Columbia University Press, New York

Gärtner K (1987) Kriterien dermateriellen Prüfung von Genehmigungsanträgen. Dtsch Tierarztl Wochenschr 94:100–102

Godlovitch R, Godlovitch S, Harris J (eds) (1971) Animals, men, and morals: an enquiry into the maltreatment of non-humans. Gollancz, London

Grimm H (2010) Das moralphilosophische Experiment. John Deweys Methode empirischer Untersuchungen als Modell der problem- und anwendungsorientierten Ethik. Mohr Siebeck, Tübingen

Grimm H (2012) Benthams Erben und ihre Probleme. Zur Selbstreflexion einer Ethik der Mensch-Tier-Beziehung. In: Zichy M, Ostheimer J, Grimm H (eds) Was ist ein moralisches Problem. Zur Frage des Gegenstandes angewandter Ethik. Alber, Freiburg

Grimm H (2013) Das Tier an sich? Auf der Suche nach dem Menschen in der Tierethik. In: Liessmann KP (ed) Tiere. Der Mensch und seine Natur. Paul Zsolnay, Wien, pp 277–232

Hirt A, Maisack C, Moritz J (eds) (2007) Tierschutzgesetz – Kommentar, 2nd edn. Vahlen, Munich
Höffe O (1977/2008) Lexikon der Ethik. C.H. Beck, München
Kunzmann P (2007) Die Würde des Tieres—zwischen Leerformel und Prinzip. Alber, Freiburg
Mand U (1995a) Über die in § 7 Abs. 3 des Tierschutzgesetzes geforderte Abwägung ethischer Vertretbarkeit von Tierversuchen. Der Tierschutzbeauftragte 3(95):229–234
Mand U (1995b) Zur Abwägung ethischer Vertretbarkeit von Tierversuchen gemäß § 7 Abs. 3 des derzeit geltenden Tierschutzgesetzes (TierSchG), dargestellt am Modell Sepsisforschung am wachen Schwein. Diss. med. vet. Univ. Gießen, Gießen.
McReynolds P (2004) Overlapping horizons of meaning. A Deweyan approach to the moral standing of nonhuman animals. In: McKenna E, Light A (eds) Animal pragmatism. Rethinking Human-nonhuman relationships. Indiana University Press, Bloomington
Narveson J (1977) Animal rights. Can J Philos 7(1):161–178
Olsson AS, Robinson P, Pritchett K, Sandøe P (2003) Animal research ethics. In: Hau J, Van Hoosier GL Jr (eds) Handbook of laboratory animal science, vol 1, 2nd edn, Essential principles and practices. CRC Press, Boca Raton, pp 13–30
Porter DG (1992) Ethical scores for animal experiments. Nature 356:101–102
Regan T (1985) The case for animal rights. In: Singer P (ed) In defense of animals. Blackwell, New York, pp 13–26
Regan T (2004/1983) The case for animal rights. University of California Press, Berkeley
Rippe KP (2009) Güterabwägungen im Tierversuchsbereich. Anmerkungen zu einem ethischen Paradigmenwechseln. ALTEXethik 2009:3–10
Rippe KP (2012) Tiere, Forscher, Experimente. Zur ethischen Vertretbarkeit von Tierversuchen. In: Grimm H, Otterstedt C (eds) Das Tier an sich. Disziplinenübergreifende Perspektiven für neue Wege im wissenschaftbasierten Tierschutz. Vandenhoeck & Rupprecht, Göttingen, pp 331–346
Rollin BE (1995) The Frankenstein syndrome. Ethical and social issues in the genetic engineering of animals. Cambridge University Press, Cambridge
Rollin B (1998) On telos and genetic engineering. In: Holland A, Johnson A (eds) Animal biotechnology and ethics. Chapman & Hall, London, pp 156–171
Rollin BE (2012) Ethics and animal research. In: Garrett JR (ed) The ethics of animal research. Exploring the controversy. The MIT Press, Cambridge, pp 19–30
Russell WMS, Burch RL (1959) The principles of humane animal experimental technique. Potters Bar, London
Rutgers B, Heeger R (1999) Inherent worth and respect for animal integrity. In: Dol M et al (eds) Recognizing the intrinsic value of animals. Van Gorcum, Assen, pp 41–51
Ryder R (1971) Experiments on animals. In: Godlovitch R, Godlovitch S, Harris J (eds) Animals, men, and morals: an enquiry into the maltreatment of non-humans. Gollancz, London, pp 41–82
SAMW/SCNAT (2007) Swiss Academy of Medical Sciences and Swiss Academy of Sciences (SAMW and SCNAT 2007): Dienstleistung der Ethikkommission für Tierversuche der SAMW (Basel) und der SCNAT (Bern): Ethische Güterabwägung bei Tierversuchen. Eine Vorlage für die Selbstprüfung. http://tki.samw.ch/ Version 1.2, December 15, 2007. The analysis and remarks refer to this version. The current version online is version 1.2.1, August 25, 2009 (last accessed 21.02.2013)
Sandøe P, Christiansen SB, Hansen AK, Olsson A (2008) The use of animals in experiments. In: Sandøe P, Christiansen SB (eds) The ethics of animal use. Blackwell Publishing, Oxford
Scharmann W, Teutsch GM (1994) Zur ethischen Abwägung von Tierversuchen. ALTEX 11 (4/94):195–198
Schmidt K (2008) Tierethische Probleme der Gentechnik. Zur moralischen Bewertung der Reduktion wesentlicher tierlicher Eigenschaften. Mentis, Paderborn
Schmidt K (2011) Concepts of animal welfare in relation to positions in animal ethics. Acta Biotheor 59:153–171

Segner H (2012) Fish. Nociception and pain—a biological perspective. In: Eidgenössische Ethikkommission für die Biotechnologie im Ausserhumanbereich und A. Willemsen (ed) Beiträge zur Ethik und Biotechnologie Bd. 9. Bern.

Singer P (1976) All animals are equal. In: Regan T, Singer P (eds) Animal rights and human obligations. Prentice-Hall, Englewood Cliffs, pp 148–162

Singer P (2011/1979) Practical ethics, 3rd edn. Cambridge University Press, New York

Singer P, Cavalieri P (1994) Deklaration über die Großen Menschenaffen. In: Singer P, Cavalieri P (eds) Menschenrechte für die Großen Menschenaffen. Das Great Ape Projekt. München, Goldmann, pp 12–16

Stafleu FR, Tramper R, Vorstenbosch JMG, Joles JA (1999) The ethical acceptability of animal experiments: a proposal for a system to support decision-making. Lab Anim 33:295–303

van den Daele W (2008) Soziologische Aufklärung und moralische Geltung: Empirische Argumente im bioethischen Diskurs. In: Zichy M, Grimm H (eds) Praxis in der Ethik. Zur Methodenreflexion in der anwendungsorientierten Moralphilosophie. de Gruyter, Berlin, pp 119–151

Verhoog H, Visser T (1997) A view of intrinsic value not based on animals consciousness. In: Dol M et al (eds) Animal consciousness and animal ethics. Van Gorcum, Assen, pp 223–232

Webster J, Bollen P, Grimm H, Jennings M (2010) Ethical implications of using the minipig in regulatory toxicology studies. J Pharmacol Toxicol Methods 62(3):160–166

Wild M (2012b) Fische. Kognition, Bewusstsein und Schmerz. Eine philosophische Perspektive. In: Eidgenössische Ethikkommission für die Biotechnologie im Ausserhumanbereich und A. Willemsen (ed) Beiträge zur Ethik und Biotechnologie Bd. 10. Bern.

Wild M (2012a) Die Relevanz der Philosophie des Geistes fürden wissenschaftsbasierten Tierschutz. In: Grimm H, Otterstedt C (eds) Das Tier an sich. Disziplinenübergreifende Perspektiven für neue Wege im wissenschaftsbasierten Tierschutz. Vandenhoeck & Ruprecht, Göttingen, pp 61–86

Wolf U (2012) Ethik der Mensch-Tier-Beziehung. Klostermann, Frankfurt am Main

Printed by Publishers' Graphics LLC